THE MASTER HANDBOOK OF TELEPHONES

THE MASTER HANDBOOK OF

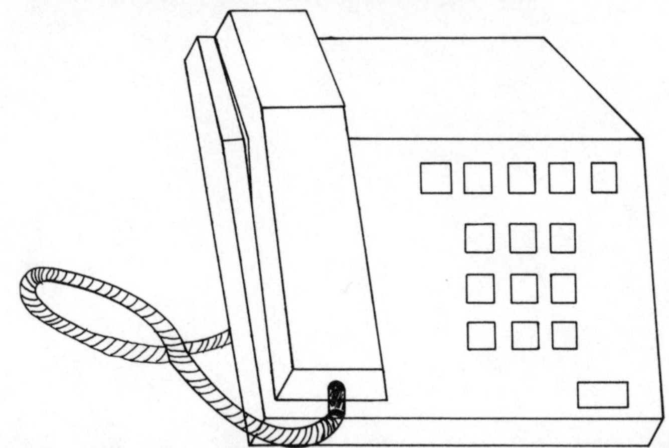

TELEPHONES

BY ROBERT J. TRAISTER

FIRST EDITION

FOURTH PRINTING

Printed in the United States of America

Library of Congress Cataloging in Publication Data

Traister, Robert J.
The master handbook of telephones.

Includes index.
1. Telephone. I. Title.
TK6165.T7 621.386 81-9092
ISBN 0-8306-0017-5 AACR2
ISBN 0-8306-1316-1 (pbk.)

Contents

Preface

Much time has come and gone since the days of Alexander Graham Bell and his invention, the telephone. This instrument of communication which almost all of us take for granted has developed to a very high state of the art. So much so, in fact, that many persons know relatively little about all that is offered to today's consumer in this area. Telephone communications are no longer limited to hardwire lines. Cordless telephones abound for the everyday consumer along with the more exotic forms of communications which may be had from the comfort of the private automobile or even aboard a pleasure boat at sea.

The phone system, too, has progressed to a point where rapid communications are available to almost any point in the world. Operator intervention is fast being phased out and replaced with computers and automated equipment. Microwave links are the rule rather than the exception that they were only a few years ago, and fiberoptic cables are now replacing the older copper conductions. In placing a long-distance phone call, your voice may be transferred through several different systems before reaching the party on the other end.

The purpose of this book is to inform the reader as to the myriad of telephones and telephone-related devices which are presently available. Most of these require no special phone company installations. Indeed, most are connected by the user to the already-existing line. Some will take and send messages while others will provide the user with a lot more versatility in taking and placing calls. All of them serve a very definite and worthwhile purpose in convenience, efficiency, and in saving time and money. Many of the devices described in the following pages may well be the answer to many of your problems. Others will serve to spark the imagination on ways they might be used around the home or office.

The material presented in these pages is arranged in a straightforward, uncomplicated manner, so regardless of your present knowledge of telephones and telephone systems, there is little chance of being disconnected.

The author would like to thank the many suppliers and manufacturers of telephone devices and accessories for supplying research information about the many products and systems available. Special assistance was provided by Michael Calabrese and Radio Shack in Front Royal, Virginia. Thank you one and all.

I wish to acknowledge the following trademarks:

Touch-Tone is a trademark of American Telephone & Telegraph Co.
Flip-Phone is a trademark of GTE Automatic Electric.
Duofone is a trademark of Radio Shack a division of Tandy Corp.
RECORDaCALL is a trademark of Avanti Inc.
The Perfect Answer is a trademark of International Telephone and Telegraph Corp.
The Easy Answer is a trademark of International Telephone and Telegraph Corp.
Call Jotter is a trademark of Quasar.
Modulus is a trademark of Tele-Devices Corp. and Tele-Devices Ltd.
Tele-Recorder is a trademark of TT Systems Corp.
Fone-A-Lert is a trademark of Floyd Bell Associates Inc.
Bel-Ringer is a trademark of Floyd Bell Associates Inc.
Tel-A-Tone is a trademark of Floyd Bell Associates Inc.
Muraphone is a trademark of Mura Corp.

Robert J. Traister, Sr.

Dedication

To my mother, Mary F. Traister, who continues to influence me, even though she may not suspect it at all times.

Chapter 1

The Telephone Instrument

Today, nearly every home in the country is equipped with at least one and probably several telephones. These instruments may appear in many different shapes and sizes. Some are in basic black although these are the exceptions rather than the rule they were twenty years ago. Through the telephone, communication with almost any part of the world is possible within seconds or minutes. The telephone companies as well as the phones themselves have taken many great strides over the past twenty years. In recent years, highly complex telephones and telephone devices have come on the market and open up a whole new scope of advantages for the consumer. The telephone company customer is no longer required to rent his instrument from the phone company. Any number of specialized telephone devices are available from mail order catalogs, electronic warehouses, and even from your corner market. Additionally, your local telephone company will probably sell you an instrument at an attractive price, again foregoing the necessity of renting.

TELEPHONE OPERATION

While operation of the telephone seems like a very simple matter, it only appears so because of the complex electronics, machinery, and technology which has gone into the establishment of the telephone system and especially the central office. In most instances, the central office is the main control point which receives input from all telephones in the system or in a specialized area of the system, decodes the input, and provides the correct answering and/or switching signals to other telephones in the circuit. The central office also transfers calls from its own system into other communication systems which is comprised of an elaborate network of electronics, computers, microwave relays, and satellites which surround the world. This highly complex function is handled through a very sophisticated automation system which has many millions of control and activation points. Your telephone is one of these points and with it a myriad of services and communication can be had from the comfort of your home or office.

The purpose of this book is not to explain the intricate workings of the entire telephone system. This topic could fill many large volumes. The following pages are devoted to phone instruments and devices, most of which can be easily purchased and installed by the reader in a home or office. But, in order to fully explain these devices, it is necessary to have a basic understanding of how the standard phone instrument operates and some of the basic functions it is able to key up at the central office.

The major function of most telephones is to provide voice communication over distances ranging from less than a mile to several thousand miles. The telephone system works on basic electrical principles whereby power from one system is transferred to power in another system and then back again. The power being referred to is communications power and, in human beings, this most often takes the form of the voice. In a standard telephone instrument, the human voice is converted into electrical energy that can be transmitted over the telephone wire. When you speak into the mouthpiece or transmitter of your telephone headset, a movable diaphragm causes a miniscule electrical current to be induced into the telephone wires and over all systems. This is not a steady current but one which fluctuates with the changing of the tones which make up the human voice. Human speech patterns are highly complex in nature but the telephone system reproduces these sounds faithfully in most cases.

When listening to a voice over the phone line, the caller on the other end has had his voice transferred into electrical current which is in turn passed on to the earpiece of receiver in your handset. An element in the receiver which is very similar to a magnetic speaker converts electrical pulses back into sound waves which are detected by the human ear. This is a very simple explanation of how voice waves are transmitted and received but serves to provide a basic understanding of the electrical nature of the telephone system.

Your telephone and phone system is made up of *current dependent* electrical and electronic circuits. Without current a phone system cannot function. This has just explained a very small portion of the operation of the telephone and it is necessary to know a bit more about your instrument and its connection to the central office.

In addition to providing the electrical means to convert speech into current and current back into speech, the instrument must be able to select any other phone in the system for a connection and to indicate when it is in use. Each set in a given area is connected in some way by a pair of copper wires called a *loop*. This loop terminates at the central office which is a switching center. Phone calls within the loop area pass through the central office. These will mostly include local phone calls. Calls to numbers outside of the loop area are routed to other offices through trunk lines.

Perhaps the operation of your phone can best be explained by picturing in your mind every step you go through when placing a telephone call. The first step is the removal of the handset from the cradle. When the handset is removed, a switch engages which provides a signal at the central office which in turn transmits a dial tone to let you know that a line is available. Once a dial tone is heard, you dial or punch in the desired number. The central office receives the electrical pulses corresponding to the dialed number and selects the correct pair of wires to engage that number. During this process a ringing signal is sent to the phone with the number which has been selected, alerting the party being called by ringing the internal bell within their instrument. When someone answers by lifting the handset, the central office connects your line with theirs. Now, the previous discussion on how voice is transmitted comes into play. When the conversation is ended by replacing the handset on the cradle, the closing of the cradle switch signals the central office that the call has been ended and an electronic command to release both lines for other calls is issued.

CENTRAL OFFICE

As has been pointed out, your telephone is a current operated device. The central office supplies the current for operation of your phone. Fifty volts DC is the standard line value which can be measured at your phone with a voltmeter. Telephone hookups rarely are cause for any concern regarding fires as the supply of current is limited to a very low value. Even if the wiring pair should be found short-circuited, only a small amount of current will be allowed to flow. This amount is so low that heating of the wire elements is not a factor and the danger of the situation is almost completely removed. The reason for this current limitation is the small size of the conductors from the telephone system. This is not always an advantage, however as small diameter wiring presents a higher resistance to current flow than does the larger diameter conductor. Since the current is supplied at the central office, the farther a phone is from this point the lower the amount of current will be that it receives. For this reason, there is a physical limitation on the distance the telephone may be from the phone company. This might better be stated by saying that there are definite limitations on the size of the loop. If you are close to the phone company, your supply current should be relatively high, but phones on the farthest point in the loop will have a much lower current value. This is the reason for the poor quality of some rural telephones.

The reason for the current drop is line resistance. Resistance is a factor which limits the flow of current. While copper wiring is considered to be a conductor, it also has a resistive element which increases with the length of the line. Telephones close to the phone company have less of a resistance factor between them and the central office. Telephones at a farther distance from the central office are connected by longer lengths of wiring and a much higher resistance factor is present.

In most telephones, there is a small electronic circuit which consists of resistors, capacitors, coils, and transformers. In more modern instruments you may also find diodes and transistors mounted on a small piece of printed circuit board. The circuit boards containing solid state components are usually compensated circuits which are designed to make up for the current variation in lines which are located at varying distances from the central office. With proper compensator circuits, the operation of a telephone on a line far removed from the central office will be almost as good as that of an instrument located in close proximity to this central switching point.

The bell or ringer which indicates someone is calling you is made up of two cup-shaped metal bowls which are struck by a clapper. When someone dials your number, the central office passes alternating current through the coils of the ringer. This causes a magnetic field which rapidly reverses itself. The clapper is attached to an armature which moves back and forth in relationship to the varying magnetic field. The ringer is an AC device and is always used in series with a capacitor to isolate the direct current from the ringer circuit while allowing alternating current to pass.

DIAL PULSERS AND TOUCH-TONE PADS

Once you are connected to the central office, the dial tone received, and you are ready to select the number you want to call, there must be some way of signaling the central office which number you desire to call. This is a more complex procedure because of the hundreds of thousands of variables involved. No single switching or indicating function can adequately convey the coding information needed for the central office switching circuit to connect your wiring pair to the desired wiring pair you are calling. The coding of the desired numbers to the central office is normally handled by either a dial pulser (rotary dial) or through a touch-tone pad.

Figure 1-1 shows a standard dial-type telephone instrument which consists of a handset used to transmit and receive communication signals and a body unit containing the dial pulser, bell, transformer, and many contact points and switches. When you lift the handset of your phone, a switch closes which completes a circuit. Current starts to flow, indicating to the central office that you are preparing to make a phone call. The electrical current flow powers a transmitter in the handset which is actually a carbon microphone and produces an alternating current signal when sound waves strike the element. The current which is supplied by the central office does not alternate. It is direct current which is required by the carbon microphone in order to operate properly and produce alternating current in response to sound waves. The alternation of the current from the carbon microphone corresponds directly to the voice or sound pattern directed across its element. The telephone receiver is a very sensitive permanent magnet loudspeaker. When the alternating current which corresponds to the voice passes through coils which interact with the permanent magnet, a diaphragm is forced to move in relationship to the varying magnetic field which is produced by the alternating current. This movement creates audio waves which are detected by the human ear and which correspond directly to the alternating current input.

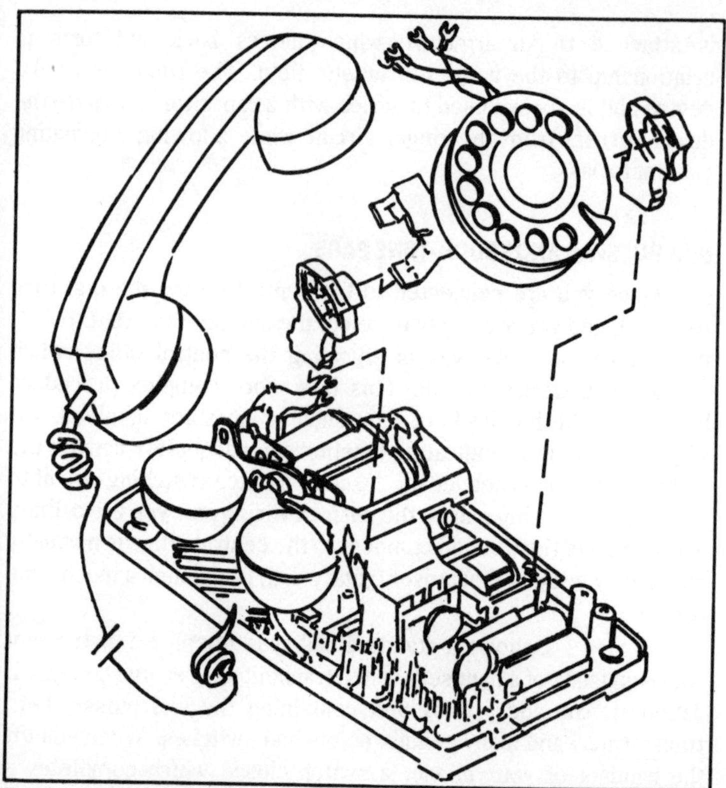

Fig. 1-1. Dial telephone instrument consisting of handset, dial pulser, and base (courtesy of GTE Automatic).

The Dial Tone

When the handset is lifted, a complex switch which contains a group of normally open and normally closed contacts is triggered and activates circuits inside the main body of the telephone instrument. Some of these switches connect the speech network (the transmitter and the receiver) to the telephone loop, and DC begins to flow. This current trips a relay in the central office. This switching action connects the loop to various other central office circuits. The dial tone is returned to the caller in the form of alternating current corresponding to the frequency of the audio tone which is converted by the receiver in the handset.

The rotary dial, although mechanically complex, is simply a switch which rapidly opens and closes the line circuits to signal the central office and transmit the number of the called party. Opening and closing a line five times signals the office that the number 5 has been dialed. Opening and closing the line eight times indicates the number 8. In many systems a number may be coded to the central office simply by pushing the plunger, which is closed by the handset resting in the cradle, up and down a number of times to correspond to the desired number. This of course is a non-exacting way of indicating a number which is why the rotary dial is used. When you turn the dial, an internal spring is compressed and auxiliary set of contacts called shunt springs are actuated. These switch some circuits which effectively remove the receiver portion of the handset from the line during the dialing process. When the dial is released, a ratchet mechanism engages and a cam rotates to open and close the contacts at an evenly spaced rate. This pulses the loop signaling the central office. A governor controls the return of the dial so that only six break pulses are conveyed to the central office at the correct rate of speed which is 10 pulses per second. An interruption of one sixteenth of a second is maintained between each pulse. This means that the contacts are open during pulsing for longer periods of time than they are closed. Figure 1-2 shows a typical rotary dial mechanism.

The ratio of open time to closed time is referred to as percent break, and the usual requirement is for the dial to produce a pulse of approximately 61% break. So, if you dial a 6, the central office will receive a coding of an open period of .061 seconds followed by a closed period of .039 seconds, then another opening of .61 seconds. This sequence will be repeated three times for the number six. Rotary dials must be able to pass rigid quality control inspections to insure proper operation. It can be seen that if the spacing and cam mechanisms should be even slightly out of order, then improper numbers will be coded through to the central office. Each dial consists of a gear train, mainspring, numeral ring, contact spring assembly, and a station number card which are mounted on a rigid metal base. Leads coming off the rotary dial are attached to the line and to the internal circuitry of the telephone. Most rotary dials have four leads coming out of their switching mechanism for these attachments.

A more modern means of central office coding is through a push button electronic switch. Often called the touch-tone pad, this is a signal system that is rapidly replacing the rotary dial. It has fewer moving parts and is mechanically simple. Instead of sending groups of pulses for the central office, touch-tone sends complex audio signals which are decoded at the central office. Each time a button is depressed, two precise tones are produced and transmitted to the central office. Again, two precise tones are transmitted simultaneously each time a button is pressed. The signaling could be handled effectively by only one audio number but this could lead to false decoding should line noise or other interference be present between your phone and the central office. For the central office decoding process to be accomplished, the two tones must be present. One portion of the decoding circuits locks onto the higher tones while another portion locks to the lower ones. If both tones are not present and exactly on frequency simultaneously, a number is not coded into the system. All of the tones used lie within the human voice frequency spectrum, so the possibility of interference to a single tone would be great. This is another reason for the dual tone system. Table 1-1 shows the corresponding tones for the various digits on the push-button network.

The high and low tones are internationally standardized so the combinations indicated in the chart will be identical with those used throughout the world. The tone frequency must be accurate to within at least 1.5% of their stipulated values under all conditions of temperature. This accuracy is achieved by the internal electronic circuitry which uses a method of digital division. A quartz crystal is often used as the central frequency reference source. Figure 1-3 shows the touch-tone pad mounted in a standard telephone unit. It has a 12 button matrix and uses highly reliable

Fig. 1-2. Inside view of a rotary dial mechanism.

Table 1-1. Corresponding Tones for Digits in the Push-Button Network.

Digit	Tone	Frequencies
0	941 and	1336 Hz
1	697	1209
2	697	1336
3	697	1477
4	770	1209
5	770	1336
6	770	1477
7	852	1209
8	852	1336
9	852	1477
*	941	1209
#	941	1477
	(Low group)	(High group)

oscillators normally controlled by integrated circuits. Only 10 of the push buttons are normally used. Two extra buttons are supplied for special future uses which may be localized to certain areas and phone companies. Some special systems may already use these spare buttons to commit certain numbers to memory circuits and to recall them upon command.

Some systems which have recently converted to the touch-tone method of number selection may not be fully through the transitional period, so the full advantage of touch-tone number activation is not fully realized. These systems can be easily detected because a clicking sound resembling that of a rotary dial will be heard after each number is punched into the system. These central offices are still using a basic dial system which is activated by the touch-tone signal arriving from each instrument. For this reason, these systems do not offer any speed advantages in connecting you to your dialed number but they do offer you the convenience of quickly punching in a number on your phone and then waiting while the dialing procedure it done electronically. Most fully touch-tone systems speed up the time it takes to ring your desired number.

While telephone instruments can come in many different sizes, shapes, and colors, almost all of them will have the basic internal parts described previously. Many may offer additional components to carry out their specialized functions; however, none can do without the basic items mentioned in this chapter. Any telephone device which is designed to provide voice communication must have a transmitter and receiver, a dial or touch-tone system for selecting the desired number, an indicator device to alert the person that someone is calling his number, and some form of equalization circuit. Without all of these components, the device will be a true telephone instrument as we know it.

Electronic Ringers

Some of the more modern instruments may combine these components and may change them to other forms. For instance, many telephone instruments have now done away with the mechanical ringers in favor of electronic circuits which usually provide an audible tone to a loudspeaker upon activation. These audio tones often resemble the ringing of the mechanical device but are more dependable because of the absence of moving parts. Moving parts always entail stress on the various components and eventual metal fatigue and inoperation. For instance, the mechanical ringer discussed earlier can only operate so many times before the device fails due to an armature break or some other mechanical difficulty. Fortunately, the lifetime of a ringer almost always outlasts the useful life of the telephone; however, dust and other foreign matter can become lodged between the clapper and the bell and cause inoperation. This is, again, due to the fact that moving parts are used in the old style ringer, and mechanically moving devices are usually not as dependable as non-moving electronic circuits. The latter are not as subject to inoperation or improper operation due to physical shocks and the collection of foreign matter. They can be damaged by improper voltages and sometimes a component will simply fail for no apparent reason. Generally speaking, however, purely electronic devices are more dependable than those of a mechanical nature.

The carbon microphone elements which have been standard for so many years in telephone handset transmitters are now being replaced, at least by the private companies, with smaller lightweight ceramic element microphones which more faithfully reproduce the human voice. Ceramic is not the only material used in these microphones, and a wide range of microphone selection is often available. Tiny condenser microphones may often be found along with others of more modern design, but the fact is that none

Fig. 1-3. Push-button dial mechanism mounted in standard telephone unit.

are quite as reliable as the old carbon element microphones. Carbon microphones are very sturdy and will withstand a great deal of abuse. Telephone handsets are normally dropped to the floor many times during their service life, and some of the more sophisticated microphone designs simply will not withstand this type of abuse. The carbon elements are generally inexpensive whereas other microphone elements may cost several times more. Carbon elements were originally used and still are today because of the many different places telephones are installed. With the coming of the personally owned telephone, the consumer had the choice of what to buy based on where it was to be located. This, of course, was and still is not very practical for the phone company who attempts to manufacture as few different instruments as possible, any of which can be used for almost every purpose. Today, even the phone companies are selling many different types of telephone instruments and have deviated from the ruggedized, carbon microphone elements in some of their models.

INSTALLATION

In many instances, it is no longer necessary to have your local phone company install a telephone device which you have purchased. This is provided that you already have a standard telephone in your home or at least the terminal block and line which has been properly installed by the phone company. Figure 1-4 shows the various types of terminal blocks and receptacles which are standard to most phone companies. Once installed, they allow for other devices to be substituted simply by removing the plug from the block and inserting the plug of the new instrument. Should the present phone company terminal block or receptacle not mate with the plug of a personally purchased unit you wish to install, you must have the phone company pay a service call to your home to modify or install the proper receptacle. It is against phone company rules for you to do this yourself. The only attachment you are allowed to make to their phone line is through a mating plug which you may insert into the receptacle. Any modifications or changes which must be made to accept another telephone instrument must be made by the phone company. You may, however, modify the plug on your purchased telephone to correspond with the existing receptacle which was formerly installed by the phone company. A source of phone connectors and associated hardware is available through most hobby outlets such as Radio Shack.

In most instances, you must notify the phone company prior to attaching any type of telephone instrument to their phone lines. This involves simply calling the business office and providing them with the information which is normally placarded on all FCC approved telephone equipment. In some areas, even the approved equipment may not be adaptable to the phone system. In these instances the phone company has the right to refuse to allow you to hook on. These instances will be very rare. Should you attach a purchased instrument to the phone line after notifying the phone company and then they find at a later date that your instrument is causing problems with the overall system, they can notify you by letter of the problem and demand that it be removed from their system.

In recent times, the leasing of telephones from the phone company has been slowly replaced by the purchase of individual instruments from various manufacturers and distributers. This can save the user money over a fairly short period of time because the phone is purchased on a one-time basis and thereafter is privately owned rather than leased. The popularity of owning your own phone would lead one to believe that this privilege had only recently come into being; however, it has always been possible to own your own phone rather than lease from the phone company, but this had not been generally publicized until recently. Additionally, the phone companies used to charge extra fees for adaptation devices which made private phone ownership uneconomical.

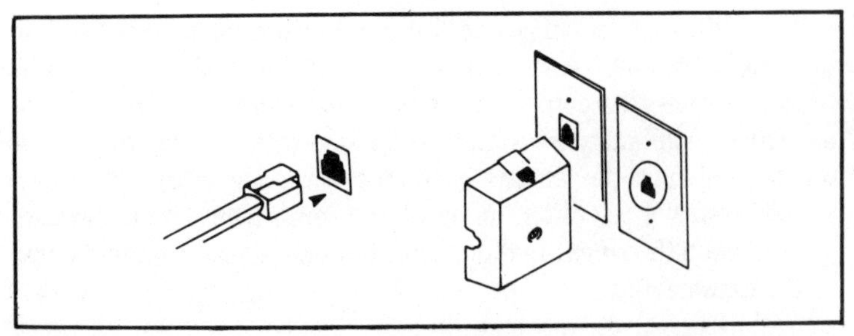

Fig. 1-4. Various terminal blocks and receptacles used for telephone connections.

The boom in private phone ownership came with the announcement of a Federal Communications Commission order that stated that anyone would be able to buy and install their own phone without having to pay a special fee as long as the instrument installed is FCC licensed and the phone company is notified before installation. All the phone companies need supply is the line and a connection jack.

The main concern with attachment of various devices to the phone line is the potential for damage to the network or the general lowering of communications quality as a whole from instruments which are not designed to operate with specific phone company installation. Under the present system, equipment which meets the FCC registration requirements has little potential for causing problems in the telephone system network. This registration does not necessarily indicate that the equipment is of high quality or that it will even work. What it does indicate is that the equipment has little or no potential for causing damage to the phone company's network.

Manufacturing companies who design and build various types of telephones and telephone related equipment must now register all of their new equipment with the Federal Communications Commission. Each company must prove through various schematic drawings and tests that their equipment is compatible with the phone system network. Once the application has been approved, the FCC issues the manufacturer a registration number which must be placarded on the device along with a ringer equivalence number. This latter number is needed to determine if a device can be connected to a specific type of line. In the United States there are over 50 possible ringer variations which vary from phone company to phone company. Variations will even occur between lines owned and operated by the same company. Improper ringer combinations connected to a noncompatible line can disrupt all service on that line and possibly cause problems at the central office. Some manufacturers overcome the problem that can result from improper ringer combinations by selling phones without ringers. These mostly fall into the extension phone category where the user simply wants a telephone in another room of the home and can depend upon the ringer in the master telephone to alert him to pick up the extension.

CENTRAL OFFICE REQUIREMENTS

Just as the FCC protects the consumer and the manufacturer by requiring certain specific tolerances and operational characteristics of telephone equipment, this government agency also protects the phone companies by placing requirements upon the consumer as well as the manufacturer. The telephone company is solely responsible for terminating the phone line in the home or office and for providing an outlet through which the line can be accessed. The customer has no authority to alter this line or termination in any way. His only legal access to the telephone line is through the receptacle. This is called the *interface* between the phone company and the customer. A stipulation is placed on the purchased equipment and on the manufacturer stating that all telephone lines must be terminated in an appropriate jack or plug for interface with the phone line. Additionally, if the instrument jack and the phone company outlet do not match, the phone company must be called to install an appropriate adapter at their outlet. A small fee is usually charged for this service.

While you as the owner of a telephone instrument have the right to use this instrument with your present phone line as long as it meets FCC requirements, if your device should be determined to be the cause of malfunction in the telephone system, the telephone company has the right to ask you to remove your telephone or they may discontinue service.

JACKS, CONNECTORS AND OUTLETS

Your local telephone company offers many different components for access to the telephone line. In recent years, the mini-modular jack has become the most popular but the older four prong plug assembly is still seen from time to time. Figure 1-5 shows some of the modular hardware which you are likely to encounter in telephone line termination. All of these plugs and connectors accomplish the same task, that of terminating the

Fig. 1-5. Modular hardware used in telephone line terminations.

telephone line in a manner which provides easy access. The ease of this access will depend on how your telephone instrument is terminated. Most instruments sold today use the mini-modular plugs which readily connect to newer telephone installations in a few minutes.

The FCC has adopted a group of standard plugs and jacks to be used in phone company installations and by the manufacturers of privately sold telephone instruments. Most telephone instruments are to be connected through 6 or 8 conductor modular plugs and jacks. If the telephone company makes any changes in its lines or switching equipment which may affect the operation of customer-provided equipment, it must notify the customer of its intentions. So, the possibility of the phone company coming to your home and changing all of the jacks is quite remote and cannot be done without prior notification. Neither can the phone company make changes in their equipment which would make your device incompatible without notification.

In some instances, there have been reports of phone companies refusing to allow connection of privately owned devices to their lines. This is often due to ignorance on the part of a few phone company employees as to proper procedures for handling these requests. Diplomacy plays an important part in these rare instances, and if permission is denied, it must be stated in writing from the company. From this point on, appeal to the FCC is possible.

TELEPHONE OWNERSHIP REQUIREMENTS

Just as the manufacturer and telephone companies must meet certain FCC requirements, the owner of a privately purchased telephone must also obey certain rules. Many older phone line installations will not contain accessible jack terminations. In other words, the line is terminated in a connector block to which separate conductors from the telephone have been attached by screw-in terminals. The owner of a private telephone is not permitted to remove wires from this termination and then connect the bared wire from his telephone instrument. In situations such as these, the customer must order the installation of a jack from the telephone company and must pay for this installation. If the customer already has a jack, he is legally required to contact the telephone company, stating his intent to connect a privately owned device and provide them with the proper information.

The owner of the private telephone is himself charged with making certain the device is properly operational. It does not take a highly skilled technician to know when most of these devices are not operating properly. The manufacturer's instruction manuals which are legally required to accompany each sale provide the information needed to identify possible problems. When an instrument you attach to the phone line is determined to be malfunctioning, you must immediately remove it from the line and not connect it again until proper repairs have been made.

Customer owned equipment may be connected only to private lines and not to party lines. These devices may not be connected to complex systems such as pay telephones, and many manufacturers prohibit connection of a wide assortment of devices to multi-line installations. These latter two systems normally involve specially wired telephones which use special apparatus to indicate to the central office which party on the line is making or receiving a phone call. Connection of a privately owned telephone could completely disrupt this encoding and decoding sequence.

UNREGISTERED EQUIPMENT

So far the discussion has centered around equipment which has been registered by the Federal Communications Commission as being suitable for connection to most phone lines. Amateur radio operators and experimenters often find it desirable to access telephone lines with specialized equipment which may not perform the same functions as a standard telephone. Amateur radio operators in the United States often run "phone patches" for servicemen and others in foreign countries, allowing them to talk to friends and relatives in the United States. Figure 1-6 shows how this is accomplished. Contact is established through radio transmitters and receivers. The output of the receiver is connected to the phone line while the output from the phone line is connected to the microphone input of the transmitter. When the party at the distant point speaks into the transmitter microphone, the received voice is fed directly to the phone line. When the party at the other end of the phone line speaks, the voice output is fed to the microphone input of the transmitter at this end of the conversation. The transmitter takes the input from the phone line and broadcasts it to the distant receiver. In this manner, communications between parties thousands of miles apart may be had for the price of a local phone call. An amateur radio operator is contacted in the same city or an adjacent city as the person with whom a conversation is desired. Once the amateur has established that receiving conditions are adequate to carry on a phone patch, he will contact the party by local phone. The connection to the phone line is made through a special circuit which matches the output of the receiver to the input of the line and the output of the line to the input of the transmitter. This circuit is usually called the phone patch.

Amateur radio operators enjoy building much of their own equipment. Phone patches often fall under this category, but most of these are not FCC registered devices, although anyone may obtain a licensing form and submit his or her circuit design for approval. Since there is a time element involved, this is rarely done by individuals, experimenters, or builders; however, you may connect non-registered devices through registered protective circuitry. The telephone company can often supply this type of device which is called a protective interface by some companies. Alternately, commercially available line connection devices can sometimes accept the input from phone patches and other specialized circuits. While the specialized circuits are not registered, the commercial equipment which attaches it to the line is and serves as a protective circuit.

If there is any doubt as to the feasibility or legality of any connection to the phone line, ask your local company officials for their advice. The author has found that most of these individuals

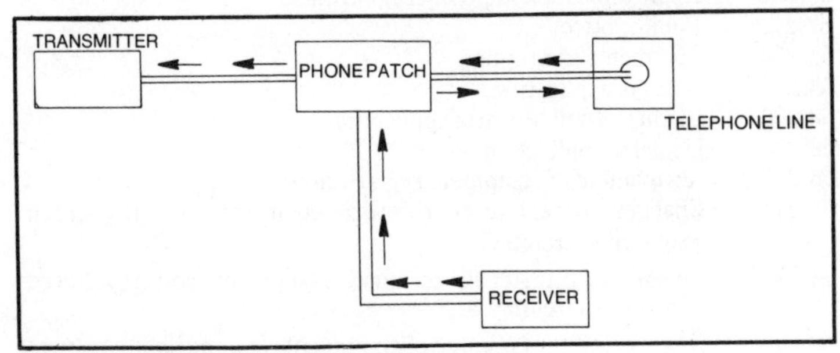

Fig. 1-6. Block diagram of a telephone patch system.

are very cooperative and with the influx of devices on today's market, they themselves are often confused as to what will work and what won't. In these instances, they will contact the proper technical source and get you an official ruling on what you can and cannot do with a specific piece of equipment.

FCC RULES AND REGULATIONS PART 68

In order to completely understand what your part as the owner of a private telephone may be in contacting your local telephone company and in staying within the law, the following is a partial reprint of FCC rules and regulations Part 68 which applies to the connection of privately owned telephone instruments and other devices to the telephone line. Each reader is requested to thoroughly familiarize him or herself with these regulations to avoid future complications and possible violations of the law.

PART 68 - CONNECTION OF TERMINAL EQUIPMENT TO THE TELEPHONE NETWORK

Subpart A - General

Sec.
68.1 Purpose.
68.2 Scope.
68.3 Definitions.

Subpart B - Conditions on Use of Terminal Equipment

68.100 General.
68.102 Registration requirement.
68.104 Means of connection.
68.106 Notification to telephone company.
68.108 Incidence of harm.
68.110 Compatibility of the telephone network and terminal equipment.

Subpart C - Registration Procedures

68.200 Application for equipment registration.
68.202 Public notice.
68.204 Comments and replies.
68.206 Grant of application.
68.208 Dismissal and return of application.
68.210 Denial of application.
68.212 Assignment of equipment registration.
68.214 Changes in registered terminal equipment and registered protective circuitry.
68.216 Repair of registered terminal equipment and registered protective circuitry.
68.218 Responsibility of grantee of equipment registration.
68.220 Cross reference.

Subpart D - Conditions for Registration

Subpart E - Complaint Procedures

Authority: Secs. 4, 201, 202, 203, 204, 205, 208, 215, 218, 313, 314, 403, 404, 410, 602, 48 Stat, as amended, 1066, 1070, 1071, 1072, 1073, 1076, 1077, 1087, 1094, 1098, 1102; 41 U.S.C. 154, 201, 202, 203, 204, 205, 208, 215, 218, 313, 314, 403, 404, 410, 602.

Source: 40 FR 53022, Nov. 14, 1975, unless otherwise noted.

Subpart A - General

Purpose. The purpose of the rules and regulations in this Part is to provide for uniform standards for the protection of the telephone network from harms caused by the connection of terminal equipment thereto.

Scope. (a) Except as provided for in paragraphs (b) and (c), the rules and regulations in this Part apply to the direct connection after May 1, 1976 of all terminal equipment other than coin telephone, and PBX and key telephone equipment to the telephone network, for use in conjunction with all services other than party line service, and to the direct connection after August 1, 1976 of all terminal equipment other than coin telephones to the telephone network, for use in conjunction with all services other than party line service.

(b) Unless otherwise ordered by the Commission, all items of equipment, other than PBX and key telephone equipment, of a type directly connected to the network as of May 1, 1976 may be connected thereafter up to January 1, 1977 and may remain connected for life - without registration, unless subsequently modified.

(c) Unless otherwise ordered by the Commission, all PBX and key telephone equipment of a type directly connected to the network as of August 1, 1976 may be connected thereafter up to January 1, 1977 - and may remain connected for life - without registration, unless subsequently modified. (41 FR 12664, Mar. 26, 1976)

Definitions. As used in this part:

(a) Direct Connection: Connection of terminal equipment to the telephone network by means other than acoustic and/or inductive coupling.

(b) Harm: Electrical hazards to telephone company personnel, damage to telephone company equipment, malfunction of telephone company billing equipment, and degradation of service to persons other than the user of the subject terminal equipment, his calling or called party.

(c) Interface: The point of interconnection between terminal equipment and telephone company communication facilities.

(d) Longitudinal Voltage: One half the sum of the potential difference between the tip connection and earth ground, and the ring connection and earth ground.

(e) Loop Simulator Circuit: A source of DC power and a load impedance for connection, in lieu of a telephone loop, to terminal equipment during testing. Figure 1-7 is a schematic drawing of the loop simulator. When used, the simulator shall be operated over the entire range of loop resistance as is indicated on the Figure, and with the indicated polarities and voltage limits. Whenever loop current is changed, sufficient time shall be allocated for the current to reach a steady-state condition before continuing testing.

(f) Metallic Voltage: The potential difference between the tip and ring connections.

(g) Multi-Port Equipment: Equipment that has more than one telephone connection with provisions internal to the equipment for establishing transmission paths among two or more telephone connections.

(h) One-Port Equipment: Equipment which has either exactly one telephone connection, or a multiplicity of telephone connections arranged so that no transmission among such telephone connections within the equipment is intended.

(i) Power Connections: The connections between commercial power and any transformer, power supply, rectifier, converter or other circuitry associated with registered terminal equipment or registered protective circuitry. The following are not power connections:

(1) Connections between registered terminal equipment or registered protective circuitry and sources of non-hazardous voltages.

(2) Conductors which distribute any power within registered terminal equipment or within registered protective circuitry.

(3) Green wire ground (the grounded conductor commercial power circuit which is UL-identified by a continuous green color).

(j) Registered Protective Circuitry: Separate, identifiable and discrete electrical circuitry designed to protect the telephone network from harm, which is registered in accordance with the rules and regulations in Subpart C of this part.

(k) Registered Terminal Equipment: Terminal equipment which is registered in accordance with the rules and regulations in Subpart C of this part.

(l) Telephone Connection: Connection to telephone tip and ring and all connections derived from telephone tip and ring. The term "derived" as used here means that the connections are not separated from telephone tip and ring by a sufficiently protective dielectric barrier.

(m) Telephone Network: The public switched telephone network.

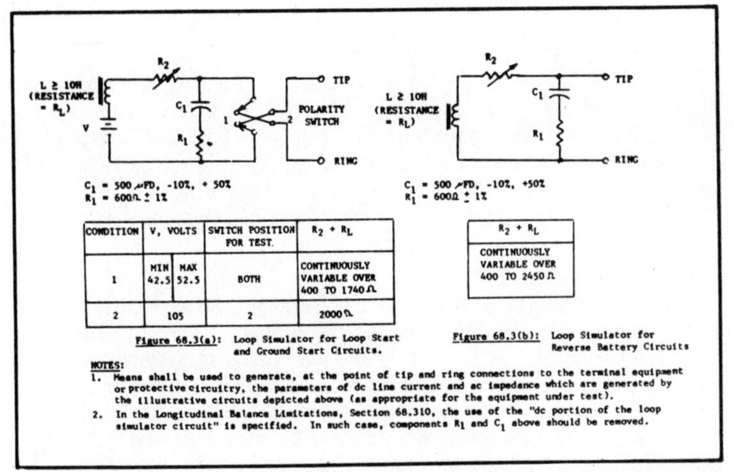

CONDITION	V, VOLTS		SWITCH POSITION FOR TEST	$R_2 + R_L$
	MIN	MAX		
1	42.5	52.5	BOTH	CONTINUOUSLY VARIABLE OVER 400 TO 1740 Ω
2	105		2	2000 Ω

$R_2 + R_L$
CONTINUOUSLY VARIABLE OVER 400 TO 2450 Ω

Figure 68.3(a): Loop Simulator for Loop Start and Ground Start Circuits.

Figure 68.3(b): Loop Simulator for Reverse Battery Circuits

NOTES:

1. Means shall be used to generate, at the point of tip and ring connections to the terminal equipment or protective circuitry, the parameters of dc line current and ac impedance which are generated by the illustrative circuits depicted above (as appropriate for the equipment under test).
2. In the Longitudinal Balance Limitations, Section 68.310, the use of the "dc portion of the loop simulator circuit" is specified. In such case, components R_1 and C_1 above should be removed.

Fig. 1-7. Loop simulator circuits as shown in Part 68 of the F.C.C. Rules and Regulations.

Subpart B - Conditions on Use of Terminal Equipment

General. Terminal equipment may be directly connected to the telephone network in accordance with the rules and regulations in Subpart B of this Part.

Registration Requirement. Terminal equipment must be registered in accordance with the rules and regulations in Subpart C of this Part, or connected through registered protective circuitry, which is registered in accordance with the rules and regulations in Subpart C of this Part.

Means of Connection. (a) General. Except for telephone company-provided ringers and except as provided in Subsection (c), all connections to the telephone network shall be made through the standard plugs and standard telephone company-provided jacks, or equivalent, described in Subpart F, in such a manner as to allow for easy and immediate disconnection of the terminal equipment. Standard jacks shall be so arranged that, if the plug connected thereto is withdrawn, no interference to the operation of equipment at the customer's premises which remains connected to the telephone network shall occur by reason of such withdrawal.

(b) Data Equipment. Where a customer desires to connect data equipment which has been registered in accordance with 68.308(a) (4) (i) or (ii), he shall notify the telephone company of each telephone line to which he intends to connect such equipment. The telephone company, after determining the attenuation of each such telephone line, between the interface and the telephone company central office, will make such connections as are necessary in each standard data jack which it will install, so as to allow the maximum signal power delivered by such data equipment to the telephone company central office to reach but not exceed the maximum allowable signal power permitted at the telephone company central office.

(c) (Reserved)

(41 FR 28600, July 12, 1976)

Notification to Telephone Company. Customers connecting terminal equipment to the telephone network shall, before such connection is made, give notice to the telephone company of the particular line(s) to which such connection is to be made and shall provide to the telephone company the F.C.C. Registration Number and the Ringer Equivalence Number of the registered terminal equipment or registered protective circuitry.

(41 FR 8048, Feb. 24, 1976)

Incidence of Harm. Should terminal equipment cause harm to the telephone network, the telephone company shall, where practicable, notify the customer that temporary discontinuance of service may be required; however, where prior notice is not practicable, the telephone company may temporarily discontinue service forthwith, if such action is reasonable in the circumstances. In case of such temporary discontinuance, the telephone company shall (1) promptly notify the customer of such temporary discontinuance, (2) afford the customer the opportunity to correct the situation which gave rise to the temporary discontinuance, and (3) inform the customer of his right to bring a complaint to the Commission pursuant to the procedures set out in Subpart E of this Part.

Compatibility of the Telephone Network and Terminal Equipment. (a) Availability of interface information. Technical information concerning interface parameters not specified in this Part, including the number of ringers which may be connected to a particular telephone line, which is needed to permit terminal equipment to operate in a manner compatible with telephone company communications facilities, shall be provided by the telephone company upon request.

(b) Changes in telephone company facilities, equipment, operations or procedures. The telephone company may make changes in its communications facilities, equipment, operations or procedures, where such action is reasonably required in the operation of its business and is not inconsistent with the rules and regulations in this Part. If such changes can be reasonably expected to render any customer's terminal equipment incompatible with telephone company communications facilities, or require modification of alteration of such terminal equipment, or otherwise materially affect its use or performance, the customer shall be given adequate notice in writing, to allow the customer an opportunity to maintain uninterrupted service.

Subpart C- Registration Procedures

Application for Equipment Registration. An original and two copies of an application for registration of terminal equipment and protective circuitry shall be submitted on FCC Form 730 to the Federal Communications Commission, Washington, D.C. 20554. An application for original approval of an equipment type directly connected to the network on May 1, 1976 may be submitted as a short form application (unless the Commission specifically requests the filing of complete information). All other applications shall have all questions on the form answered and include the following information:

(a) Identification, technical description and purpose of the equipment for which registration is sought.

(b) The means, if any, employed to limit signal power into interface.

(c) A description of all circuitry employed in assuring compliance with Part 68 including the following:

(1) Specifications, including voltage or current ratings, of all circuit elements, whether active or passive, in that part of the device which provides the isolation means at the interface.

(2) A circuit diagram containing the complete circuit of that part of the device which provides the isolation means at the interface. If this portion of the device is subject to factory or field adjustment by the manufacturer's service agency, instructions for these adjustments shall be included. In addition, if the device is designed to operate from power supplied by electric utility lines, the circuit diagram shall also include that portion of the device connected to such lines, including the power supply to the internal circuits and whatever means are employed to isolate such utility lines from the internal circuits.

(3) If a service manual is submitted, and any of these items are covered therein, it will be sufficient to list the pages in the manual on which the information specified in the item(s) appear.

(d) A Statement that the terminal equipment or protective circuitry complies with and will continue to comply with the rules and regulations in Subpart D of this Part accompanied by such test results, description of test procedures, analyses, evaluations, quality control standards and quality assurance standards as are necessary to demonstrate that such terminal equipment or protective circuitry complies with and will continue to comply with all the rules and regulations in Subpart D of this Part. The Office of Chief Engineer may issue an OCE Bulletin describing acceptable test methods; other test methods may be employed provided they are fully described in the application and are found acceptable by the Commission.

(e) A photograph, sample or drawing of the equipment label showing the information to be placed thereon.

(f) Photographs, 8″ × 10″ of the equipment of sufficient clarity to reveal equipment construction and layout and labels for controls, with sufficient views of the internal construction to define component placement and chassis assembly. Photographs smaller than 8″ × 10″ will be acceptable if mounted on paper 8″ × 10″ and of sufficient clarity for the purpose. Insofar as these requirements are met by photographs or drawings contained in service manual or instruction manual included with the application, additional photographs are required only to complete the required showing.

(g) If the device covered by the application is designed to operate in conjunction with other terminal devices, to couple their signals from or to the interface, the application shall list the class(es) of such other devices, together with their pertinent specification details.

(41 FR8048, Feb. 24, 1976, as amended at 41 FR 10224, Mar. 10, 1976)

Public Notice. (a) The Commission will issue public notices of the filing of applications for equipment registration and the grants thereof. No grant will issue before 20 days from the date of the public notice of the filing of the application.

(b) The Commission will maintain lists of equipment for which it has granted registration and for which it has revoked registration.
(40 FR 53023, Nov. 11, 1975, as amended at 41 FR 8049, Feb. 24, 1976)

Comments and Replies. Comments may be filed as to any application for equipment registration within 20 days of the date of the public notice of its filing. Replies to such comments may be filed within 10 days of the date of filing of such comments. All comments must be served on the applicant and all replies must be served on all parties filing comments. An original and three copies of all comments and replies must be filed.
(41 FR 8049, Feb. 24, 1976)

Grant of Application. (a) The commission will grant an application for equipment registration if it finds from an examination of such application and other matter which it may officially notice, that the equipment will comply with the rules and regulations in Subpart D of this Part, or that such grant will otherwise serve the public interest.

(b) Grants will be made in writing showing the effective date of the grant and any special conditions(s) attaching to the grant.

(c) Equipment registration shall not attach to any equipment, nor shall any equipment registration be deemed effective, until the application has been granted.

Dismissal and Return of Application. (a) An application which is not filed in accordance with the requirements of this part or which is defective with respect to completeness of answers to questions, execution or other matters of a formal character, may not be accepted for filing by the Commission and may be returned as unacceptable for filing unless accompanied by a fully supported request for waiver.

(b) Any application, upon written request, may be dismissed prior to a determination granting or denying the equipment registration requested.

(c) If an applicant is requested by the Commission to furnish any additional documents, information or equipment nor specifically required by this Subpart, a failure to comply with the request within the time, if any, specified by the Commission will result in the dismissal of such application.
(40 FR 53023, Nov. 14, 1975, as amended at 41 FR 8049, Feb. 24, 1976)

Denial of Application. If the Commission is unable to make the finding specified in 68.206 it will deny the application. Notification of the denial will include a statement of the reasons for the denial.

Assignment of Equipment Registration. Commission equipment registration may not be assigned, exchanged or in any other way transferred to another party, without prior written notice to the Commission.

Changes in Registered Terminal Equipment and Registered Protective Circuitry. No changes in registered terminal equipment and registered protective circuitry shall be made without prior written notice to the Commission. If such change in registered terminal equipment or registered protective circuitry results in any change in the information furnished the Commission pursuant to 68.200, the grantee shall submit a revised application for equipment registration in accordance with 68.200.

Repair of Registered Terminal Equipment and Registered Protective Circuitry. Repair of registered terminal equipment and registered protective circuitry shall be accomplished only by the manufacturer or assembler thereof or by their authorized agent; however, routine repairs may be performed by a user, in accordance with the instruction manual if the applicant certifies that such routine repairs will not result in noncompliance with the rules and regulations in Subpart D of this Part.

Responsibility of Grantee of Equipment Registration. (a) In applying for a grant for an equipment registration, the grantee warrants that each unit of equipment marketed under such grant will comply with all the applicable rules and regulations in Subpart D of this part.

(b) The grantee or its agent shall provide to the user of the registered equipment the following:

(1) Instructions concerning installation, operational and repair procedures, where applicable.

(2) Instructions that registered terminal equipment or protective circuitry may not be used with party lines or coin lines.

(3) Instructions that when trouble is experienced the customer shall disconnect the registered equipment from the telephone line to determine if the registered equipment is malfunctioning, and that if the registered equipment is determined to be malfunctioning, the use of such equipment shall be discontinued until the problem has been corrected.

(4) Instructions that the user must give notice to the telephone company in accordance with the requirements of 68.106.

(c) When registration is revoked for any item of equipment, the grantee is responsible to take all reasonable steps to ensure that purchasers and users of such equipment are notified of such revocation and are notified to discontinue use of such equipment. (41 FR 8049, Feb. 24, 1976)

Cross Reference. Applications for registration of terminal equipment shall, in addition to the requirements of this subpart, comply with the provisions of Subpart L of Part 2 of this chapter.
(41 FR8049, Feb. 24,1976)

Subpart D - Conditions for Registration

Source: 41 FR 12673, Mar. 26, 1976, unless otherwise noted.

Labelling Requirements. (a) Registered terminal equipment and registered protective circuitry shall have prominently displayed on an outside surface the following information in the following format:

Complies with Part 68, FCC Rules
FCC Registration Number:
Ringer Equivalence:

(b) Registered terminal equipment and registered protective circuitry shall also have the following identifying information permanently affixed thereto:

(1) Grantee's name.

(2) Model number, as specified in the registration application.

(3) Serial number of date of manufacture.

Environment Simulation. Registered terminal equipment and registered protective circuitry shall comply with all the criteria contained in the rules and regulations in this Subpart, both prior to and after the application of each of the mechanical and electrical stresses specified in this section, not withstanding that certain of these stresses may result in partial or total destruction of equipment.

SUMMARY

Many telephones and telephone accessories are now available to the average consumer which were only dreamed of a few years ago. The connection of these devices to the already installed telephone line and terminal block is not nearly as taboo a subject as it was previously. In order to compete, even the phone companies are now selling telephones to their customers who desire this, and many are of the decorator type which offer many advantages over the telephone normally installed on a rental basis.

This chapter has endeavored to inform the reader of the basic operation of standard telephone systems. This understanding will be of more than theoretical usefulness if you delve heavily into the many aspects of telephone ownership and operation. Compliance with the federal rules and regulations along with telephone company rules has also been stressed because this is a most important aspect of this subject. Without this strict compliance, the public's right to purchase and connect privately-owned telephones to the public utility might soon be terminated.

While telephones and associated devices will differ in style, features, and performance, all of them should have several basic things in common. Each should be licensed by the FCC and contain the registration and ringer equivalence number if a ringer is used. Each should be terminated in a plug which can be directly attached to the phone company's receptacle, and each requires notification of the phone company *before* installation is attempted. By following all of these legally required steps, your privately owned phone will do exactly what you expect of it and what you purchased it for. All of this added convenience will be had with the permission of, if not blessing, of your local company. Don't try to cut any of these procedures. This would represent a blatant disregard for the rules and could result in the termination of your phone service or at worst the disruption of an entire portion of the communication capabilities in your area. Most of the FCC registered devices on the market are compatible with the majority of phone systems. Those which may not interface in certain areas can be modified or connected through a special company-supplied interface device. Always play fair with the company which has provided you with communication capabilities. I think you will find your local phone company will meet you more than halfway in every instance.

Chapter 2
Standard Telephones For The Home And Office

Having discussed the standard telephone instrument, the logical progression would be to identify and describe just what's available to the average consumer for use in the home. Many of these same instruments will also find varied uses in an office environment. Today, the telephone need not be a detriment, and the shiny black instruments, while still available, are often discarded and replaced with streamlined, appealing substitutes. Many of these more modern instruments offer decided advantages over their predecessors. Built-in electronic circuits may be included which provide many convenience features unheard of until only a few years ago.

ITT ROTARY DESK PHONE

ITT Personal Communications has long been regarded as one of the top manufacturers of telephone instrumentation in the world. Their products seem to cover the range from mundane to highly exotic. Starting at the bottom end of the scale, the instrument shown in Fig. 2-1 is typical of the majority of phone system instruments placed in the average home or office.

The rotary desk phone depicted is the familiar and popular standard desk model with a rotary dial. This instrument, like most, provides a control on the bottom which allows the owner to adjust the volume of the ringer. Often, high volumes will be used for office purposes and in homes which generate a higher than normal ambient noise level. More sedate environments will usually dictate the ringer volume being set to near minimum level.

An advantage of volume adjustment which may not often be realized is sometimes utilized in office situations where two or more separate telephone instruments are installed. While many offices may have a multi-line phone with selection switches to engage two or more lines (with maximums approaching ten or more), separate instruments of single-line construction are often advantageous for certain applications and purposes. A difficulty experienced in this latter situation is encountered when two or more phones are placed side by side on a desk. It's nearly impossible to tell which phone is ringing if all of the ringer volumes are set to the same level. True, if each phone is separated from the other by several feet, the selectivity of the human auditory senses can easily determine the correct phone to answer, but this amount of space is not always available and is rarely so when four or five separate phones must be used. The standard method of determining which phone is ringing is to place the hand on each instrument to detect the sound vibrations. Obviously, if you're busy enough to require four or five telephones, you probably don't have the time to go through the identification process as described. Some single-line phones can be fitted with lights which indicate a calling party, but this is often impractical if the answering secretary is on the other side of the room and must answer the phones by reaching across the back of her desk.

The author has a system which uses three single-line phones. One instrument has its ringer set for maximum volume, another medium volume, and the third is set for minimum volume. After a few days of becoming accustomed to the various ringer volumes, it is now possible to immediately identify which line is being called. This can even be done while in adjacent offices to the one containing the telephone instruments. This "staging" of ringer volumes is quick, convenient, inexpensive, and highly efficient.

The ITT rotary desk phone is available in several decorator colors, including white, beige and cocoa brown. It is interesting to note that this instrument is not even available in "basic black", which was a standard of not too many years ago. Installation is handled in the conventional manner, using the mini-plug receptacle which the phone company installs. If this set-up is already present at the installation site, disconnection of the original phone and connection of the ITT model should be accomplished in less than one minute.

Fig. 2-1. Typical telephone instrument using a rotary dial (courtesy of ITT Personal Communications).

ITT PUSH-BUTTON DESK PHONE

Very similar to the previous desk phone discussed, the ITT Push-button Desk Phone offers the convenience, speed, and flexibility which can only be had with push-button number selection. Whereas the rotary dial phone is mostly mechanical in internal design, push-button phones utilize solid-state circuitry to "punch-up" the desired number to be rung. Rather than being built from discrete components such as transistors, diodes, etc., the push-button phone uses one or more integrated circuits which contain many of these discrete device equivalents on a single, crystalline chip.

Pushing one of the buttons on the control panel causes a multi-tone signal to be generated and passed down the line to the phone company control point. These tones must be accurate to a very high degree. The transmission of simultaneous tones is a surer means of control than a method using only one tone. Devices at the phone company are activated only upon proper reception of two accurate tone signals. This prevents false triggering from line noises which lie within the audio spectrum. These might accidentally cause extraneous numbers to be incorporated into the number selection sequence.

Of course, speed in calling an outside number is the main advantage of push-button telephone systems. The correct number can quickly and easily be punched in, cutting about half the time off the dialing process. In systems set up to respond immediately to the generated tone sequences, the speed at which the selected number rings the phone being called is also substantially increased.

The ITT Push-button Desk Phone is shown in Fig. 2-2. This model is representative of most push-button desk phones in use in the United States. It closely resembles the rotary dial type of telephone previously discussed in style and appearance. Due to the replacement of the rotary dial, which is a mechanically-derived switching device with a more compact solid-state replace circuit, push-button phones often weigh considerably less than their rotary dial counterparts.

Fig. 2-2. ITT push-button desk phone (courtsy of ITT Personal Communications).

The same three color choices are available for the push-button phone from ITT, which are white, beige, and cocoa brown. The adjustable ringer volume feature is also included.

DIAL-IN-HANDSET TELEPHONES

Another type of telephone device which can be classified as a "standard" for the industry today is the popular dial-in-handset model. The ITT rotary dial version is shown in Fig. 2-3 and offers the owner the convenience of dialing the desired number from the same portion of the device which is used for transmitting and receiving voice communications.

It is quite possible that the trend toward installing long coiled cords on the handsets of telephones similar to the models previously discussed was the impetus to design the dial-in-handset models. As cords lengthened, it became possible for a homeowner to answer a phone in, for instance, the kitchen and then to carry on a conversation while going about the household routine of preparing dinner. This added a great deal of flexibility to telephone usage, but it was still necessary to return to the main body of the instrument to hang up or to dial another number. This was considered to be a bit of a nuisance, so the dial-in-handset models were developed to make telephone use even more flexible.

Fig. 2-3. Dial-in-handset telephone instrument (courtesy of ITT Personal Communications).

With these new models, it no longer is necessary for the caller to dial from the fixed or resting location of the telephone instrument. The long cord is still of full advantage. The owner can now pick up the handset, carry it to a comfortable easy chair or a work area and still have the full convenience of being able to complete one phone call, hang up, and then dial another. Of course, it is still necessary to return the handset to its permanently mounted cradle when all calling is completed.

Hanging the phone up from a remote point is accomplished through a push-button switch located just below the rotary dial mechanism. This switch automatically returns to the *on* position when released. This switch breaks the line, serving the same purpose as hanging up the handset with the previous stationary systems. Once the new dial tone has been established, the caller is free to dial another number without ever having to leave the original calling position.

Since the dial-in-handset models were developed, they have received excellent response from the consumer market, so the next logical step for the manufacturers was to offer push-button models as well. The ITT Push-button phone is shown in Fig. 2-4. A comparison can be made between the dial and push-button dial-in-handset phones in the same manner as was done with the earlier desk phones. Electronic circuits have replaced the mechanical rotary dial. The same basic integrated circuit design is used for replacement in this model. This adds more convenience and speed to the number selection process and often reduces the weight of the handset.

It should be pointed out that electronic circuits are generally much more reliable than their mechanical counterparts. Mechanical devices involve moving parts. Whenever objects move, especially when they are in direct contact with each other, friction is created which results in component wear. A mechanical part will eventually wear out. Electronic circuits do not usually incur friction wear. They generally have no moving parts so wear, as it is applied to mechanical components, is not a factor. The life expectancy of an electronic circuit is usually longer than the anticipated life of its housing. In other words, the electronic circuits used in push-button telephones will usually outlast the usefulness of the entire instrument. Breakage of the plastic case will probably occur before the electronic circuit fails, and the entire instrument will be replaced.

Fig. 2-4. Push-button version of previous telephone model (courtesy of ITT Personal Communications).

Rotary dial phones often last past case breakage, but the dialing mechanisms have been known to fail due to surfaces which have been worn to a point where the establishment of switching contacts is no longer reliable. Here, the rotary dial may have to be replaced, as repair is not usually practical, especially in older models. Another selling point for electronic telephone components is the replacement cost factor. Generally, it is far less expensive to replace an entire electronic circuit, which is usually housed in one tiny chip, than to replace a complicated rotary dial mechanism. The replacement will also be less expensive, as the labor time is substantially reduced.

While electronics is depended upon to deliver the tones required for push-button systems, the author has purposely made a distinction between standard push-button instruments and what are to be later referred to as electronic telephone instruments. The latter will be thoroughly discussed in another chapter and generally depend upon electronic circuitry to provide many more functions than dialing a number in the manner already described. It should be understood, however, that all touch key or push-button phones are electronic devices or, more appropriately, combinations of mechanical and electronic devices.

While only a few models of standard home and office telephone instruments have been pictured and discussed so far, many similar models are available from different manufacturers. Some may be slight modifications of the styles shown, but all of them have been tailored around the same physical design configuration.

It is interesting to note the added versatility of various model sequences. For instance, the desk phones discussed in this chapter can be used or mounted in one position only. They were designed for the sole purpose of resting on the surface of a desk or other similar piece of furniture. But it can be seen from the figures depicting the dial-in-handset models that they, being newer in design, boast more versatility in mounting configurations. Since the handset contains ninety percent of the working parts, the cradle simply serves as a convenient resting place when the handset is not in use. Placement of the handset in its cradle also disengages the line, arming the telephone to receive an incoming phone call. The cradle is the element of increased mounting versatility in this case, as it is often designed to rest on a desk surface, as were the previous models, but can also be wall mounted. A convenient tap or hole which enables the cradle to be hung, picture fashion, on a flat wall surface is located through the plastic base. Should the phone be originally purchased for desk use, the same instrument could later serve its owner as an out of the way wall-mounted telephone. Previously, telephones were made only for desk mounting or for wall mounting, but not both.

This move toward versatility, both in telephone uses and services and in physical or mechanical positioning and mounting, is a major cause of the many exciting telephone instruments and accessories available on today's market. It is stressed that the phones discussed so far in this chapter do not differ by extreme amounts from the phones of twenty years ago, with push-button dialing being the most notable improvement. Style is another, and, by comparison, those twenty-year old phones are very ugly, an interior decorator's nightmare.

FURTHER CHANGES AND MODIFICATIONS

Lest the reader believe that these changes under discussion are the only ones that have been put into effect for the standard telephone instrument as it was originally designed, it is important to note still further innovations in style and utility which are found in instruments offering basically the same functions as those already discussed.

Figure 2-5 shows a rotary dial telephone instrument which differs greatly in style from other models. This instrument does not closely resemble earlier telephones, as do the previously discussed models. Named the *Ultraphone* by its manufacturer and distributor, ITT Personal Communications, this modern phone features an angled, low-profile appearance and a handset that fits in the base instead of being perched atop. A recessed handle in the base allows it to be carried in the wall-mounted position. The recess can be used to hold the handset when the owner does not wish to hang up. Measuring 12″ by 7 ¾″, the Ultraphone is available in white and ivory finishes. Another optional model of this same design is offered in brown handcrafted leather covering a beige finish.

This instrument has the advantage of a desk phone, while still providing the versatility of the handset-enclosed dial model which might also be wall mounted. It is light enough to be carried from calling position to calling position, so the ability to dial from a

Fig. 2-5. Ultraphone desk model (courtesy of ITT Personal Communications).

Fig. 2-6. Ultra contemporary styling in the viva phone (courtesy of ITT Personal Communications).

remote location is still available. It also offers streamlined styling, which many consider to be more modern than even the dial-in-handset models.

It is very interesting to learn what some manufacturers have done with the standard desk-mounted only telephone while still retaining much of its original features. The picture of the ITT *Viva* phone in Fig. 2-6 shows that a little exterior modification can make an old design look as new as tomorrow.

ITT advertises this model as a desk phone with contemporary styling and dramatic color accents in earth-tone shades. The interior components are probably exactly like those of the earlier desk phones and may even be identical to models the reader may presently have in his or her home.

While it is true that the Viva phone would not blend well with an Early American home or office decor, its sleek lines and brash styling would certainly compliment a contemporary living room or office. Whereas early telephones were devices to be hidden away or, at the most, hopefully blend with the room decor, these newer models are expressions unto themselves and are styled to add to their surroundings instead of simply blending with them.

The ITT Viva phone is available as a rotary dial instrument only. One model uses a brown faceplate and base stripes, while another model is available with a red faceplate and base with multiple red side stripes. As is the case with all telephones mentioned in this book, installation usually involves the removal of the mini-plug attached to your old phone and to the mini-plug wall jack. After this step is completed, the owner merely plugs the jack of the new model into the present wall outlet. This process usually takes less than a minute, but the results in room styling can be tremendous.

Push-button systems, too, may take many different forms. Figure 2-7 shows a model which is similar to the Ultraphone. This is a push-button, multiline version which offers the same convenience features and styling as its predecessor. Intended mainly for business office uses, this model does not weigh appreciably more than the Ultraphone and can be placed on a desk or hung on the side of a wall. It is available in rotary dial versions as well for areas where push-button number selection is not available.

Figure 2-8 shows the ITT *Rotary-Dial Delta Phone*, which takes streamlined styling a step further. Much of the excess material in the plastic handset has been cut away to arrive at a very slim package. This style of phone is very similar to telephones depicted in several futuristic science fiction movies and television programs. Following other trends, the Delta model may be installed on the top of a desk or table, or it may be mounted vertically on the surface of a wall. This multi-line phone will accept up to four telephone lines which are selected by four push-button switches. A fifth switch is used to place the various lines on hold. This latter mode allows the owner to engage other lines while keeping the hold line open. When the other call or calls have been

Fig. 2-7. Push-button version of the Ultraphone (courtesy of ITT Personal Communications).

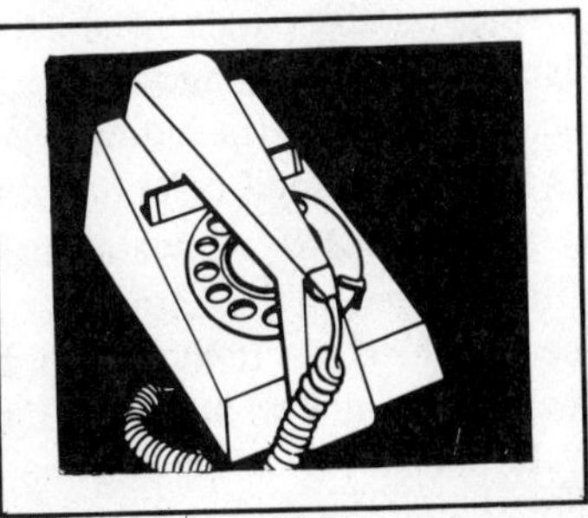

Fig. 2-8. Rotary dial Deltaphone (courtesy of ITT Personal Communications).

handled, the first line may be re-engaged to the handset by simply depressing the desired line button again.

Getting back to the more conventional designs, the ITT *Domino* model shown in Fig. 2-9 is more representative of the modification of more standard telephone design. So named because of its black-on-white finish, the Domino's physical appearance is somewhere between the standard desk model and the Viva phone discussed earlier. Intended for contemporary home and office uses, this instrument can be considered a specialty item to some, mainly because of its two-tone finish. It is interesting to note that even the handset is finished in black and white and blends with the base when in the cradled position.

PERSONALIZED TOUCHES

While many variations in telephone styles, colors, and tonal blends are available from the many telephone instrument manufacturers, it is a relatively simple task to customize the telephones which you own. Again, this should only be attempted with instruments which have been purchased by you and not with rented devices from your local or area company. Spray-on paints and enamels are available from local hardware and painting supplies stores at very reasonable prices. These can be used to re-color your present telephone to make it accentuate or blend with a change in your interior decorating. Also, large savings may be had by purchasing telephone instruments from surplus distributors. These items are usually serviced and replacement parts added where needed, but there are often scratches and blemishes on the plastic cases which is all the average person ever sees. Many of these purchases will be the old, basic black desk phones, which are perfectly usable but which do little for a room's decor. Any of these surplus purchases can be quickly transformed into very decorative, useful items by adding a coat of paint. The total process should take only a few hours (including drying time) and should cost less than two dollars if single color modifications are attempted.

Figure 2-10 shows a desk phone with the cover removed. Most models in this category will have two bolts in the base which secure the cover to the frame. By removing these bolts, the cover will easily slip off and painting processes can be started. Figure 2-11 shows the bottom of the desk phone. Notice the mounting position of the two bolts. One is usually located toward the right side of the top or back of the base, while the other will be found near the left side at the opposite end. Usually, these bolts cannot be

Fig. 2-9. The Domino model exhibits a stark appearance with a black-on-white finish (courtesy of ITT Personal Communications).

Fig. 2-10. Push-button desk phone with cover removed.

completely removed but will unscrew to a point where the cover will slide off. This prevents the bolts from being lost.

The author recommends acrylic enamel spray paint, which is durable, inexpensive, and very easy to apply. It also dries quickly and sets up into a hard finish. Place the case on several sheets of newspaper to prevent accidental staining of the floor or other surface. It is also a good idea to set the case up on tiny wood blocks, raising the edges off the newspaper surface. Remove all grease, dirt and other foreign matter from the surface of the case by using a household cleaner. Wipe the surface clean with a lint-free cloth or towel and allow adequate time for the case to dry. Should moisture be present when the painting process begins, the finished surface may crack or run. Wet areas can also cause the paint to dry unevenly and create a blotchy finish that can never be properly matched. When this occurs, repainting is the most practical cure.

The proper use of spray paints can best be assured by carefully reading the container label. Most will tell you to shake the can until a mixing ball inside breaks loose and rattles. Once this has occurred, the container should be vigorously shaken for at least one minute. This agitation procedure should be repeated for about fifteen seconds after each minute of spraying time.

Make sure that the temperature of the room is at least 60 degrees when painting the phone case. If the case has been stored outside, allow it to reach room temperature before beginning. Hold the can about 12 inches from the surface of the case and depress the spray nozzle. Move the container back and forth over the surface to prevent heavy concentrations of paint from being applied to a limited area. When painting with a spray can, a large quantity of the paint is going to fall on the paper or spray over or under your intended target. This cannot be helped. Any attempts to place the

Fig. 2-11. Bottom of desk phone showing bell volume adjust.

spray nozzle closer to the case to concentrate more paint on the work surface will result in uneven paint distribution. Most manufacturers recommend two thin coats of paint on the work surface rather than a single heavy coat. The number of coats necessary for your case will depend on the original color and the color of the spray paint. Obviously, it will take many more coats to change an original dark-toned case to a lighter color than it would to switch from one light color to another. While it can be done, it is very difficult to convert a black phone to a snow white color, but the same black phone can easily be switched to dark red, green, blue, etc. Figure 2-12 shows the proper positioning of the spray can in relation to the telephone case.

Once the base portion has been painted, it may be set aside to dry. The next work area will involve the handset. In more modern phones, this part can be removed from the base frame and its coiled cord by disconnecting the mini-plug at the transmitter base (the

Fig. 2-12. Proper positioning of spray can for applying paint.

lower end of the handset). Older models may require some soldering to properly remove the cord, although many also offer set-screw contacts which can be removed with a small flathead screwdriver. It is absolutely necessary to remove the handset from the cord to prevent the painting of undesired surfaces. It will probably be necessary to replace the cord anyway with one of a matching color. Figure 2-13 shows how the mini-plug is removed from the handset.

When the handset has been freed, unscrew the earpiece and mouthpiece which protect the receiver and transmitter elements. The bottom (transmitter) element should simply fall out of its housing once the cover has been removed, as it usually is connected to the phone line through pressure contacts. Figure 2-14 shows the freed element and the pressure contacts which still lie within the handset housing. Remove the element and set it aside for the moment. Leave the mouthpiece cover off as well.

The earpiece element will require a little more effort for proper removal. After the cover is removed, shake the element loose from the well. When it comes free, you will see two wires attached to screw contacts on the element back (Fig. 2-15). Using a tiny jeweler's screwdriver, loosen the two screws just enough to allow the contacts to slip free. Do not pull too hard on the wires which are attached to the contacts, as they may separate. Set the disconnected element aside. Do not replace either cover.

Returning to the other end of the handset, remove the circular case which contains the pressure contacts for the transmitter

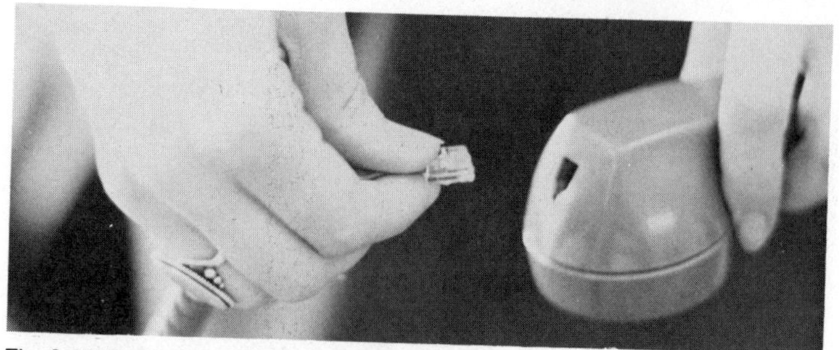

Fig. 2-13. Removal of modular plug from handset.

element. This process is shown in Fig. 2-16. The wires at the back of this part attach to the earpiece element when in use, but the removal of this element will allow the wires and the transmitter well to slip completely from the handset. Set this assembly aside until the painting is finished and you're ready to reassemble the telephone.

Using the same spraying techniques as before, cover all visible surfaces of the handset. It is emphasized that the element covers are not to be screwed back into place during the part of the procedure under discussion. These elements are to be painted separately after the handset is finished. Be certain to cover all areas of the handset. This part of the telephone usually has more curved surfaces than the base. Direct only a very light coat of paint to the threaded areas around the ear and mouthpiece.

When the handset is completed, set it aside for drying. If excess paint is noticed around the threaded areas, remove most of it with a slightly dampened cloth. Touch up if necessary. Now, paint the two element covers in the same general manner. Again, be careful not to overpaint the inside, threaded areas of these pieces. After their surfaces have been completely covered, take a tapered toothpick and remove any paint globs which may have covered the many holes through which sound is transferred. This process may also be done after the paint has thoroughly dried, should you miss any residue in the first attempt.

Depending on the phone you are refinishing, no further work may be required, but many types have the standard base coloring also applied to sections of the rotary dial exterior or to the faceplate which covers the bottom of the push buttons in models with this type of number selection. Rotary dial phones may require the touching up of the dial face by using a small paint brush. Spray an adequate amount of paint from the can on a non-absorbent surface, then dip the brush in the paint and apply it to the smaller surfaces of

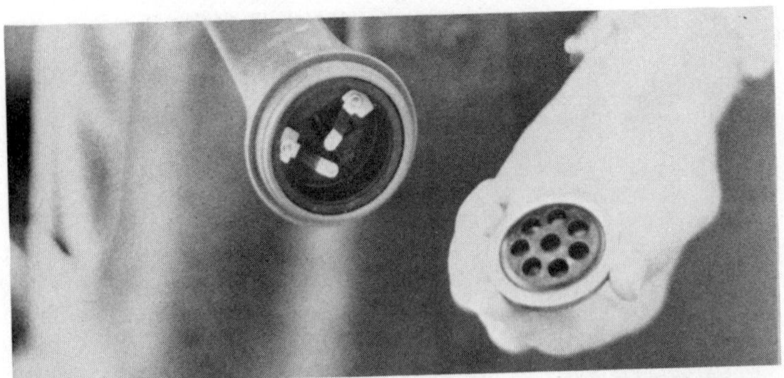

Fig. 2-14. Transmitter element removed from handset. Note the pressure contacts which still lie within the handset housing.

the dial face. Masking tape may be applied in small strips to the metallic surfaces which are not to be painted to protect them from accidental contact with the brush.

Modern push-button telephone instruments have a plastic, removable faceplate which slips off with the rest of the case. This can be easily unsnapped when the case is removed or may be left in place during the base painting procedure. Figure 2-17 shows how the faceplate is most easily removed. A screw-in mounting bracket inside the case is first removed which releases the holding clamp on the outside of the faceplate. Brush painting of the faceplate is recommended, because most models have a special seamed design. These seams will tend to fill with paint when spraying, although the excess may be easily wiped away. If the spraying method is opted for, make certain that a very light coat is initially applied.

Fig. 2-15. Removal of receiver element.

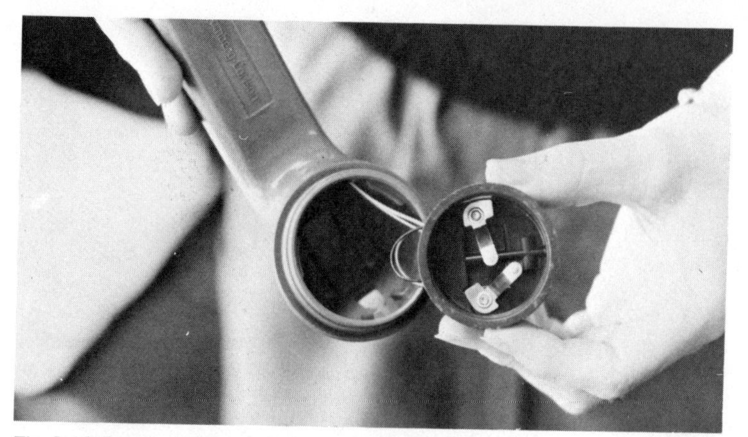

Fig. 2-16. Removal of transmitter element well from handset.

While painting of telephone instrument cases is very simple and involves a minimum of time, effort, and the taping of areas where paint is not to be applied, care must be exercised to arrive at a professional appearing final product. Some readers may choose to further disassemble the case in order to avoid taping of areas which are not to be painted. The cradle switch may fall into this category. This switch is the one which is closed when the handset is hung up on the base. Usually, two clear plastic nipples protrude from the base wells and are pushed downward by the weight of the handset when hung up. These engage the switch arm which is metallic and mounted in the base. For a truly professional job, the plastic nipple assembly should be removed. Fortunately, this takes only a minute or so and requires only a medium-sized flathead screwdriver. Figure 2-18 shows the screw position in the case. The entire activating mechanism should slip easily from the case when the screw is removed and can be set aside until after the main painting has been completed. Figure 2-19 shows the assembly sliding from the case. While this last step is not essential to the appearance of the painted case, it does make the job a bit easier. Alternately, small pieces of cotton or newspaper may be pressed down over the nipples and into their housing wells. However, a close inspection of this area after painting will reveal the presence of the former finish.

The line cord and handset cord, as was previously mentioned, cannot be recolored in a practical manner. Most attempts at painting will result in uneven finishes which crack as the cord insulation is flexed. The only means of matching a cord to the new finish is to buy a new one. For this reason, customized colors or color schemes should be thought out ahead of the actual painting procedure. Colors should be chosen to match or blend with the colors of easily obtainable, commerical line cords. These can be purchased from many different hobby stores, telephone outlets, and directly from most phone companies. It may be possible to dye an old line cord to closely match a customized case coloration. This will depend upon the original color of the cord and the color to which it is to be changed. Check with your local hardware store for advice in this area.

REASSEMBLY

Once the painting has been completed and the finished work allowed to dry, the instrument may be reassembled. Small areas of

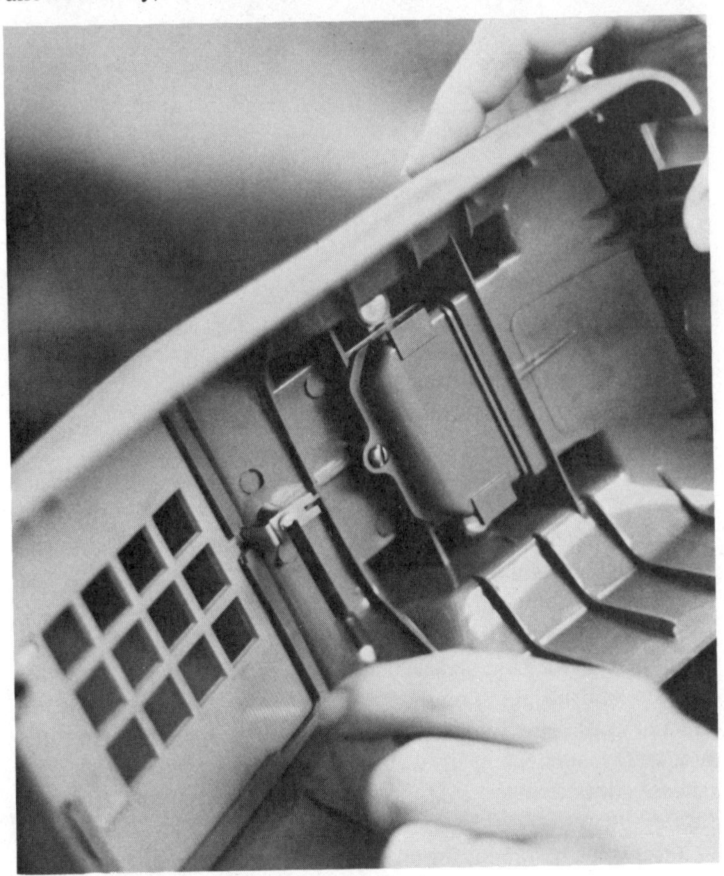

Fig. 2-17. Removal of bracket holding faceplate.

Fig. 2-18. Location of cradle switch set screw.

paint along the mating surfaces may have to be lightly scraped in order to facilitate assembly. The fit may be rather tight on some models, but careful cleanup of heavy coats of paint will aid in this stage.

Start first with the handset. Push the receiver element wiring down through the case and reconnect this element by inserting the contacts and re-tightening the set screws. Reseat the transmitter element well. Now, screw on the receiver earpiece. Place the transmitter element in its well and attach the cover. Insert the handset cord connector in its receptacle. Set the handset aside for the moment.

Replace the cradle switch contactor assembly and the face-plate if these items were removed. Align the cord receptacles and place the case down over the frame. Move this case from side to side and back and forth until it rests into place. Tighten the base screws.

The line cord may now be reinserted along with the handset cord. Test the cradle switch to make certain it moves unhindered and activates the switch arm properly. For push-button instruments, make certain the buttons move easily within their plastic faceplate grids. If all operates normally at this point, the job is complete. Recheck your work in a few weeks, looking for any signs of peeling enamel or scratched finish. These small blemishes can be touched up with a small brush and the same color or colors of paint used in the original painting scheme.

Your phone may now be attached directly to the line and used exactly as before, or you may wish to go even further and purchase

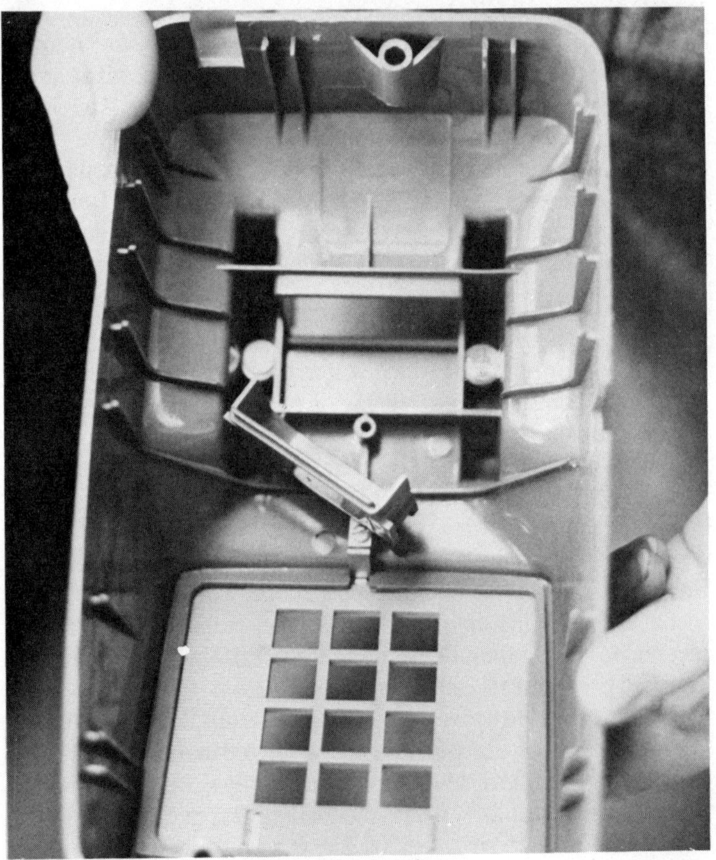

Fig. 2-19. Cradle switch assembly sliding from case.

plastic applique with adhesive backings which may be applied to certain portions of the case. These come in a variety of patterns and color combinations. Some of the forms include flowers, butterflies, bubbles and complex kaleidoscope patterns. The sky is the limit. Today, almost anyone can custom paint and design the telephone instrument to be an attractive addition to any room decor while still preserving the original purpose of the device. If the finished product turns out to be less appealing than imagined, one has only to start from scratch and repeat the entire process using different colors and patterns.

SUMMARY

It can be seen from this discussion that redoing the color scheme of your home or office telephone instrument is no more complicated (less so, in fact) than varnishing a piece of furniture. It is a job that can be expertly accomplished by almost anyone who knows how to use spray paint and can do touch-up work with a small paint brush. The reader is cautioned that these customizing procedures can only be carried out on telephone instruments which have been purchased from a retail store or from your local phone company. You must own the telephone. To perform this work on a rented phone which your local office merely supplies for your use is against telephone company rules. No modifications can be made to these latter devices.

Always choose a color scheme which allows you to purchase a handset and/or line cord which will match, and use a paint which will not flake easily. Again, enamel is to be preferred for its hard, lustrous finish. Take special precautions to remove all of the telephone circuitry from the direct area where the painting is taking place. While any electronic components will not be harmed by accidentally covering with paint, mechanical parts may be completely ruined or, at least, rendered temporarily inoperable. Be professional in your painting operations. The sole purpose for changing the original color is to make the instrument conform to or accentuate your other home or office furnishings and fixtures. Finished instruments which show tell-tale signs of splotched finishes, uncovered areas, portions where paint has been allowed to run and surfaces which should not be painted immediately cancel out all of your intentional efforts. It takes only a little longer to do the job correctly, and even this amount of time is microscopic when compared to the many years of use you will receive from your project.

Chapter 3
Decorator Phones

It has already been stated that the modern telephone has been redesigned by many manufacturers and has reached a stage of development where it adds to the decor of its surroundings rather than being an unattractive necessity of modern life. It is important to note that the decorator telephone has been with us for a long time.

It is unusual to note that the decorator phones of years gone by, often called French phones, are coming back. On the outside, they resemble most accurately the phones of years gone by, while on the inside, modern mechanical and/or electronic components assure completely modern operational advantages.

Most of the decorator phones in this chapter are supplied by many different manufacturers. While some will have slight mechanical differences, they will closely resemble the same model line from manufacturer to manufacturer. To stress availability, our local Radio Shack store was chosen in order to see just what could be obtained by the average citizen in the way of classically-styled decorator phones. Many, many different models were found, along with this company's usual line of highly sophisticated electronic telephone instruments and attachments. There are certainly many other suppliers of these models under different product names, but these are representative of what is generally available on the market today.

These phones are compatible with most telephone systems throughout the United States. As usual, it is necessary to contact your local company and notify them of your intent to install a privately-owned instrument and to provide them with the proper information about the device to be installed. With very few exceptions, these phones will interface with almost every system.

CLASSIC TELEPHONES

Following traditional styling, most of the classic telephone instruments will be available only in rotary dial models. There are a few exceptions, however, which are able to blend the very old with the very modern and still present a very distinctive and tasteful appearance.

From a styling standpoint, it will take a very special person to properly design a room decor which will best be enhanced by these phones. To be quite honest, these types will not fit in well with many modern or traditional furnishings. They are attractive when properly placed, but they will stand out like sore thumbs when no forethought has gone into their selection for specialized situations. Recently, the author was visiting a friend who had purchased one for his den. The highly embossed gold-appearing exterior did absolutely nothing for the room, which was done up in a very macho style, featuring a lot of leather furniture and dark, masculine colors. He liked it, however, and this is what really counts to many people. Those who like French Provincial styling will find most of these classic decorator phones to be excellent as finishing touches for expertly decorated interiors.

At the Radio Shack we visited, the selection was excellent. Classic phones were found in a variety of styles and colors even within the framework of a single model type. Multiply this diversity with five or six classic model types, and one begins to understand just what kind of selection there really is.

Figure 3-1 shows the *Cutie*, which Radio Shack uses as the model name for this miniature French-style phone. It is very tiny as these phones go and almost resembles a child's toy phone upon first inspection. This model is available with a white base and handset grip. All metal work is gold colored. This is an ideal model for feminine styling and will go well on a small night table, bedside stand, or when placed on a vanity. The base is heavy enough to solidly lock the device into place on its resting surface, although the handset is extremely light and delicate in appearance. A closer inspection, however, reveals that this instrument, delicate though it may appear to be, is quite rugged and should easily withstand the normal abuses which telephone instruments must endure on an

Fig. 3-1. Cutie model of French-style phone from Radio Shack.

almost daily basis. Priced in the seventy dollar class, this phone would be a welcome addition to the teenaged girl's bedroom. It has been reported that this model is especially popular with this segment of the American population. The line connection cord is terminated in a standard modular plug which will immediately connect to most modern wall outlets installed by your local phone company.

Figure 3-2 shows a telephone instrument which is similar in appearance to the previous model discussed, but this phone is of more standard size. Called simply a *French-Style Phone*, Radio Shack sells it for about sixty dollars for the ivory model with gold trim. This is one of the few classic styled telephones which is available in several colors. For another ten dollars, the customer may purchase a model which is gold colored and trimmed in ivory.

Fig. 3-2. Radio Shack French-style phone of more standard size.

Both models are an attempt to capture the elegance and charm of old Europe. This is especially true when you note that instead of ringing, these phones contain continental-style buzzers which indicate an incoming call. This should take many readers back to the Charles Boyer movies, while others will only remember Tony Curtis in one of his European adventure-romances. Everyone has seen them at sometime, but most likely it was on television, at a movie theatre, or in a travel magazine.

A bit of modernization on the outside, as well as to the internal workings, is witnessed by the phone in Fig. 3-3. This is the same basic design as the previous example, except the rotary dial has been replaced with a push-button keyboard. This model will work with most rotary dial and touch-tone systems and still retains its elegant French Provincial styling. Its specifications are exactly the same as the rotary dial models and the internal buzzer is still present in place of the conventional ringer. For those persons who wish to combine just a bit of modern technology with provincial styling, this instrument may be a good choice. It is a bit more expensive due to the added internal components for the push-button dialing and sells for about seventy-five dollars from Radio Shack. It is available only in ivory with gold trim styling and weighs approximately the same as the previous model discussed. Persons who live in areas which offer phone systems requiring push-button dialing will find this model especially attractive because it is one of the few classic phones with push-button number selection.

The types of French phones discussed to this point may be more appropriate for the bedroom or dressing room areas of the

Fig. 3-3. French provincial styling is shown in this Radio Shack model.

Fig. 3-4. French continental-style phone in white and gold from Radio Shack.

home. They are ideal for placement on stands which offer very little room. In a large living room, however, they may be a little too small to add the distinction desired, so the French Continental style telephones are offered. Pictured in Fig. 3-4 is one style of this phone. It is rather massive when compared to the dainty, previous model. It has a very large base and a most prominent handset. This particular model is ivory colored and trimmed in gold. To many it would seem to be a combination of Old World styling with modern overtones. Note the rubber feet which support the base on the resting surface. Figure 3-5 is completely Old World in styling from handset to metal base feet. Simulated gold filigree makes this model especially distinctive for special placement applications. The handset cord is ivory, as are the earpiece and mouthpiece on

Fig. 3-5. Same model phone as before but with simulated gold filigree finish.

the handset. Price varies on these models depending on how much trim you elect to include in your purchase. Most of them, however, are somewhere in the eighty dollar category.

While not exactly a classic design, the phone shown in Fig. 3-6 is certainly antique and takes one back to the Elliot Ness days of the twenties and thirties. Named the *Bonnie and Clyde Special* by Radio Shack, this instrument is finished in black and gold trim and is one of the most useful models tried out. Figure 3-7 shows the correct usage, and it will be noted that you don't have to get all that close to the transmitter or mouthpiece to establish effective communications. This is an interesting phone and still offers certain advantages over our more modern versions. Since the earpiece and transmitter/mouthpiece are not contained in the same assembly, the portion which is held by the caller is not as heavy and tiresome. The author uses one of these phones in his office and finds it most convenient and, certainly, an excellent conversation

Fig. 3-6. The Bonnie and Clyde Special from Radio Shack.

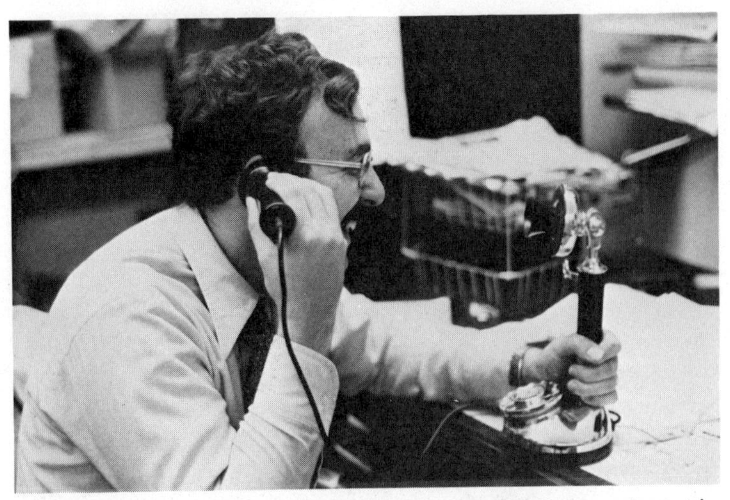

Fig. 3-7. Correct use of the 1930's style telephone. Microphone element is sensitive, so the user does not have to speak closely.

stimulator when visitors happen to see it. Radio Shack sells this model for about eighty dollars.

The models presented so far in this chapter are good representations of the Classic-style decorator phones available to the average consumer on the general market. There are many other models available, but most will duplicate or closely resemble the models featured here.

CONTEMPORARY STYLED PHONES

For those individuals who are more attuned to the contemporary styles in homes and offices today, there is a wealth to choose from in telephones whose outside appearances have been modified or changed completely to blend and add to this style of interior decorating. Some of these will bear only slight resemblance to a standard phone. Others will not stray so much from standard design practices. All of them will find uses, special and otherwise, in a great many homes and offices.

Figure 3-8 shows the Radio Shack ET-100, which is a miniature telephone instrument offering push-button dialing and auto-redial. This feature allows the last number punched up to be automatically redialed by simply pushing one button on the push-button panel. This can save the user a lot of time and movement when redialing a number which proved busy a few minutes before. Upon receiving the busy signal, the caller hangs

Fig. 3-8. Radio Shack ET-100 contemporary telephone.

up, waits an appropriate time and then depresses the numbers (#) button on the push-button panel. The number is automatically dialed by the ET-100.

Redial is accomplished by means of a simple, electronic memory circuit within the instrument's interior which automatically stores and holds the last number punched up. When another number is punched, the first memory is erased and the last number is committed to memory. Punching the numbers (#) button on the panel triggers the memory circuit, which releases its stored information into the normal dialing circuit. The stored number is dialed, but the caller doesn't have to go through the motions of punching out each digit in the number. This and other electronic advantages will be more fully explained in a later chapter on electronic telephone instruments and accessories.

The Radio Shack ET-100 offers push-button operation on most types of telephone lines, whether they are rotary dial or tone dial types. A convenient muting switch allows you to talk to someone in the room you are calling from without having to cup your hand over the mouthpiece element to prevent this conversation from going down the line to the other party. A partially coiled 16-foot cord allows for greater flexibility in the user's movements around a small room, and the phone is hung up simply by allowing it to rest, face down, on a flat surface. This modern device requires no separate cradle and base like most other telephone instruments do. The ET-100 sells for about forty dollars. For persons with true tone-dial systems, the model ET-100T is also available for about fifty-five dollars and offers a true tone-dial output.

Figure 3-9 shows the Radio Shack ET-200, which can be desk or wall mounted. This instrument offers pulse dialing which is compatible with both rotary and tone dialing telephone systems. A dial tone button or line switch is located directly beneath the pushbutton keyboard. Should you accidentally dial incorrectly, depressing this button is equivalent to hanging up. A quick press of this button cancels the portion of the number you just dialed and allows you to start over again without having to return to the base switch. This can save a lot of time for the busy executive or

Fig. 3-9. ET-200 from Radio Shack with push-button dialing.

housewife who must be doing other things while conversing on the telephone.

The ET-200 comes equipped with a pleasant-sounding tone trigger. A three-position switch on the base allows the volume level of the ring to be set. Three positions of this switch marked *high, low, off* provide good versatility. The off position is used when you are asleep or any other times when you do not wish to receive a phone call.

Like the ET-100, the Model 200 also offers the automatic redial feature. It works in exactly the same manner as with the other instrument. This circuit requires a standard 9-volt battery in order to properly store the last number. If you do not desire *last number redial*, you may eliminate this battery. Insertion of the battery is very simple and is accomplished without the use of tools of any kind. Figure 3-10 shows how the compartment cover slides back to expose the terminal connections. Once this assembly has been snapped into place, the user simply inserts the battery into its slot and slides the cover back into place. The battery maintains the memory of the last number redial circuit and must be replaced at least once a year for proper operation. For longest life, use an alkaline battery. Low battery power is indicated when the last number redial feature refuses to work or operates erratically. Since the battery powers only this circuit, all other telephone operations will remain the same.

TeleConcepts, Inc. in West Hartford, Connecticut offers a tremendous variety of highly unusual, beautiful and interesting telephone instruments. Their many designs are unique, highly useful, and an interior decorator's dream. To say that many of this company's products are apparent works of art would not be much of an overstatement. From a novelty phone, which looks like a toy

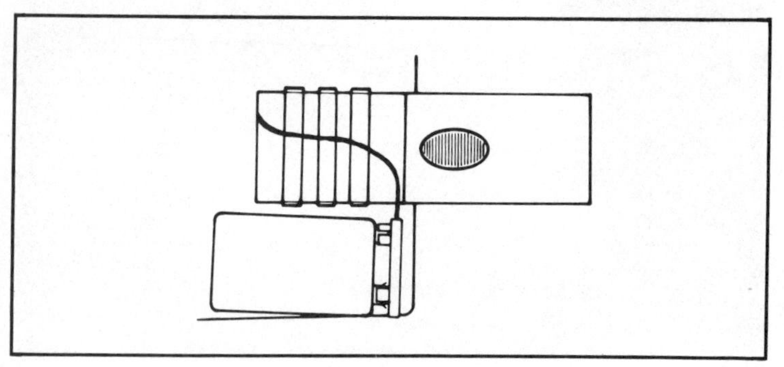

Fig. 3-10. Removal of compartment cover for insertion of battery.

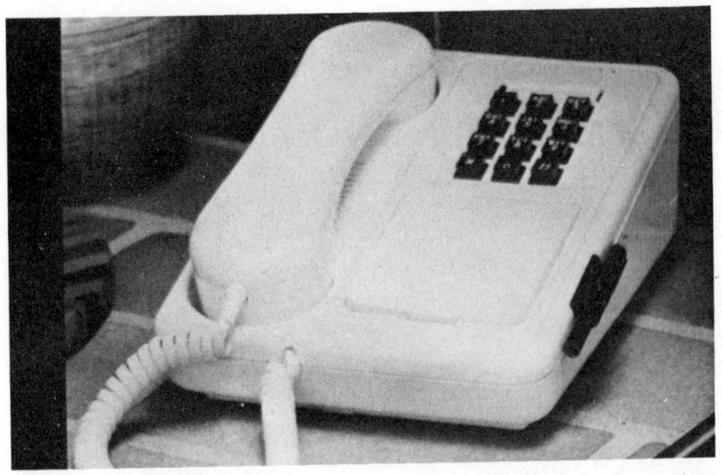

Fig. 3-11. Northern Telecom's Doodle model with built-in message pad (courtesy of TeleConcepts, Inc.).

airplane, to a sculptured designed pedestal, TeleConcepts is recognized as a leader in creative telephone instrument design.

One line of products marketed by TeleConcepts is made by Northern Telecom. Figure 3-11 shows the model called the *doodle* because it contains its own built-in doodle pad, which is a great convenience for those folks who never seem to have paper and pencil handy when asked to take an important message by phone. The pad is located on the front panel just below the push-button keyboard, while a pencil clips to the side of the base. This instrument can be used with almost every type of phone system in the United States and is available in rotary and tone dial models. Its universal styling makes it applicable to both home and office uses, and it comes in a variety of colors.

A take-off on the previous model is the *Kangaroo* shown in Fig. 3-12. This is a rotary dial version, but it too comes with push-button dialing if desired. This model adds a soft touch to any room or decor with leatherette, corduroy, or denim bean bag pouches in a wide variety of colors. Teenagers will especially like this design and, in later years, they can remove the instrument from the bean bag if a more formal look is desired. Conversely, those persons who have already purchased the Doodle model can order bean bag pouches to make the conversion to the Kangaroo model. Here is a good example of a company which builds versatility into their instruments and allows the owner to have the

Fig. 3-12. The Kangaroo model from Northern Telecom is the previous model in a special pouch (courtesy of TeleConcepts, Inc.).

advantage of two models and two appearances available from a single unit.

Novelty Telephones

In the novelty phone line, the instrument in Fig. 3-13 is highly popular. The name is even more creative than the phone. It's called

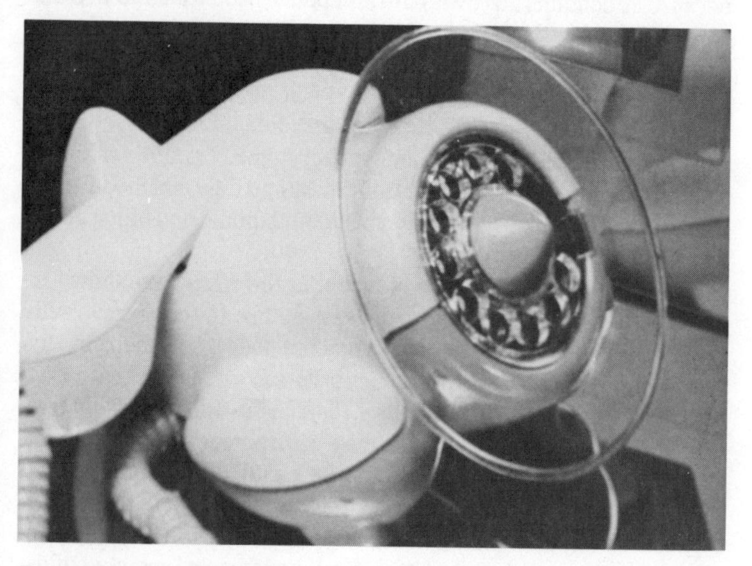

Fig. 3-13. A truly original telephone instrument called the Alexander Graham Plane from Northern Telecom (courtesy of TeleConcepts, Inc.).

Fig. 3-14. Executive office phone called the Diplomat (courtesy of TeleConcepts, Inc.).

Alexander Graham Plane and features an unusually and highly attractive design that is guaranteed to please youngsters and adults with an interest in airplanes or, at least, the unusual. The rotary dial is surrounded by a circular plastic shield which resembles a spinning propeller. This phone is available in only one color combination, which is white, orange and pumpkin. This device is perfectly compatible with most phone systems, is a true telephone instrument and not a toy, and will certainly be the cause of many conversations when guests or business associates see it for the first time. This model is available in rotary dial only and is designed for mounting on a desk, table, or other flat surface.

For the office executive, the *Diplomat* in Fig. 3-14 commands attention wherever it's placed. The most striking feature of this desk-mounted instrument is the distinctive faceplate which contains the push-button dialing keyboard. This particular model is available only in tone dial and will not be compatible with rotary dial only exchanges. The unit has a very low-profile styling and will be best suited to business decors.

For the home, the Northern Telecom model shown in Fig. 3-15 has unisex styling and is as masculine as it is feminine. This model is called *Dawn* and is available only in rotary dial models.

Fig. 3-15. Rotary dial telephone model called Dawn may be used in many settings (courtesy of TeleConcepts, Inc.).

The base is a most distinctive cone which houses the handset with its smooth lines when not in use. The author thinks of this phone a bit differently than does the company which manufactures it. While Dawn may be an appropriate name for this instrument, "Chameleon" might be a better one. Northern Telecom points out its elegance and sophisticated features in advertisements, and while this is certainly true, the same phone blends well with an informal den. It is one of the few designs on today's market which will add to, or at least blend with, almost any type of contemporary room or office decor. The buyer has a choice of many colors and may opt for a gold or chrome-plated decorator base ring. This model is designed for desk or table top mounting only.

The next two decorator phones can be considered to be marketed toward the female consumer. Figure 3-16 shows the *Rendez-Vous* which is light, easy to carry, and adds an immediate feminine touch to most mounting sites. Note that the handset is very similar to other models carried by this same company. Figure 3-17 shows a similar instrument. Called the *Athena*, this model offers a flatter base and the handset is perched atop a carved pedestal when not in use. Both of these Northern Telecom instruments are available in a variety of pastel and decorator colors, and both are available only in tone dial models.

No discussion on decorator and novelty telephones would be complete unless an instrument such as the one shown in Fig. 3-18 is included. Marketed by TeleConcepts, the *Pay Station* is a working telephone, which admittedly, will not fit in with almost any decor. A working pay phone which can also serve as a bank, this model would go well in a home rec room or bar or possibly in the bedroom of a child or teenager. It is authentic looking and works in

Fig. 3-16. A feminine appearance is presented by the push-button Rendez-Vous model (courtesy of TeleConcepts, Inc.).

a similar manner to those counterparts found in phone booths across the United States. It is available only in rotary dial models, and the basic black color seems to be the most popular.

Fig. 3-17. The Athena from Northern Telecom features a pedestal-mounted handset (courtesy of TeleConcepts, Inc.).

Fig. 3-18. Specialized room decors may take advantage of the Pay Station (courtesy of TeleConcepts, Inc.).

Resembling a hand-held microphone, the TeleConcepts *Erica* model (not shown) is not as new a style as a first glance might indicate. Similar telephone instruments were marketed back in the early sixties and were available to most telephone subscribers in the United States. This model is available in either tone or rotary dial models. The phone is hung up by simply setting it down on a flat surface, resting on its base. The dial or push-button keyboard is found at the bottom of its resting base. This instrument must be held the whole time it is in use, as setting it down will immediately disconnect the line. The transmitter element is housed in the top portion of the base. A myriad of colors are available when ordering this model. They include green, white, red, ivory, brown, yellow and beige.

It is possible to go even further in designing special decorator telephones for a general or specialized purpose. From our earlier discussion of basic phone instruments, it was learned that the telephone is really a quite simple device consisting of some minor electronics, a rotary dial or push-button keyboard ringer, and a few other items. It is possible to locate these items in almost any type of case, as long as its size is adequate to house the various components. The dial or push-button keyboard may even be located at a distance from the rest of the circuitry and connected by electrical wires. Telephones do not have to look like telephones is the gist of this conversation and the idea behind many of the unique products now being offered by TeleConcepts, Inc.

Mod Art Telephones

Figure 3-19 pictorially illustrates the idea of a telephone not looking like a telephone. Called the *Disco*, this instrument contains all of the electronics in a circular can or cylinder. The dial mechanism is centrally located, while a band is fitted around the center forming the base support. A cradle arm extends from the left side of the cylinder and holds the handset, which is ivory and chrome colored. This is a very compact instrument and will easily rest on the tiniest of stands and tables. It is available with the rotary dial, as shown, or in push-button models for tone or outpulse dialing. The Disco will attach to the present phone company hookup in a few seconds and is compatible with most systems in the United States. This series is also available in a clear, plastic casing which displays the internal components. This model is shown in Fig. 3-20.

Taking miniaturization one step further, the *Lido* is arrived at. Actually, this model, shown in Figure 3-21, uses the same, basic components as the former model discussed, but different packaging gives it the appearance of being even smaller. The sculptured base houses most of the circuitry, while a perfect ball with a flattened end houses the rotary dial or push-button keyboard. The Lido model in Fig. 3-22 is the rotary dial version. Both models are

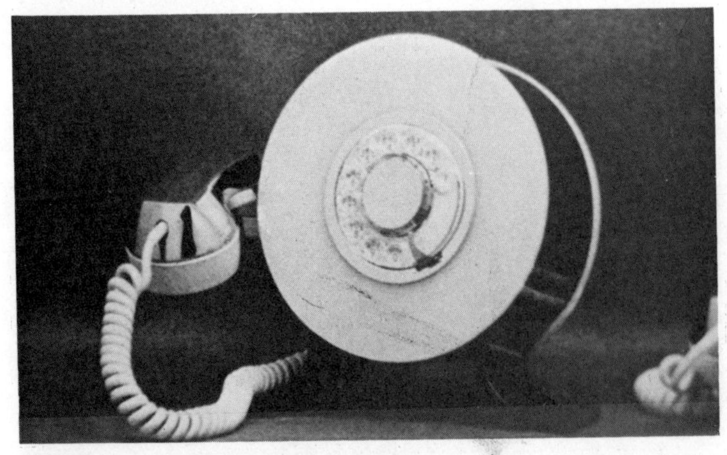

Fig. 3-19. Unusual circular styling with a side-arm cradle is presented by the Disco Mo.Jel (courtesy of TeleConcepts, Inc.).

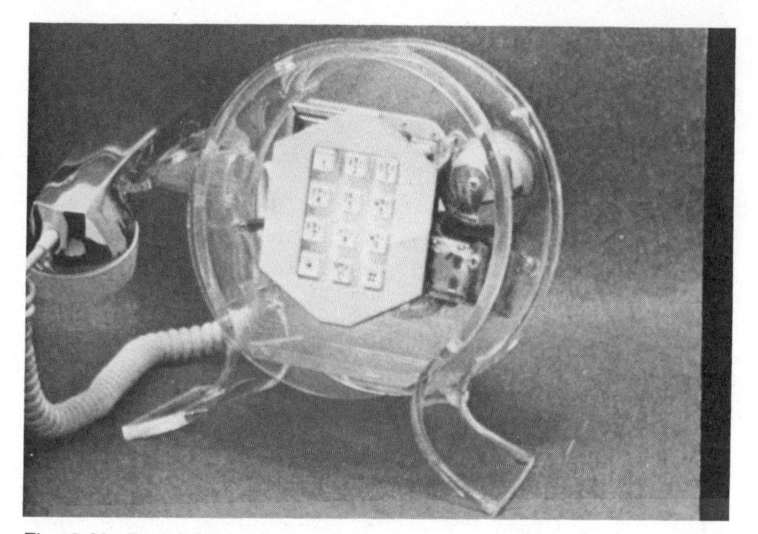

Fig. 3-20. Push-button version of the Disco in a clear case (courtesy of TeleConcepts, Inc.).

available in abalone, ivory, pearl and brown colors. The screw-on ear and mouthpieces match the finish of the base, while the remainder of the handset is finished in chrome which copies the finish of the spherical dial casing. The Lido is a very dainty instrument in appearance and is popular as an extension in bedrooms, dressing rooms, etc.

More modern in case design, the *El Rondo* shown in Fig. 3-23 contains the unused handset horizontally and on the front of the instrument. A sculptured octagonal case holds the push-button

Fig. 3-21. The Lido model features a sculptured base and a push-button mechanism mounted in a spherical case (courtesy of TeleConcepts, Inc.).

Fig. 3-22. Rotary dial version of the Lido available in abalone, ivory, pearl, and brown finish (courtesy of TeleConcepts, Inc.).

keyboard and all other circuitry. The base is really a sort of stand made from a single, curved sheet of plastic material. The cradle switch enters the octagonal case horizontally. The design of this model makes it a very versatile instrument for extension bedroom phones or for use in the most elaborate den or living area. Available in brown, silver sparkle and abalone, the latter two colors are often chosen for a swankier appearance, while the brown adds a more informal touch. The rotary dial model with a sparkle finish is shown in Fig. 3-24.

Sometimes a more or less standard telephone instrument can be transformed into an apparent work of art by placing it inside a

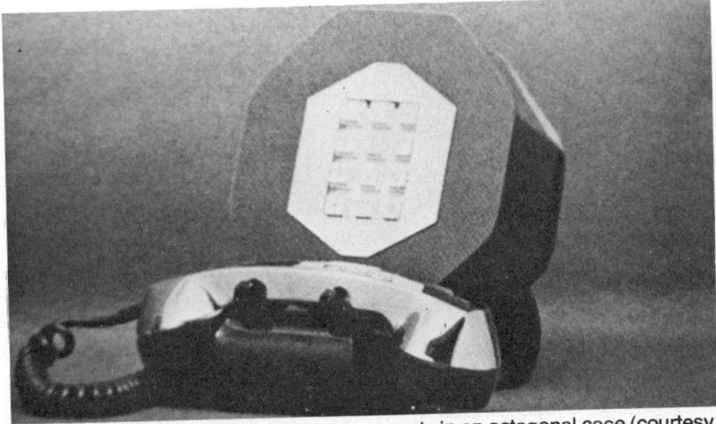

Fig. 3-23. El Rondo model houses components in an octagonal case (courtesy of TeleConcepts, Inc.).

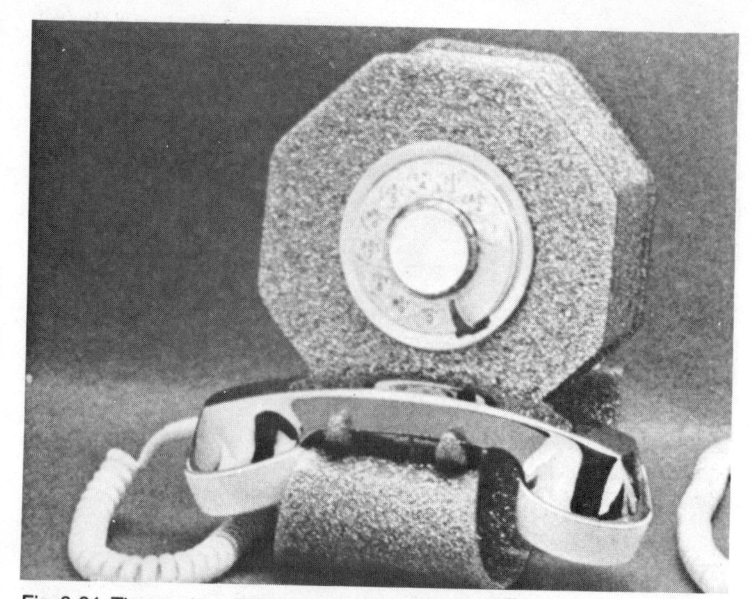

Fig. 3-24. The previous El Rondo model is also available in a sparkle finish for a more elegant appearance (courtesy of TeleConcepts, Inc.).

special stand or holder. This is the case with the *Shellmar* design shown in Fig. 3-25. Here, a telephone handset very similar to the Radio Shack ET-200 or the Northern Telecom Contempra is housed in a container which resembles a large, exotic sea shell. This container contains no electronics as such. Only the cradle switch is found in this section, which opens the phone line when the handset is not in use. Since the handset is really the self-contained

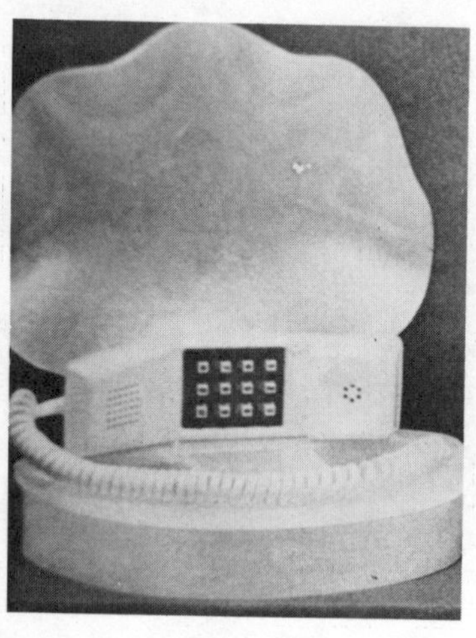

Fig. 3-25. A very elegant appearance is presented by the Shellamar which features a case resembling a large seashell (courtesy of TeleConcepts, Inc.).

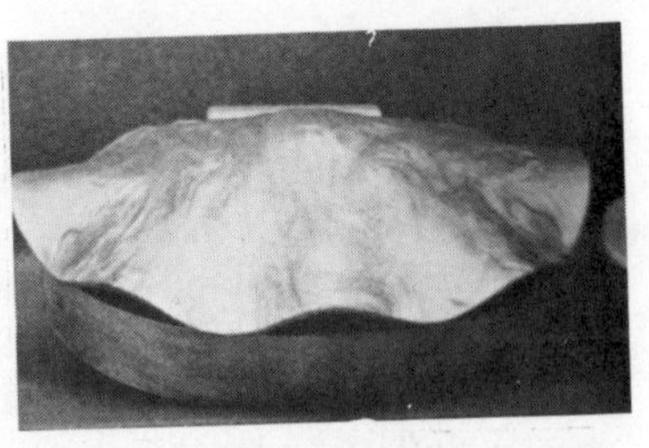
Fig. 3-26. Shellamar phone with simulated shell cover in closed position (courtesy of TeleConcepts, Inc.).

instrument, there is a hang-up switch mounted directly below the rotary dial or push-button keyboard which will allow the user to break the line in order to regain the dial tone after completing a call or dialing a wrong number. This switch is of the momentary variety and automatically returns to the open position when released. This prevents the line from being accidentally disconnected and left in this mode without the user being aware that the phone is not armed and ready to receive any incoming calls. This model is available only in rotary dial and tone dial models. Pulse dialing mechanisms are not available, which allows the convenience of push-button dialing to be had when connected to rotary systems. Figure 3-26 shows this model in the covered mode. The phone is completely hidden from view and the container appears to be an ornamental dish or jewelry box. This model is available only in two color combinations, abalone and camel.

Similar in appearance to some of the earlier decorator phones discussed in this section, the *Chromephone* in Fig. 3-27 resorts to

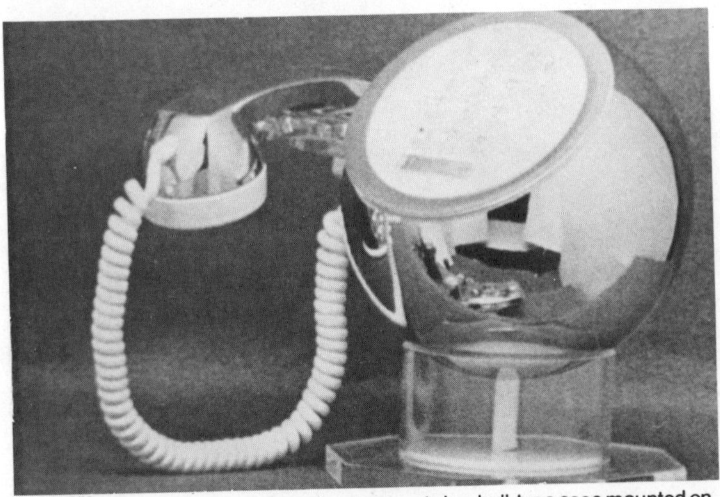
Fig. 3-27. Chromephone houses all components in a ball-type case mounted on a clear pedestal (courtesy of TeleConcepts, Inc.).

the ball type of case which houses all of the electrical components. A chrome finish is the only type available, while the handset cord and mouth and earpieces are ivory. The ball container is supported by a mini-pedestal consisting of an octagonal, flat baseplate and short, cylindrical stand. Both of these are made from clear plexiglass. A white insert within the stand completes the total look of this model. Even the cradle arm for the handset is moulded from clear plexiglass. Available in rotary dial and push-button models, the Chromephone may also be purchased for outpulse dialing using a push-button keyboard which is compatible with rotary dial systems.

A variation on the Chromephone design is the *Chrome Nouveau* in Fig. 3-28. The case design is that of a slanted cylinder with a flared top where the dial or push-button face is found. A ringed decoration is used on the faceplate which accentuates the contemporary look. A multi-layered base provides a firm support and weighs down the entire instrument while combining with the look of the main body. A plexiglass cradle arm extends from the left side of the cylinder and the handset which, like the cylinder component case, is chrome with ivory trim. This model is available in both push-button and rotary dial versions.

Futuristic Telephones

Clear plexiglass design is becoming very popular. The *Apollo* model in Fig. 3-29 provides a rectangular design with a flat base all

Fig. 3-28. Chrome Nouveau is an ultra-modern design consisting of a slanted cylinder and flared top (courtesy of TeleConcepts, Inc.).

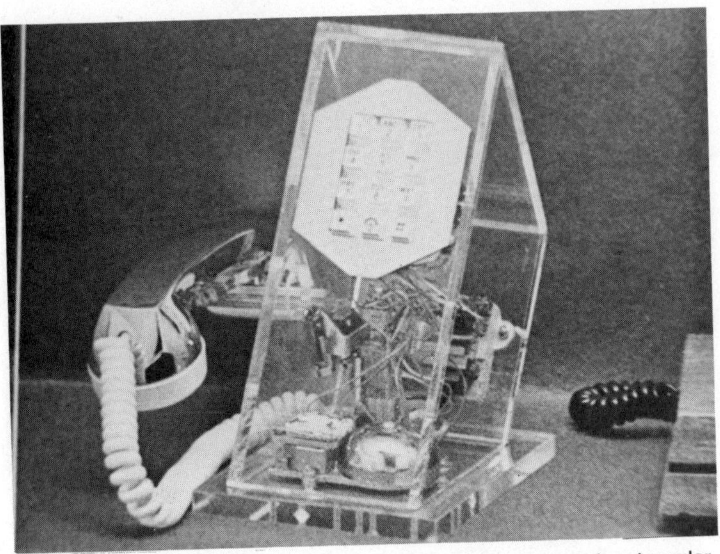

Fig. 3-29. See-Through Apollo model with a flat base and semi-rectangular design (courtesy of TeleConcepts, Inc.).

made from clear plexiglass. The internal instrument mechanism is clearly visible and is polished to be attractive and unusual in the manner with which it combines with the clear case. Either the push-button keyboard or a conventional rotary dial model may be chosen. The only part of this instrument which is not of see-through material is the handset and cord, which are ivory and chrome. While the author is not aware of any present instruments designed with a see-through plexiglass handset, this might further enhance an unusual design. The clear handset case would allow the internal transmitter and receiver elements to be clearly seen along with the interconnecting wiring. Even a transparent line cord could be used to display the conductors. One reason this has not been done so far is probably due to the fact that plexiglass scratches rather easily and the clear material prominently displays this scratching. While the main body of the telephone instrument is not handled all that much in normal use, the handset is constantly being picked up, bumped, and possibly abused to a point where, if made of the clear material, severe scratching would be visible after only a short time. Perhaps a clear lacquer coating will be developed, allowing for a total see-through design which will hold its appearance for the life of the instrument.

Looking at other clear designs, the reader will find that there is much to choose from. The model known as the *Diamente* has a

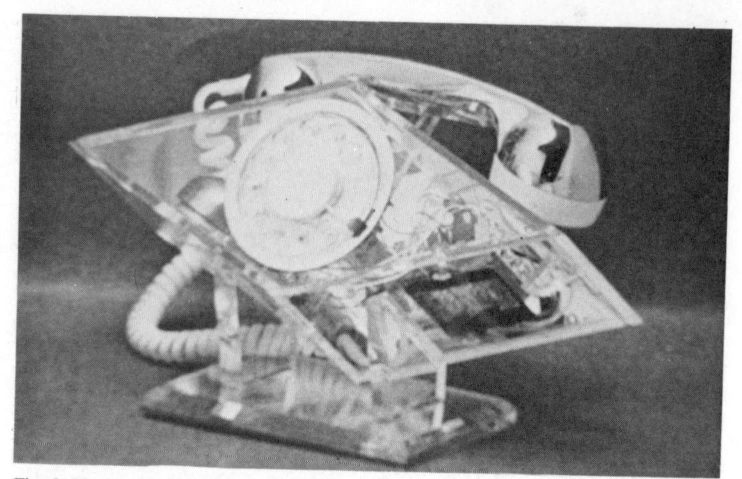

Fig. 3-30. Space-Age styling has been applied to the Diamente which features a handset mounted atop a parallelogram case (courtesy of TeleConcepts, Inc.).

complex-looking parallelogram case which is difficult to comprehend upon first glance due to the unusual angles involved in the mounting of the handset cradle and the dial face. The entire housing is supported from a clear plastic baseplate by solid rectangles. Figure 3-30 shows this model in the rotary dial version, although tone dial is also available. This model is quite a conversation starter and adds a touch of elegant sophistication when inserted into certain decors. It will also blend well with ultra-modern office

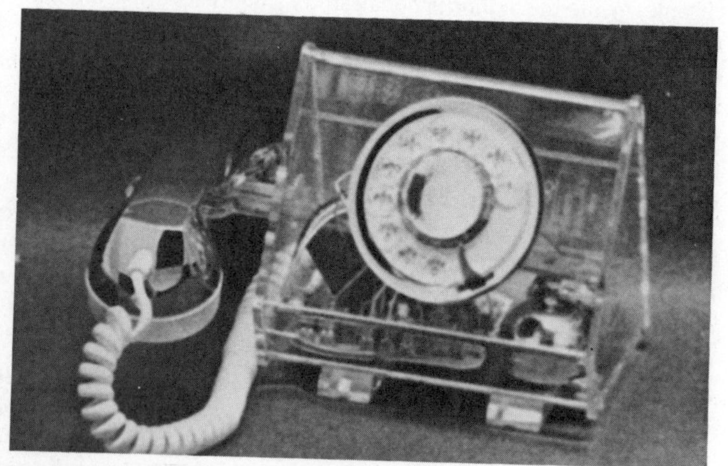

Fig. 3-31. The La Belle clear case telephone is also available in brown and pearl finishes (courtesy of TeleConcepts, Inc.).

Fig. 3-32. The Petite model features a very simple design and is small enough to be placed almost anywhere (courtesy of TeleConcepts, Inc.).

layouts, but don't try to use it with any other styling, whether Mediterranean or Early American. It just won't go.

The *La Belle* in Fig. 3-31 is a little more down to earth and may be used in a variety of settings. While the clear model is shown, this instrument is also available in brown and pearl. The brown version might even be fitted into an Early American decor without clashing too much, although there is really nothing traditional about its design. It is one of those looks which might fit anywhere under the right circumstances. This design is a very simple design which is functional and attractive. Its small size is an immediate attention-getter, and it is not as brash as many of the previous models pictured and discussed. The Petite is also available in a pearl finish, adding a more feminine appearance to the instrument, and is also available in a rotary dial model.

The last clear table model telephone in this series is the *Cube* from TeleConcepts, which is just what its name implies, a cube of clear plexiglass mounted on one edge on a clear plexiglass plate. See Fig. 3-33. The chrome and ivory handset rests on a plexiglass cradle and the entire case slants backward for ease of viewing from a seated or standing position. The push-button keyboard may be chosen for true tone dial systems or in outpulse for rotary systems, or the standard rotary dial model may be ordered.

The basic cube can be used to many good advantages in designing cases for telephones. Depending on its mounting configuration, color, and general trim, the cube-designed telephone can conform to informal decors or to the most elegant. The *Tempo* shown in Fig. 3-34 is a basic cube design with a bit of edge

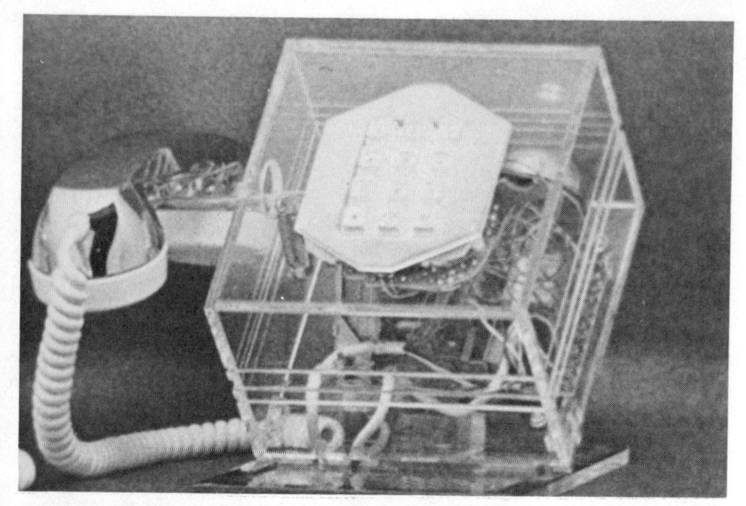

Fig. 3-33. The Cube telephone is shown here in a see-through case and is mounted on edge to a clear plate (courtesy of TeleConcepts, Inc.).

trim. It is mounted on a simple stand which tilts the main body and makes the dial or push-button keyboard more accessible. This is the pearl version, which is very elegant. For a more informal style, the clear version might be ordered. Both come with a chrome and ivory telephone handset. The stand is of the same color style as the main case.

Figure 3-35 shows the same basic compartment, but here the cube is turned slightly to provide a four-pointed star affect.

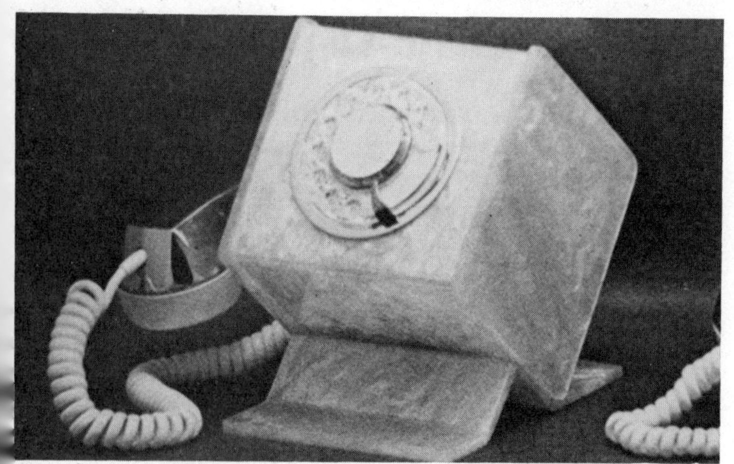

Fig. 3-34. The Tempo design is shown here in a pearl finish and mounted atop a special base in the same finish (courtesy of TeleConcepts, Inc.).

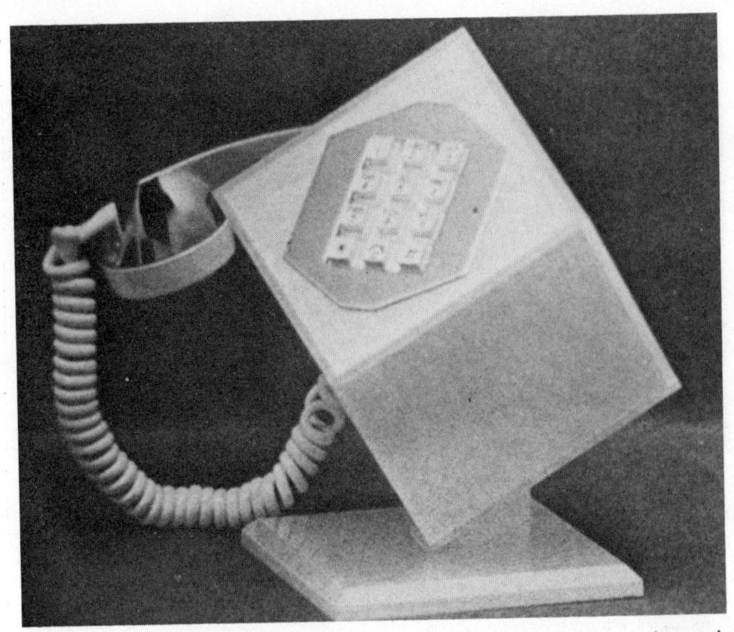

Fig. 3-35. Another cube type of housing but with the compartment turned on end for a diamond effect (courtesy of TeleConcepts, Inc.).

Attached to its pedestal base at the back, this model is also available in a clear design and in push-button or rotary dial versions.

Still another look is obtained in Fig. 3-36 using a design which is structurally similar to the Tempo discussed previously. A functional, contemporary appearance is given off by this cube which is available only in combination colors of brown and ivory, red, white and blue, and in yellow and white. The rotary dial model here may be replaced with a tone dial version also available in the same color combinations. The flat base is ordered in the same combination of colors as the main case unit. The handset for all three color combinations is ivory or white, as is the handset cord.

An unusual design for a desk phone is shown in Fig. 3-37 and uses a combination of black and clear plexiglass with some chrome trim on the handset to arrive at a very scientific-looking instrument. This model is called the Teledome and features a black, rectangular case in the middle of which a clear, plastic dome rises, revealing the internal components of the instrument. A rotary dial or keyboard rests in the center of the dome, while the handset, which is chrome with black trim, rests in a plexiglass cradle on the

Fig. 3-36. Americana model is similar to some previous designs but with a more informal, contemporary styling (courtesy of TeleConcepts, Inc.).

left side of the black base. This combination presents a very stark appearance.

While most of the instruments discussed throughout this book will be of plexiglass or plastic construction, TeleConcepts does

Fig. 3-37. A scientific-looking instrument, the Teledome is mounted in a rectangular case with a clear plastic dome for the dial face (courtesy of TeleConcepts, Inc.).

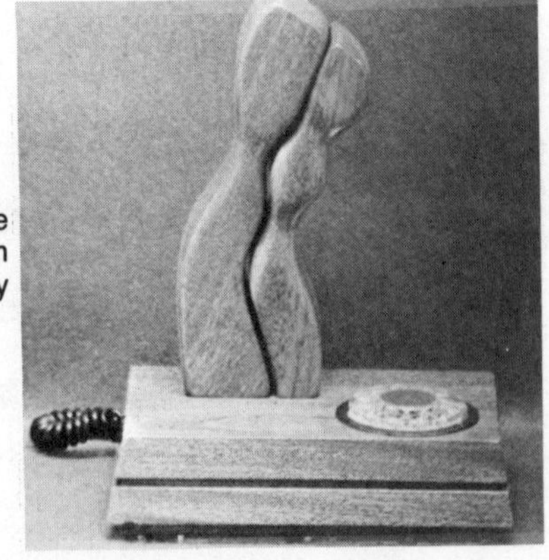

Fig. 3-38. Unique wood sculpture appearance is provided by the Adam and Eve mahogany model (courtesy of TeleConcepts, Inc.).

offer at least two models which house the electronic and electrical components in wood cabinets. The *Adam and Eve* model shown in Fig. 3-38 is available in rotary or tone dial and features a unique wood sculpture in contemporary art as a case for the instrument. The modern figure on the left is actually the handset. If not for the rotary dial, this might appear to be a fine work of art and nothing more. The same might be said of the *Eve* model which is shown in Fig. 3-39. This case has a walnut appearance and the kneeling representation of a figure is the handset. Both models should highlight any room with a lot of wood trim.

While the phones detailed here are highly unusual, it is possible to go even one step further, as is shown by the *Star Blossom* model of Fig. 3-40. This is very similar to some of the earlier cube designs discussed, but a hand-painted marble finished appearance is presented. The phone almost has the look of fine

Fig. 3-39. The Eve walnut-sculptured instrument is available in dial and push-button models (courtesy of TeleConcepts, Inc.).

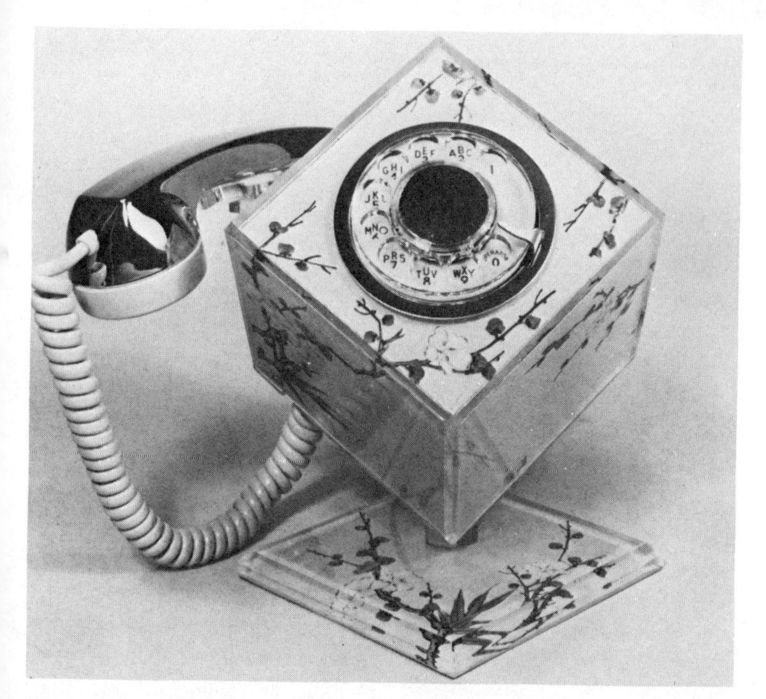

Fig. 3-40. The Star Blossom is similar to cube designs of contemporary models but is classically styled with a hand-painted porcelain case appearance (courtesy of TeleConcepts, Inc.).

porcelain. The rotary dial seems to blend better with this design than does and rotary systems alike. This same finish is carried over to the *Apollo Nikko* which is simply the clear Apollo model discussed earlier with an artwork exterior depicting traditional Japanese culture. The push-button model is shown in Fig. 3-41.

Some Japanese heritage is carried over to the *Pagoda* line of telephone instruments, although there is really nothing traditional about this ultra-modern design. Called the Pagoda because of the stacked layer appearance of the base, this model is available in many different colors and color combinations. Shown in Fig. 3-42, this is the single line model which, when made for multi-line business uses is called the *Executive* (Fig. 3-43). Rotary or push-button dialing is available, along with cases in silver sparkle, chrome and gold, black, ivory and even clear plexiglass. The handsets are usually gold with white or black trim, although some models offer chrome and white combinations with white handset cords.

Fig. 3-41. Apollo Nikko which is the previous Apollo model with an artwork exterior depicting traditional Japanese culture (courtesy of TeleConcepts, Inc.).

Once you have chosen the desk phone or phones to suit your needs, you may want to go a step further and purchase an elegant stand to place your masterpiece upon. These are also available from TeleConcepts in designs which match some of the phones already discussed. Shown in Fig. 3-44 are three stands which match the two Shellamar models and the Chrome Nouveau phone. Notice that these pedestals are designed to perfectly match the exterior appearance of each phone, with the abalone and camel

Fig. 3-42. Pagoda model desk phone features Japanese traditional styling (courtesy of TeleConcepts, Inc.).

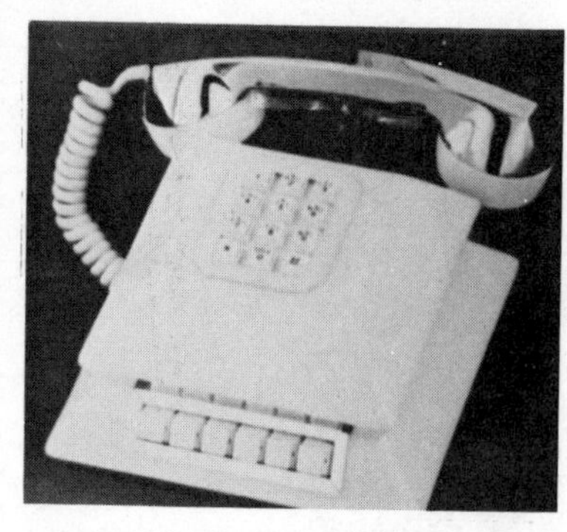

Fig. 3-43. Executive desk phone has multi-line capabilities (courtesy of TeleConcepts, Inc.).

shell look for the Shellamar phones and an ultra-modern cylinder and frame for the Nouveau.

WALL PHONES

So far, most of the telephone instruments discussed have been designed for desk mounting. While some of the designs already covered can be mounted to the side of a wall as well as placed on a flat surface, these have not been numerous. With the design innovations which take place on an almost daily basis to meet consumer needs today, even the wall phone is becoming a decorator item. Traditionally, the wall phone was a purely practical device which was designed for out-of-the-way utility and little else. While many inroads were being made into desk style telephone cases and designs, the wall-mounted phone stayed much the same as its predecessors of ten or even twenty years ago.

Figure 3-45 shows the *Mark II* wall phone from TeleConcepts, which is available in abalone, pearl, green, and yellow finishes. A clear dome emerges from the center of the unit and displays the interior, which includes a double bell ringer. Note that the chrome-finished handset hangs vertically from the cradle mount. This model may also be had in a rotary dial design for customers who subscribe to this type of system.

A total see-through wall phone design is pictured in Fig. 3-46. This is the *Mark I* and is also available in solid colors. This is the wall-mounted equivalent of the Apollo desk model described earlier in this chapter.

CUSTOM DESIGN

It is important to point out that almost all of the phones pictured in this chapter are basically identical regarding internal components. To simplify matters, it would be nearly accurate to say all of these phones use the same instrument, but mount and display it in many different forms. Most of the phones featured were available in rotary dial, tone dial, or with pulse dialing, which uses push buttons to actuate a dial phone system. They may have been packaged in many different ways, but they all were designed to initiate and receive phone calls.

If you own your own telephone instrument or instruments, you may now be thinking of ways you could build your own case or cabinet in which to mount the components from your present phone. This is very easily accomplished with the only problem area being the cradle switch mechanism, which may have to be specially fabricated.

Many unusual, beautiful and novel telephones can be arrived at using the homebuilding method. Remember, though, do not alter any of the internal circuitry or components. To do so might void the licensing of that unit.

No specific projects will be covered in this text because the subject of customized telephone design could easily fill many books. However, the author once built an extension phone into his IBM typewriter. Only the basic components were included for this

Fig. 3-44. Three custom stands which are designed to match specific telephone models (courtesy of TeleConcepts, Inc.).

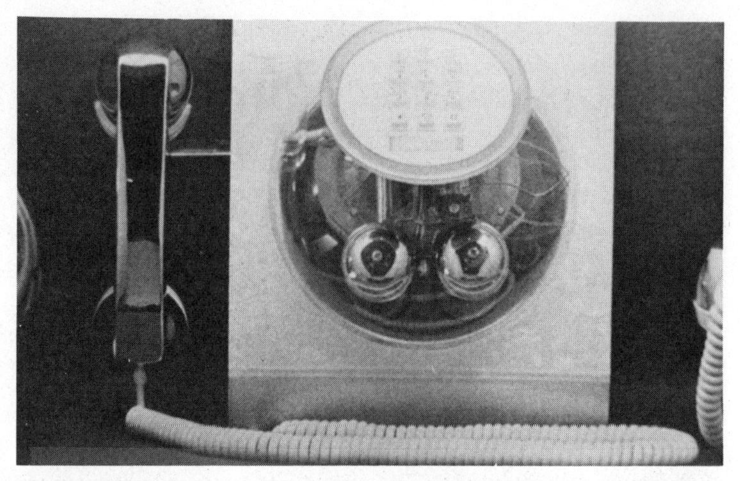

Fig. 3-45. Mark II wall phone available in a variety of finishes, each featuring a clear plastic dome (courtesy of TeleConcepts, Inc.).

answer-only set-up, which included a compact handset, a line switch and little else. The phone handset rested on the right side of the typewriter when not in use. The only component which was located inside the typewriter case was the microswitch to disconnect the line upon hanging up. It was quite simple, really, and involved only a few hours work. Of course, if a dial, ringer, and other components were added, the mechanical complexities of the job would have tripled the difficulty factor. While this typewriter phone was built as a novelty, it came in quite handy, as the author could answer the phone while completing an assignment and continue working during the conversation. It was also an indication of the author's dedication to work when visitors to the office noticed this unusual arrangement. This may not have always been a *true* indication, however. Seriously, many homebuilders have been able to install telephone components in clocks, adding machines, bottles, etc. These are often practical designs but, as often, may be novelty arrangements which are of immediate interest to others. A friend of the author even has a part-time business of installing telephone components in anything the customer wants. Usually, the customer supplies the surplus telephone, which must be licensed for operation by the FCC. A bid is given on the installation in whatever case the customer desires. The most unusual installation which has been made to date was on the side of a marble toilet. This latter arrangement, obviously, offers only limited decorating applications.

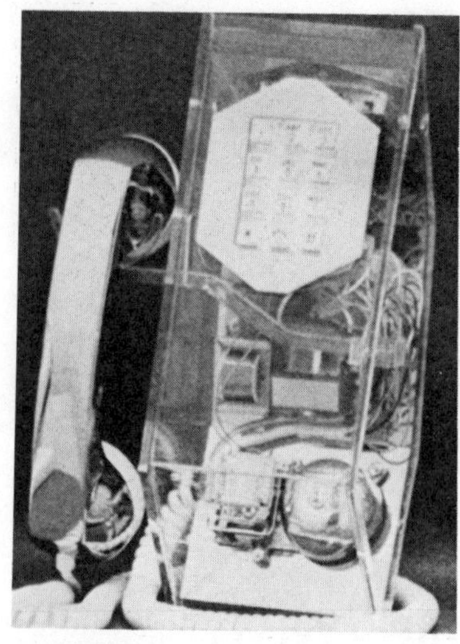

Fig. 3-46. Mark I Wall phone featuring a total see-through design (courtesy of TeleConcepts, Inc.).

SUMMARY

With the amount and variety of decorator phones on today's market, it is possible to outfit any home, office, or room with telephone instruments which match and accentuate decors. True, in many instances, the telephones supplied by your phone company will be completely appropriate, but should they prove undesirable, at least there is a practical alternative in the decorator phones which may be privately owned. As is the case with all other merchandise, the more attractive or intricately designed instruments will be priced higher than their simpler counterparts. It is possible to purchase decorator phones with 24 Karat gold trim or plating, which costs in the many thousands of dollars. More to the point, some of the decorator phones of interest to the average consumer will cost well over one hundred dollars, but many will be priced between forty and one hundred dollars complete.

It should be apparent by now that decorator phones are really furniture. They can be likened to console televisions or stereo systems which are purchased in styles to match the rest of the furniture in the rooms where they are to be located. Like the telephone, they have specific purposes, home entertainment in this case, whereas the telephone is used for personal communications. When a stereo is not in use, it then becomes furniture. The same is true of the decorator phones. And just like fine furniture, the finer the styling of the decorator phone, the higher the cost, even though all telephones in this series accomplish the same basic function of dependable communications.

Chapter 4

Telephone Answering Devices

The home and office telephone has become such a vital link with the rest of the world that it has become necessary for many persons to monitor their phone calls on a 24-hour basis. This may be done by a telephone answering service which automatically takes all phone calls to a specific number after a certain time or when the owner is out of the office. This is an expensive service, comparatively speaking, and can be eliminated through an electronic device that will perform the same function for only a fraction of the long-term cost. This electronic device is known as a *TAD* or telephone answering device.

TADs are available in many price ranges and so designed to perform many different functions. Some will only deliver a message saying that the person dialed is not in and will return at a certain time. Others will supply a brief message to the caller and allow him to leave a message which is recorded on magnetic tape. When the owner returns to his telephone location, he may then rewind the tape and play back his messages. Still others will do all of the above and allow the owner to have his messages played back to him on the telephone. These systems are usually tone activated. The owner dials his own number, hears his recorded message to callers, and then injects a preset tone from an oscillator, which is supplied with the answering device, into the phone he is calling from. The answering system in his home or office decodes the tone and immediately plays back all of the messages into the phone line. In this manner, a businessman may leave his office for a trip to another state and still be able to receive his messages without having to be physically present at the location of the answering device.

All TADs operate in basically the same manner. Most of them use magnetic tape on which the owner's message is recorded. The caller's messages are usually recorded on the same loop of tape although on a different track than the one which contains the owner's message. When the telephone rings, the induced current trips an internal relay which attached the line directly to the output of the tape recorder. The pre-recorded message is played down the line to the caller. At the end of the pre-recorded message, an audio tone is automatically inserted alerting the caller that when the tone stops, he may leave a brief message which will be recorded. This tone usually activates another internal switching device which connects the input of the tape recorder to the output of the phone line through a matching transformer which allows the caller's voice to be recorded. At the end of this sequence, the finer answering systems may automatically recycle and indicate that one message has been received. The next call is handled in exactly the same manner with the indicator showing that a second call has been received. After each sequence, the TAD automatically rearms itself and is ready for another incoming call.

TADs are not especially new devices. They appeared in the late 1940's and early 1950's but were very large and heavy devices because they used vacuum tubes instead of solid-state components. These early TADs were designed primarily for businesses, as they were much too expensive to be considered as practical home-use devices. The better units cost between $600 and $1000, which may not seem like a tremendous amount today, but in the later forties and fifties, this represented quite an expenditure.

With the practical development of the transistor and other solid-state devices, TADs became much smaller in size and less expensive. Many now use integrated circuits to miniaturize the devices even more and to make them ultra-reliable. They are portable and most can be easily held in one hand.

THE BASIC TAD

Almost all telephone answering devices are simply tape recorders which record incoming messages directly from the phone line and play back prerecorded messages into the phone line. A microphone is necessary for the user to record his message onto magnetic tape. Many of the newer answering devices appearing on today's market even use standard cassette tapes which may be purchased in any hobby store. This tape recorder device must also have the capability of playing back its audio messages into a standard speaker or headphone. This allows the user to hear the received messages. With a little additional circuitry, almost any standard tape recorder could easily be turned into a TAD.

Magnetic tape is sensitive to the alignment of tiny particles along a north-south pole. Recording tape is coded with a microscopically thin layer of iron oxide which comes into contact with the electromagnetic head of the tape recording device. The head has a north and south pole which are located in close proximity to each other. The space between the north and south pole is known as the *gap*. This is where the modulation process takes place through the alignment of the magnetic iron particles. This process is graphically illustrated in Fig. 4-1.

Modulation is the term which means to change. If a steady tone were applied to the magnetic head, then the iron oxide particles on the magnetic tape would all be aligned identically. However, as the tone changes, as is the case with the complex waveforms making up the human speech pattern, the modulation process aligns the magnetic particles in a symmetrically coded pattern which corresponds to the fluctuation of the audio input to the head.

An electronic process is used to deliver the input from the telephone line or from the spoken voice at a microphone input to the heads. Usually a microphone by itself has insufficient power output to properly modulate the audio tape. In this case, amplifying circuits are placed between the microphone and the tape heads. These electronic amplifying circuits are fairly simple in nature and are often composed of single transistor circuits. The low level electrical output from the microphone is multiplied or amplified in the circuit. In many instances, there will be two or more stages of amplifications before the signal is transferred onto audio tape. These amplifier circuits faithfully reproduce what was originally obtained at the output of the microphone. So, the modulation pattern which is applied to the tape is identical in waveform with the applied speech at the microphone.

Most TADs contain an integral loop of audio tape which cycles continuously. It is not necessary to rewind the tape to a preset point to hear a message, as this is usually done automatically. When a call comes into the number, the current supplied by the central phone office triggers a relay which causes the tape recorder

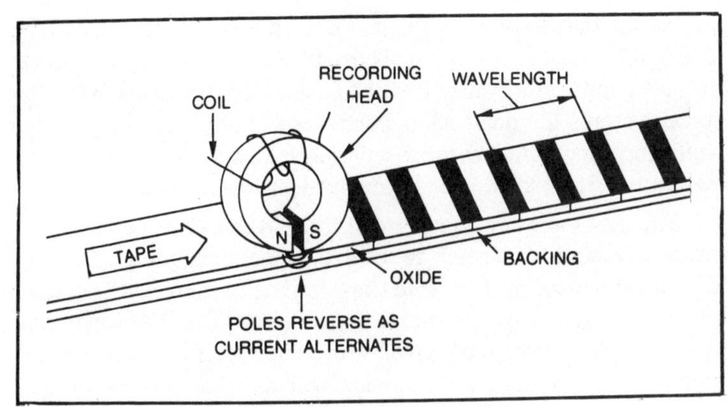

Fig. 4-1. Electromagnetic recording process.

to be attached to the phone line. In this mode, the audio output of the tape recorder is first switched into the line to play its message to the caller. This is usually accomplished in the following manner. The first ring of the phone closes a relay and sets up the electronic connection. Some systems allow the adjustment of ring selection so that the machine may be triggered on the second, third, or other ring.

The outgoing message which is played to the caller must give the required information as to what number has been reached and whether or not a message can be left. This is previously recorded through a microphone. When the outgoing message reaches the end of its cycle, it is cued to activate a beep on the phone line and to switch to the recorder mode. This is sometimes done with a separate recorder circuit and tape drive but may also be accomplished by using another portion of the audio tape. These latter systems will use the top half of the audio tape for recording the user's message, while the bottom half of the tape receives the incoming messages.

Once the beep has been heard by the caller, he will have anywhere from 20 seconds to 3 minutes or more, depending on the machine, to complete his message. When the message is complete, the machine will still continue to record for the maximum period of time allotted for each message. At the end of this time, it automatically rearms itself, awaiting another phone call.

Many answering devices have made provisions for indicating whether or not a message or messages have been received while the user was away. Some even indicate exactly how many messages were received. Depending on the machine, to hear a playback of the recorded messages necessitates moving a switch or lever to the playback position. This will usually rewind the tape to the start, and another button may be pushed to activate playback. In some systems the playback is continuous to the end of the tape, while in others, the machine will automatically stop after each message and is started again by identical push of the button.

After all calls have been noted, the TAD may be returned to the *answer mode* to be ready to receive the next incoming call. The TAD is often used as a "middle man" for business executives who are often plagued by unwanted phone calls. The TAD may be placed in the answer mode even while the owner is in the office. When the phone rings, the device will indicate that he is not present and request that the caller leave a message. Meanwhile, the executive can be eavesdropping by listening to the incoming call on a separate telephone or on the handset of the TADs which are telephone instruments and answering devices combined. If the call is an important one, as is indicated by the message which the caller is in the process of giving, the executive can immediately interrupt the sequence and indicate that he has just arrived.

With most answering devices, it is necessary to be physically present at the recorder in order to hear a playback of messages. Ninety percent of the time this presents no problem, but for the traveling executive or the businessman on vacation, the wait of several days or weeks to hear the various messages could be costly. Fortunately, specialized TADs now offer a method of overcoming this problem by featuring a remote actuation circuit through which all recorded messages may be played over the telephone upon the command of a remote audio encoder.

To receive all messages, the owner simply calls the number guarded by the TAD. When the answering device takes over, an audio tone is sent down the phone line from the remote calling point. This is often a subaudible tone or one which does not occur within the normal human speech pattern. The TAD contains a decoder circuit which acts upon this specific tone. Upon receiving the command from the remote encoder, the TAD rewinds its message tape and plays all messages back into the telephone line.

Using remote actuation, a TAD owner may monitor his messages from the next town or in a foreign country. Some of the newer devices even allow for erasure of previous messages and the resetting of the TAD on a daily basis. One unit even responds to the owner's voice and no other. This system requires no external remote encoder to broadcast its recorded messages. The electronic pattern of the owner's voice accomplishes the encoding task.

As can be seen, all telephone answering devices have many likenesses. Each, however, is also different, and different manufacturers have found some very unusual ways of solving the technical problems or at least making them less of a bother. Audio tape is always the heart of the unit and in many devices, two or more audio tapes will be used. The audio tape could be called the "memory" if we were describing TADs in computerized terminology.

The manufacturers of modern TADs all use audio tape, but the means of distributing this tape across the tape heads and collecting it will vary from manufacturer to manufacturer. Some types still use reel-to-reel recording and playback, as shown in Fig. 4-2, rather than packaged cassettes. Both of these systems have a master reel and a takeup reel, and both systems require the rewinding of the tape onto the master reel before playback is attempted. Other systems will use a continuous loop which is shown in Fig. 4-3. This is similar to the 8-track tape cartridge which is used with stereo systems in many homes. Here, there is only one reel. The tape is fed from the inside up across the tape heads and is wound around the outside of the coiled tape reel. The TADs which use cassette tapes have the advantage of easily changeable tape cartridges which can be replaced by the owner. Most of the systems which contain endless loops and reel-to-reel tape distribution systems are not owner serviceable and must be returned to the manufacturer for repair. Audio tape is a very sturdy substance as long as it is not abused, but dirt on the capstan and heads can cause abrasive damage and a weak tension spring may

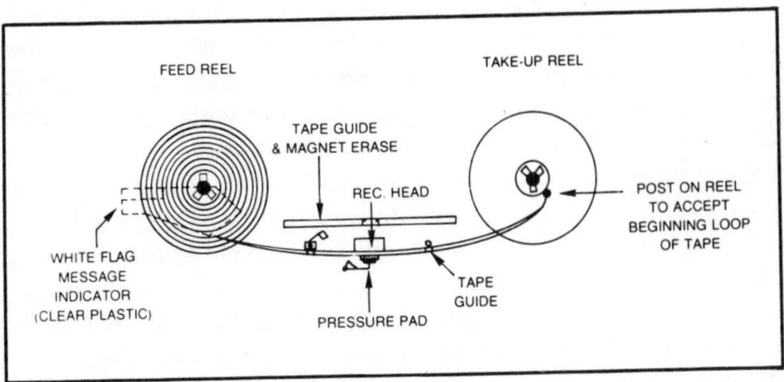

Fig. 4-2. Reel-to-reel TADs are still used in many applications (courtesy of Phone-Mate).

cause the tape to foul and become hopelessly entangled in the drive mechanism. The older a tape is, the more susceptible it is to failure. For this reason, the changing of the tape periodically should keep the TAD in a good state of repair and highly operational.

USE AND OPERATION

Once a TAD has been chosen, the phone company notified, and the installation made, one would think that no difficulties would be encountered after this point. This is true from an electronic and mechanical standpoint in most cases. With just a little preventive maintenance, the TAD should provide many years of adequate performance without having to be returned to the factory for a major overhaul. No one knows just how long a modern TAD will last, although there is a limitation on the amount of wear and tear the magnetic tape can withstand. TADs which are used in business locations may receive a large number of calls on a daily or nightly basis, while at-home units will receive comparatively few calls. The life of TADs is determined in hours of operation rather than duration of ownership. Under-use of the device could lead to failure if dust is allowed to build up and infiltrate the delicate motor drive mechanism. Therefore, when a TAD is to be stored for long periods of time, it is best to insert the entire unit in a plastic bag and seal it tightly. Some of the commercial moisture absorption packets might also be included in the bag to prevent any condensation buildup.

Since the electronic and mechanical operation is not a major problem, you might be wondering what other problems may occur. According to many owners, the immediate problems which are encountered after the unit is properly operational involve the content and wording of the messages which are heard by the callers. This may seem like a very minor operational point, but it is probably the most important aspect of the entire device. The reason for purchasing a TAD is to be able to take messages from callers when you're not there. In order for them to leave their message, an explanation is recorded for play down the line to them.

A great deal of psychology must be applied at this point, as most people do not appreciate being spoken to by a machine when they want to talk to a human being. About the only thing they dislike more than this is being required to talk back to a machine. True, this drawn-out mental conclusion does not usually go through their minds; rather it is more of a instinctive aversion. For

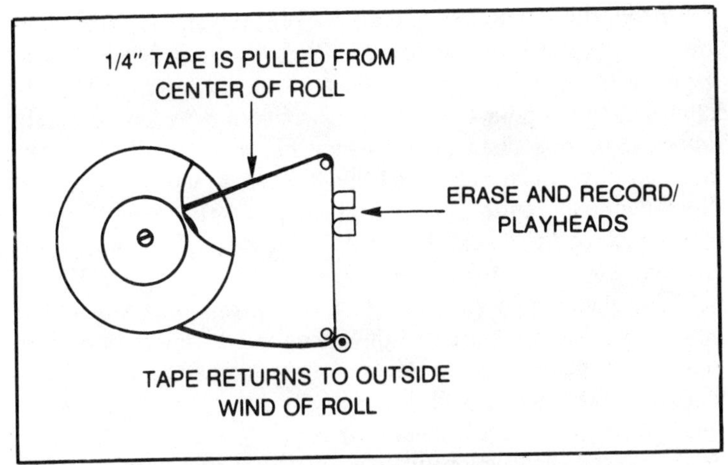

Fig. 4-3. Continuous loop TAD tape carriage (courtesy of Phone-Mate).

this reason, what you put on the recording plays a most important role in making the caller feel at ease and willing to communicate with you via a recorded message.

Almost anyone who has ever owned a TAD is aware of the number of people who simply hang up rather than leave a message. This is often because they are in a hurry or feel they must talk directly to you. Some persons may feel, and rightly so, that someone else such as a secretary or aide may listen to the messages and jot them down for you, the owner. In the author's opinion the reason for over 50% of the hang-ups is the completely unimaginative and boring messages which are heard by the caller. Often, a hasty message is unprofessionally recorded by the TAD owner before leaving the home or office. Little thought is given to the content and no thought is readily given regarding what the caller hears when the outgoing message is played.

First of all, in order to record a message which is as audible as possible requires microphone technique. Most TADs are equipped with inexpensive microphones which present varying audio characteristics. One device might require that the person recording the message be very close to the pickup. Other such devices may have microphones which will insert distortion into the tape if worked as closely as the previous device. It is necessary to experiment with your particular machine. Record a message and then listen to the playback. Better yet, call your number on another telephone to hear exactly what it will sound like to the caller. This is the best way to check the audio quality and to determine what recording procedures should be used in making other tapes.

During the process of listening for audio quality, you may even learn a bit about radio announcing. You may notice that you're speaking too rapidly or too softly or even incorrectly. This is important to the person calling your answering device, as many times noisy long distance lines may be involved. While your messages may sound understandable, if not clear, when checked on a local phone, the entire contents of the tape may be totally useless to a person on the distant end of an out-of-town phone line or to someone who may be hard of hearing.

The importance you wish to place in this area will often depend on what you purchase the phone for. Business purposes may dictate more professionalism in the recorded message than will casual at-home use. If you find you do not have a pleasant speaking voice and you expect to receive many phone calls from persons you do not personally know, you may want someone else to record the message for you. Some businesses find that a woman's voice is more preferable to a man's regarding their calling clientele. Other businesses may find just the opposite to be true.

All of these factors must be considered before making the recording which is to be used in the answering mode. One should never hastily record a message without consideration and forethought. Each message should be considered with the caller in mind and should be written down on paper and possibly edited for a minute or so. This assures that what you want to say is expressed clearly and concisely. This is the reason for your purchase in the first place, the dissemination of information and the collecting of same when you are not present to directly do so yourself.

Once the audio recording technique and the proper management of the voice have been worked out, the actual message content must be decided upon. This phase of TAD operation will often depend on the type of device you have. Some allow as little as ten seconds for the recorded message, while others may allow up to a minute. Ten seconds is a little too short, while a sixty-second message is much too long. Several years ago, the author conducted an intense research project for a radio station to determine just how long the average attention span of a radio listener is. The average length of time that a person will listen to a radio commercial attentively is between 12 and 14 seconds. Amazingly, the length of the commerical had little impact on the normal attention span length. In other words, people heard about as much of a 30-second commercial as they did of a 60. Fortunately, telephone answering devices hold a captive audience, at least for a short period of time.

Unlike the radio commercial, a person who calls a number is normally listening intently to the information coming from the other end of the line. They are not as susceptible to outside distractions as would be the case with most radio listeners who may be driving automobiles, ironing clothes, etc. However, the maximum length of a recorded telephone message should not exceed 30 seconds, and 15 to 20 seconds is even more desirable.

Message Style

You will have to try many different combinations, and test them for trial periods in order to determine which message style works best for you. A business which takes orders by telephone may find a simple, direct approach most advantageous. For example:

"Good evening, this is Smith Communciations. In order to provide you with 24-hour service, you may place your order with this electronic device. When you hear the beep, state your name, address, and phone number and then give your order along with the catalog number, quantity, size, and color, if applicable. Your order will be processed during regular store hours tomorrow. Thank you for your business. And now, here is the tone."

This is rather unimaginative, but this message does clearly state the intention of the answering device and explains carefully and concisely what the caller is to do. After-hours ordering, although highly used in the business world, is not the chief use of TAD's. Most of the time these devices are purchased by persons who are often away from the home or office and cannot afford to hire someone to take their calls or who find the hiring of another staff member impractical. The following is an example of an informal message which might be used in a small insurance business, one where the agent personally knows most of his clients.

"Hello. This is Bob Smith. I'm sorry, but I'm not able to answer the phone at the present time, so I am using this electronic device to answer my calls for a little while. When you hear the tone, please leave your name and number, and I'll call you back right away. Thank you for your call, and I'll look forward to hearing your message, which you may leave right after the tone which will occur in two seconds."

Again, this is an informal message and it lets the caller know right away what he originally called for will be taken care of as soon as possible. He originally called to talk to his insurance man who is not in, but who had taken the courtesy of explaining the situation to the caller by TAD. You will notice that two references were made to the tone which indicates to the caller that he may deliver his message. In this message and the previous one, the caller was actually led directly into the tone by introducing it with words.

To those persons not used to leaving a message with a TAD, the process can be a little confusing upon the first few attempts. If the message is correctly worded, all of the guesswork and hesitation are removed. This is good from a psychological standpoint because a person who is not sure of what to do in order to leave a message will most often hang up rather than risk appearing ignorant by botching the message.

Most of the time it is not a good idea to indicate that you are away from the home or office, especially for an extended period of time. One TAD owner left a message for callers stating that he was out of town for a few weeks on vacation and left a Florida phone number for those persons who found it necessary to get in touch with him before he returned home. Upon his return, he discovered that thieves had completely cleaned out his home and even taken his answering device. Although never proven, thieves could have potentially found out that he was gone for an extended period of time by hearing the message from the TAD. This particular home was even fitted with a device which turned various lights on and off automatically during times of the day and night in order to give the appearance that someone was home. Incidentally, this device was also stolen.

Rather than indicate that you are away from the home or office, it is much better to say that you are temporarily out or cannot answer the phone at the present time. This is a much safer message to leave because you never know who may be calling at any time. Always give the impression that you will be returning shortly and, at the same time, indicate that you will get back to the caller as soon as you are near the telephone.

If the direct approach in phone messages is not successful, it may be necessary to add a bit of humor to your taped messages. This can be done in good taste in most home situations and businesses and may serve to put a smile on a caller's face who might otherwise be disgruntled with the answering device. The following message was used successfully at the offices of a small journalistic business in Virginia:

"Greetings. This is Jack Smith of Jack Smith Writing Associates. Alas, you have cut me to the quick for I have momentarily departed my office. Be of stout heart for I shall return posthaste and when I do, I shall harken to your behest. Now, if you don't understand any of that, don't be alarmed because I didn't either, but you've got to admit it's different from the plain old recorded messages you normally hear. If you leave your name and phone number after the beep, I'll get back to you right away and will do away with all of the Shakespearean theatrics. Thank you. The beep should be occurring shortly. To beep or not to beep. That is the question."

For the particular business stated, this worked very well, and very few persons (comparatively speaking) hung up without leaving a name and phone number. Due to the nature of the business, this comedic message was found to be appropriate. Another business might not fare as well.

Analyzing what was said in the above message, we find that all of the criteria which was previously given for determining message content is there. The caller knows that you are "temporarily" out and that you will return the call upon your return. The caller is also given elementary instructions on what to do, "leave your name and number after the beep." All the proper information is disseminated but in a way which is entertaining, original, and effective. The author has found that if you can hold the caller's attention until the tone occurs, you have a much better chance of getting him to leave his name and number.

Once a suitable message has been found, don't stop there. If you have many of the same people calling you on a more or less regular basis, you will find it necessary to change the message at regular intervals. Some TAD owners even find it necessary to change the message at regular intervals. Some TAD owners even find that people will call just to hear the "message of the week," if each is entertaining and original.

Humorous Messages

Another way of holding the caller's attention while not resorting to a humorous slant on your message is through the use of music. The following message was used to excellent advantage in a tax consulting business located in a city of about thirty thousand population:

Background: Soft string music.

Female voice: Contralto.

"Thank you for thinking of us. We appreciate your call and would like to show our appreciation by calling you back as soon as possible. (Pause) (Music increases in volume) (Music fades and voice returns) Following our musical selection of the week, *Michelle* by The Living Strings Orchestra, you will hear a short tone. When this stops, please leave your name and phone number so we may return your call immediately. Thank you, and the tone will sound within five seconds."

The message is nothing spectacular, although the softly hoarse contralto voice is rather staggering upon first listen. What really makes this message something special is the musical background. For a truly professional effect, a local radio station may be able to produce several messages for you which can be inserted at different time periods. Warning: Nothing sounds worse than musical messages recorded on inexpensive tape recorders or with inferior microphones. The TAD's which will accept standard cassette tapes will probably be best suited to high-production messages. Alternately, TAD's which have built-in tape and tape drives but which provide an external input jack may also be used. Here, it is necessary to record the messages on a separate tape recorder. The output of the tape recorded message is then fed to the input of the TAD. Most of these inputs will be for a very low level drive such as the kind which is present at the output of a microphone. It may be necessary to "pad" the output of the tape recorder a bit to prevent distortion. It may also be necessary to install a matching transformer between the output of the tape recorder and the input of the TAD. Both a matching transformer and a variable pad are usually used, but the pad may be used without the matching transformer where the tape recorder output impedance closely matches the TAD input. Also, if the TAD should have a high-level input, the pad could possibly be done away with.

It should be pointed out that most TAD's are not really intended for the accurate reproduction of music. They are designed for voice-only messages. Many, however, can be used for messages with musical backgrounds. These models will usually be higher in price than the others. The TAD's which will do a reasonably good job on musical recordings do so, because their tape drive systems are more stable and the tape passes evenly and smoothly over the recording/playback heads. Speed, too, is held very constant. When these criteria are not met, tape *flutter* occurs. This can be detected by listening to the playback. If flutter is present, the musical content of the message will sound erratic. Tones will change, and the sound will resemble that of a turntable which is spun by hand while the stylus is resting on the record and the amplifier is operating.

Flutter is most often caused by the tape being jerked across the tape heads by the tape drive. When the tape is allowed to travel in an up and down motion to excess, the same audio condition will exist. In the former situation, the tape is actually travelling across the heads at a variable rate of speed. For a fraction of a second it reaches one speed and then instantly speeds up or slows down in response to the uneven pressure applied by the tape drive assembly.

When using voice messages only, flutter is not as easily detectable. This is due to the intermittent pattern tendencies of the human voice, whereas music is usually an almost continuous series of complex tones which do not have significant pauses as is true of the human voice. While flutter is actually the result of inconsistent speed characteristics in the tape mechanism, this term is most-often used to describe a condition where the speed change is constant and periodically repeats the speeding and slowing operation in very short periods of time. Another condition which is also speed oriented is gradual in nature and lasts for far longer periods of time without repeating. This is called *off-speed* recording or playback and describes the tendancy of certain tape drives to gradually slow up as the tape is advanced. This is often difficult to determine from a casual listening standpoint. The obviousness of this condition will be dependent upon how much the speed is slowed or speeded up.Battery powered TAD's are most subject to this condition. As the internal batteries begin to weaken, the motor begins to slow up. If the recorder portion of the circuit has recently been used to record a message, the condition may go unnoticed for a long time, but if a previously recorded tape is used for playback, the speed change is usually very apparent. The reason for this seeming oddity can be seen when we realize that the playback speed is relative to the recording speed. If a TAD with weak batteries is running slower than normal, then it will run slower during both the record and playback modes. This means that a message which is recorded at a speed which is ten per cent slower than normal will sound perfectly normal if played back at 10 per cent below the normal speed. But, if a recording is made at normal speed and played back at a lower speed, the effect will be immediately discernible. When batteries are replaced in a TAD requiring them, it is usually necessary to re-record the taped message. The chances are, this message was recorded at a lower than normal speed due to the weakened condition of the batteries. When fresh batteries are supplied, the playback speed will return to normal, and the previously recorded tape will be played back at a faster rate. The effect will be immediately noticeable as well.

Generally, whenever battery operated TAD's are in use and a speed malfunction is detected, the batteries should be the first possible trouble source examined. If replacement causes a message to return to normal speed (if running slow) or causes a message to speed up to above normal playback, this has been at least part of the source of the trouble. Other conditions can cause speed deficiences to occur. These are defective or aged tape, dirt on the capstan or record/playback heads, foreign matter entering and caking on the gears or other portion of the drive mechanism, and slipping belts between the motor and capstan if they are used. A general cleaning of the drive mechanism, if accessible, should be performed every month, or more often if the TAD is subject to heavy use. Some TAD's have sealed tape drive mechanisms which are intended to be serviced by the manufacturer. If you attempt to gain access to the internal workings, this may void your guarantee. The latter units do not need the periodic cleaning as the non-sealed devices do. The sealed TAD's are very well protected from dirt and dust as well as internal moisture build-up.

Complete care and maintenance instructions are almost always included with each TAD purchased. Of as much importance, these manuals also include what *not* to do. Follow the manufacturer's instructions as closely as possible, and your TAD should perform up to its design specifications. While many of these devices appear to be rather simple in design, the tape drive mechanisms are often dependent upon fairly critical adjustments. Even slight unknowledgeable tampering can cause mayhem and necessitate a costly, unwarranted repair job.

PUBLIC INFORMATION USES

In recent years TAD's have been heavily incorporated into the dissemination of public information. In these applications the outgoing message feature is used alone. The ability of these devices to record messages from the caller is not used and many of the TAD's designed for outgoing purposes only do not offer this latter function. Commercial radio stations are often plagued by phone calls requesting information which is regularly aired through broadcasting. During periods of snow or inclement weather, requests for school closing information can create difficulties in staffing the telephones. The TAD may play an important role at these facilities. Using the outgoing message capabilities of the device, complete school closing information can be recorded in a minute or so. When the closing information is given over the air, the announcer will mention the phone number to call for a repeat of the information. It takes a few weeks to get the listeners accustomed to this system, but after the initial breaking-in period, many man-hours have been saved using the TAD.

Weather information is also often recorded onto the TAD tapes. Some radio and television stations even sell brief commercial time on these recordings to local businesses. The "Weatherline" is highly publicised on the air so listeners who have missed the on-the-air forecast can call for a full update. An announcement on the TAD might read:

"The WXYZ Weatherline is sponsored by RJT Enterprises, 100 Papermill road, your supplier for all winter sporting needs. Stop in today and look over the large selection of skiing equipment and accessories now on sale at 40% off. And now the weather . . . Partly cloudy today with a high in the lower forties. Clear and much colder tonight, low in the mid twenties. Chance of snow 20%. The WXYZ Weatherline has been presented by RJT Enterprises, home of winter sports."

The versatility in public information services is tremendous using these inexpensive TAD's. Another radio station, as part of a Christmas promotion, urged the children of listeners to call a certain number to talk directly with Santa Claus. A typical Christmas greeting was recorded by a Kris Kringle sound-alike. Additionally, fifteen seconds were alloted for the children to list their gift desires. In this instance, the response was so large that the local phone company called the station asking that this promotion no longer be continued. Due to the large number of people, all calling the same number simultaneously, major problems had started to develop at the central switching office.

A minister friend of the author's reports that he uses the church's TAD to provide community information for all callers. The phone number is listed in the local newspaper and is aired over the local broadcast media. This is done as a public service for churches. Contained on the tape is a brief devotional message followed by a listing of hospitalized individuals in this small, rural community.

The list of other uses for the *outgoing message only* capabilities of the TAD is almost endless. From an informational standpoint, the TAD does not replace but is an extension of the human being whose voice is heard on the tape. It allows him the ability to be several places at once, informationally. The TAD allows businesses to take advantage of a cost-effective method of offering services around the clock, seven days a week. The expenses involved are diminutive compared with the potential for customer trust and loyalty.

SUMMARY

The telephone answering device (TAD), once an expensive tool of the larger commercial businesses, has now been developed to a point where almost any homeowner can readily use and afford this device. Its possibilities encompass a broad range of services which, before its development, were accomplished by direct human effort, resulting in a waste of time, money and manpower. Today, every segment of the U.S. population is taking advantage of the many conveniences TAD's have to offer. From the ultra-sophisticated units to the simple, battery-powered models, TAD's are no longer a dream but a dream come true for the average consumer. Small businesses can now afford to offer the customer convenience offered only by large corporations previously. Even the teenager with his or her own telephone can buy a TAD out of carefully saved allowance money. The housewife need no longer miss important phone calls while she is out of the home, and the bachelor on the go can spend less time around the home awaiting that call from a "maybe" date.

There is no question about it. The TAD has directly affected the lifestyles of literally millions of Americans. These have been mostly pleasing affects resulting in more freedom of movement for all of those who have taken the TAD plunge. As development continues, the price of TAD's should fall even lower, and more sophisticated devices will continue to present themselves on the market.

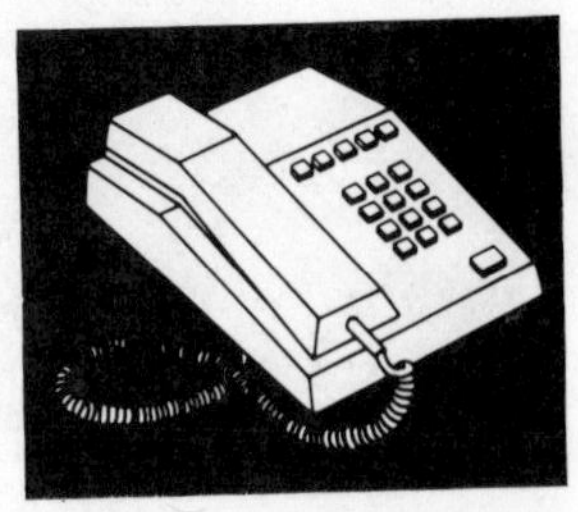

Chapter 5

The Telephone Answering Device Market

Having discussed the principles of the TAD elsewhere in this book, it's now important to know the various types of TADs available to the consumer and the characteristics of each. You will find a large array of these devices available from many different manufacturers, prices range from around $100 upward, and each offering special features, some of which may be of direct interest to you.

This chapter will provide information and specifics on a wide range of telephone answering devices that should serve as an excellent reference source for potential buyers. In selecting such a device, you will more than likely see many similarities between the models offered by several different manufacturers. Price is one factor, but more important is service and also the unit's warranty. Other features tend to remain consistent from manufacturer to manufacturer, but some slight differences may prove important to you, depending upon the area in which you live or the way you intend to use the unit, and be the deciding factor in making your purchase, Don't be afraid to shop around and compare prices. By doing so, you'll be sure of obtaining equipment that best suits your current needs at a price that will fit your wallet.

PANASONIC KX-T1510/KX-T1520

Panasonic, a name which has long been respected in the home electronics field, offers two telephone answering devices which would seem to be attractive buys for many consumers. Figure 5-1

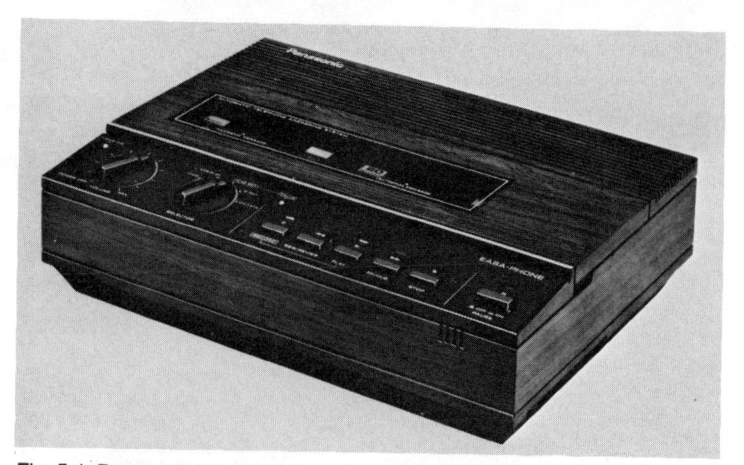

Fig. 5-1. Panasonic KX-T1510 Telephone answering device uses dual cassettes for recording incoming messages and playing back the outgoing messages or announcement.

shows the Model 1510, which offers many features for its owner. It offers almost every TAD service, with the exception of remote playback activation. The Model 1520 (Fig. 5-2) is almost identical to the 1510, except that it offers the remote playback activation feature. Both models use two cassette tapes for recording incoming messages and the playback of the outgoing one. This is a convenient feature, as incoming calls can be saved if necessary and this tape replaced with a new one. TADs that do not have removable cassette tapes must depend on internal tapes which are not readily replaced by the owner in case of breakage. On the other hand, cassettes are readily available almost everywhere, and new tapes will help guarantee the quality of both incoming and outgoing recordings through a regular tape replacement regimen.

When the telephone rings, the Model 1510 will automatically play the pre-recorded outgoing message to the caller. The Model 1520 has the added feature of playing up to *two* outgoing messages. The second message is automatically keyed when the caller finishes his incoming message that is recorded for later playback to the TAD owner. This second-message feature is not found on many TADs at this time, and it's a shame because it allows for an even more personal touch. Most callers who are answered by TAD devices hear only a long tone after leaving his name, number, etc. With the Model 1520, the tone will still be heard telling the caller to leave his or her message; but after this message is delivered, the second tape engages and, for instance, thanks the caller for leaving the message and wishes him or her a nice day, good evening, etc. Using the Model 1520, the last thing the caller hears before hanging up is a personal message from the owner or a member of his company or household. This is an excellent feature for businesses in maintaining good customer relations.

Another feature that is found on both models is your choice to answer the phone or not, even when you're in. If you're not sure who is calling or you might not care to speak to the caller, merely let the TAD take over and then listen to the person leaving the recorded message. If it's someone you wish to speak with, you may override the TAD and begin a direct conversation.

The remote playback feature of the Model 1520 can be of great assistance to individuals who are away from the phone, as it can be activated to play all incoming messages down the phone line by using the remote controller. The TAD user simply calls his phone number from the remote location and holds the remote controller up to the receiver of the phone he is calling from. The remote control circuitry transmits an audio tone which is transferred to the TAD. Upon receiving this tone, the TAD begins playing down the line all of the stored incoming messages it has received and recorded during the owner's absence. Skip and repeat operations are also available and are selected by the proper keying sequence established with the remote control module.

Another excellent feature of both models is found in their ability to record all telephone conversations on the line to which the TAD has been connected. When calling from the home or office, the TAD may be converted to the taping mode and both sides of the conversation recorded. This can be important to businessmen who may need to prepare typed transcripts of various

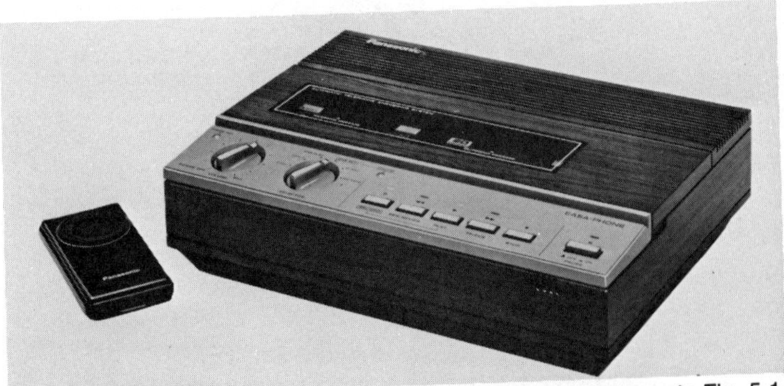

Fig. 5-2. The Panasonic KX-T1520 is similar to the model shown in Fig. 5-1 except it offers the remote-playback-activation feature.

conversations. Both models may also be used as a dictaphone by converting to this mode. Here, the phone line is internally disconnected from the circuitry, and the TAD is used in much the same manner as a common tape recorder or dictaphone.

For speed and convenience, the Panasonic Models 1510 and 1520 offer high-speed erasure of the incoming message tape. After all incoming messages have been monitored and noted, this mode is selected and all messages are erased in a few seconds. This feature can save the operator long periods of erasure time which are necessary when this latter feature is not available. The erased tape is then ready for replay to receive incoming calls again.

Figure 5-3 provides an explanation of the many controls of these models. It should be note that jacks are provided for an external speaker, earphones, and microphone. These TADs offer

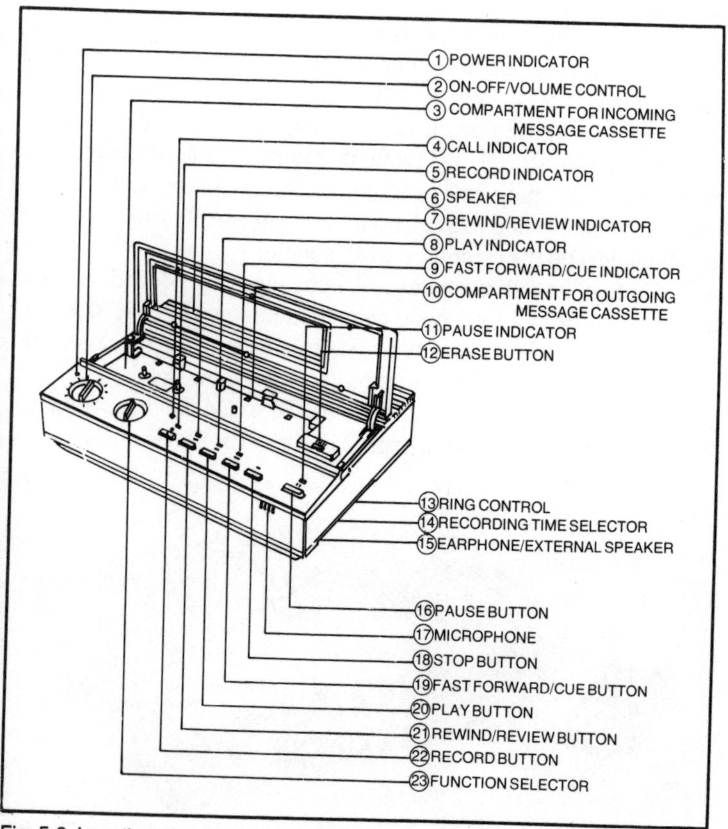

Fig. 5-3. Location of various controls on the Panasonic answering devices. Note that most of the operating controls consist of push-buttons.

most of the features of a good tape recorder along with the automatic answering capabilities.

Figure 5-4 shows the installation arrangements found on the back of each unit. The standard telephone is connected through the modular telephone *in* jack while the modular telephone plug connects to the line jack which is installed by the telephone company. The AC power cord is inserted into a standard 115-volt receptacle and provides operating power for the TAD's internal electronic circuitry and tape motors. Installation should take only a few minutes when the proper line jack has been installed by the telephone company. Older installations may require a call to your local company to have the present receptacle changed to the type

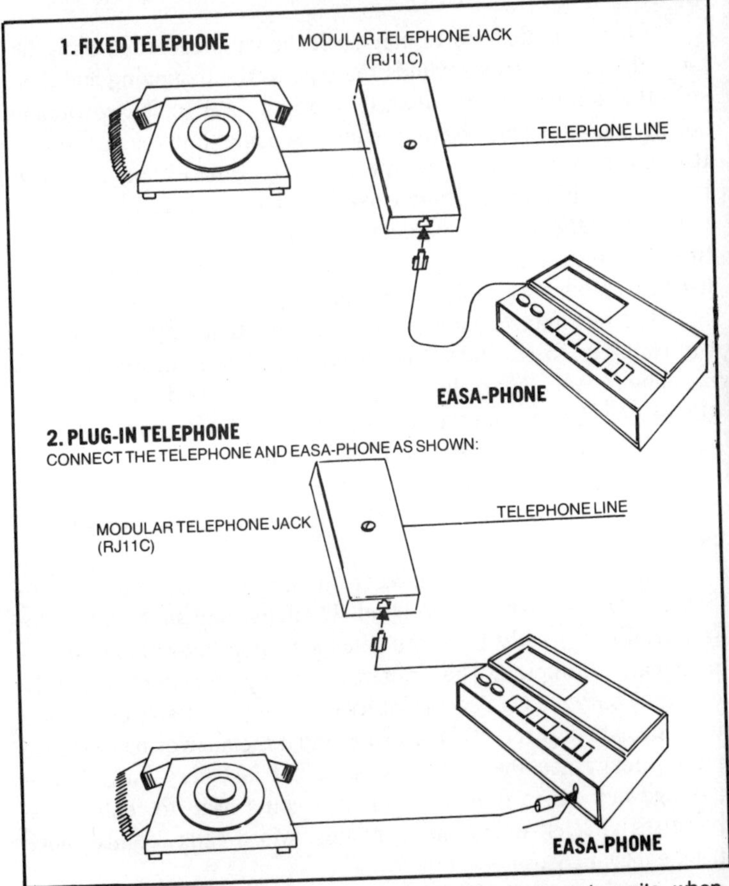

Fig. 5-4. Many variations are possible with the Panasonic units when connecting to a telephone line.

which will accept the modular telephone plug provided with the TAD.

RADIO SHACK DUOFONE TAD-10

The Radio Shack Duofone TAD-10 is shown in Fig. 5-5 and is relatively inexpensive and easy to use. Due to its battery powered circuit, it is portable and may be carried from one location to another and set up in a matter of seconds. Connection to telephone lines is made through a mini-modular phone plug designed to mate with existing jacks in office or home. All messages are recorded on an internal tape which must be manually rewound for playback. For automatic answering, all that is required is to rewind the tape (Fig. 5-6) and switch the device to the *answer* position, as shown in Fig. 5-7.

When it is desired to play back the calls recorded onto the tape, the user simply rewinds the tape to the beginning and then sets the control to the forward position. Due to the rewinding ability, messages may be played over as many times as necessary. It is not necessary to erase the tape each time either. Any new messages will automatically record over the old, erasing the old message as the new one is recorded. Since there is a considerable time lag between messages, the fast forward feature should be used to rapidly advance the tape when playing back the messages.

The maximum length of each message that may be played to the caller is 15 seconds. The caller then has a maximum of 25 seconds to leave his or her message. This is one disadvantage of the device, since it only has capabilities of playing and recording very short messages. However, due to the relatively low price of this device, it might be a good choice for non-critical applications where only short messages, such as name and phone number, are necessary.

The care and maintenance of this unit is also simple. Radio Shack advises that if a problem develops, the unit should be removed completely from the telephone line before any adjustments are attempted. The main maintenance procedure is merely changing batteries at regular intervals. The power source for this unit consists of three "D" batteries, and if the answering device is put to heavy use they will obviously not last very long. If you average more than five calls per day, you'll have to replace the batteries at least every three months. More calls require more frequent changes.

This device can be recommended for home use but for most office situations, other more elaborate TADs will probably be a

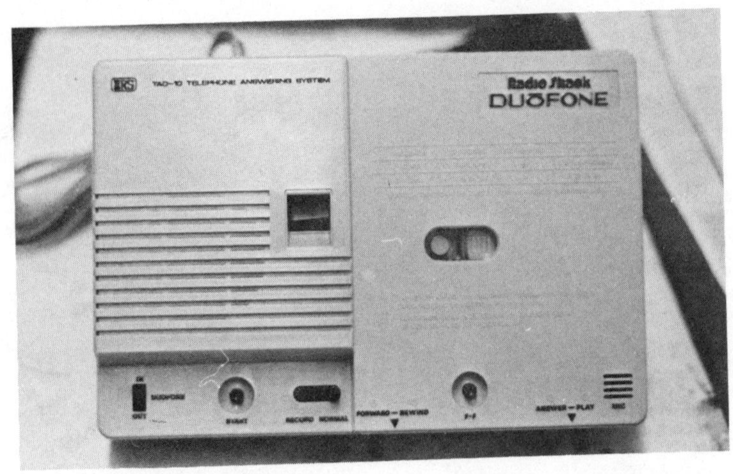

Fig. 5-5. The inexpensive Radio Shack Duofone TAD-10 is a portable unit that may be carried from one location to another and set up in a matter of seconds.

better choice, especially those that offer power supplies derived from conventional 120-volt AC circuits.

Figure 5-8 shows a schematic drawing of this unit using two integrated circuits and two transistors. The rest of the components are standard electronic parts and all may be replaced easily. The small amount of electronic components in this unit makes the Radio Shack Duofone TAD-10 reasonably trouble-free.

RADIO SHACK DUOFONE TAD-20

The Radio Shack telephone answering device shown in Fig. 5-9 offers many excellent features in a low-priced unit. It takes only a few minutes to set the device up and have it ready for use. It will automatically answer phone calls, play your answering message, and accept up to 30 messages from callers.

One feature of the TAD-20 is the remote control circuitry, permitting the user to receive messages from a remote location. All you do is dial your number and then use the portable remote control unit (carried with the user) to signal the TAD-20 to play back all messages.

The TAD-20 is powered by 120-volt house current and attaches to telephone jacks through a mini-modular plug. A built-in condenser microphone allows easy recording, and all of the features in the TAD-10 (previously described) are also incorporated into this unit.

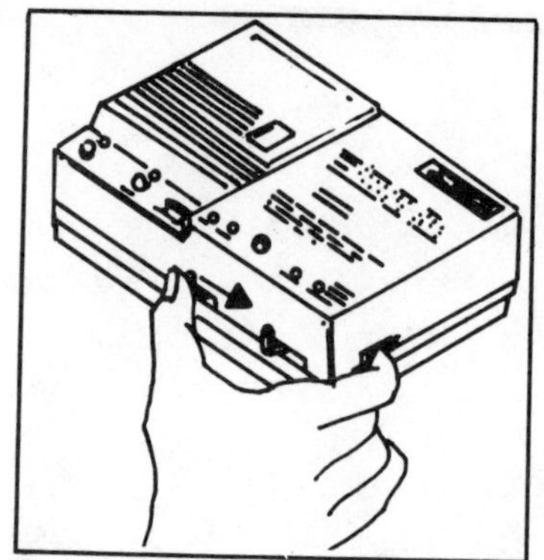
Fig. 5-6. Method of rewinding the tape on the TAD-10 for automatic answering.

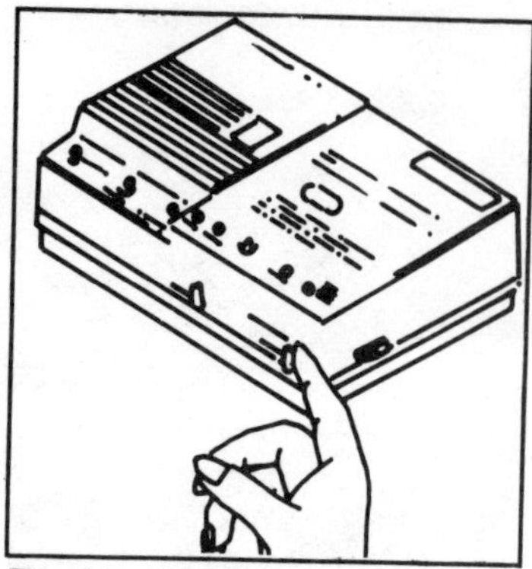
Fig. 5-7. Method of switching the TAD-10 to ANSWER mode.

To use the remote control device from a remote location, dial the number where the TAD-20 is connected and while the answering message is being played, press the button on the remote control device by holding it to the mouthpiece of the telephone. The base unit will then automatically rewind to the beginning of the tape and start playing all messages which have been recorded from callers. After listening to all of the messages, the remote control unit is once again held to the mouthpiece of the phone and the same button is pressed again. This resets the TAD-20, giving an indication that the reset function has taken place by continuing the remainder of the "answering" message from the point at which you first interrupted it. When this message is finished, the unit will shut off and be ready to answer the phone again.

Should you desire to put a message of your own on the device once you have heard all of the playbacks, just wait until the outgoing message ceases and then speak into the telephone. This feature can let someone at either your home or office know that you have checked in and received all of your messages. This might even serve as a reminder to yourself when you return.

You may call back, using the same procedure as before, and hear your messages as many times as necessary. Should you want to erase those messages that you have already heard (since they are taking up space on the tape), the TAD-20 can be remotely set to return the tape to the beginning of the reel and start recording again. However, the messages presently on the tape will be erased as new messages are recorded. To accomplish this, return the TAD-20 to normal function after hearing all messages as described previously. During the time the answering message is finished up,

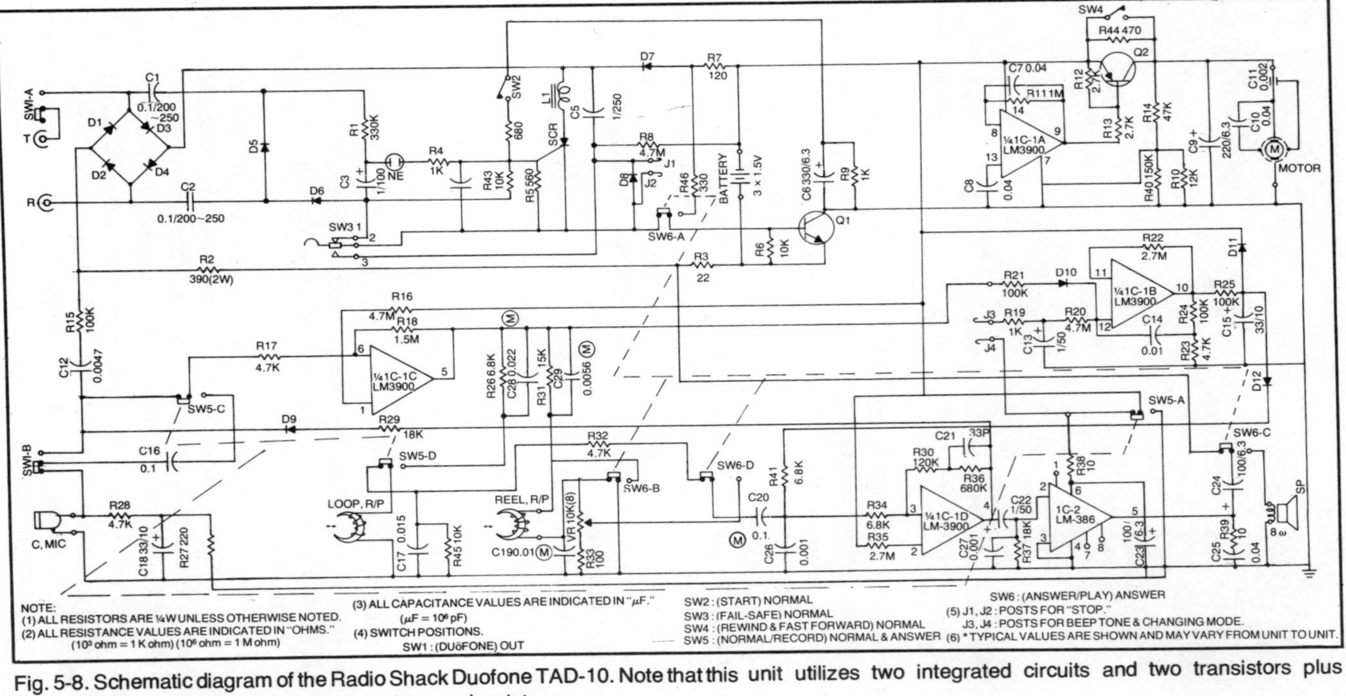

Fig. 5-8. Schematic diagram of the Radio Shack Duofone TAD-10. Note that this unit utilizes two integrated circuits and two transistors plus conventional components such as capacitors and resistors.

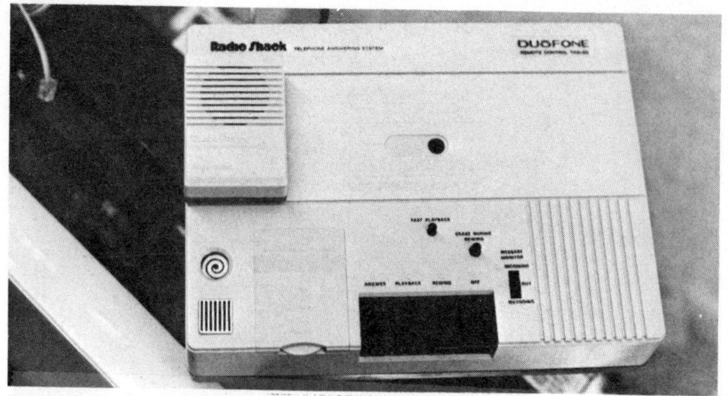

Fig. 5-9. Radio Shack telephone answering device 20 offers many excellent features in a relatively low-priced unit.

press the remote control button again. At this point, the unit will rewind and start playing back from the first incoming message. Press the button again. This action resets the unit and it will be ready to accept any new messages at any point desired. To illustrate, if you have received ten messages and only want to save the first five, you would play through message number five, then press the remote control button immediately upon completion of this message. At this point, the first five messages have been saved, but all others beyond this point will be erased as new messages come in.

The manufacturer cautions not to hang up the phone until you have reset the TAD-20 properly while using the remote control feature. If the TAD-20 is not reset properly, it will remain in the play mode, running to the end of its tape and then shutting off. This will cause no harm to the device, but no further messages can be taken.

A schematic diagram of the TAD-20 is shown in Fig. 5-10. Note that it uses two integrated circuits and about a dozen transistors. This obviously is more complex than the TAD-10 but is still relatively trouble-free and reliable.

Figure 5-11 shows a schematic diagram of the remote oscillator, which is an audio device that injects a tone into the telephone line at some remote location. This tone is sensed by a decoder circuit within the master unit and since the output of the oscillator does not fall within the range of normal speech patterns, accidental triggering of the remote playback circuit is not likely.

Care and maintenance of the Duofone TAD-20 is an easy task that involves only keeping the exterior clean and free of dust and

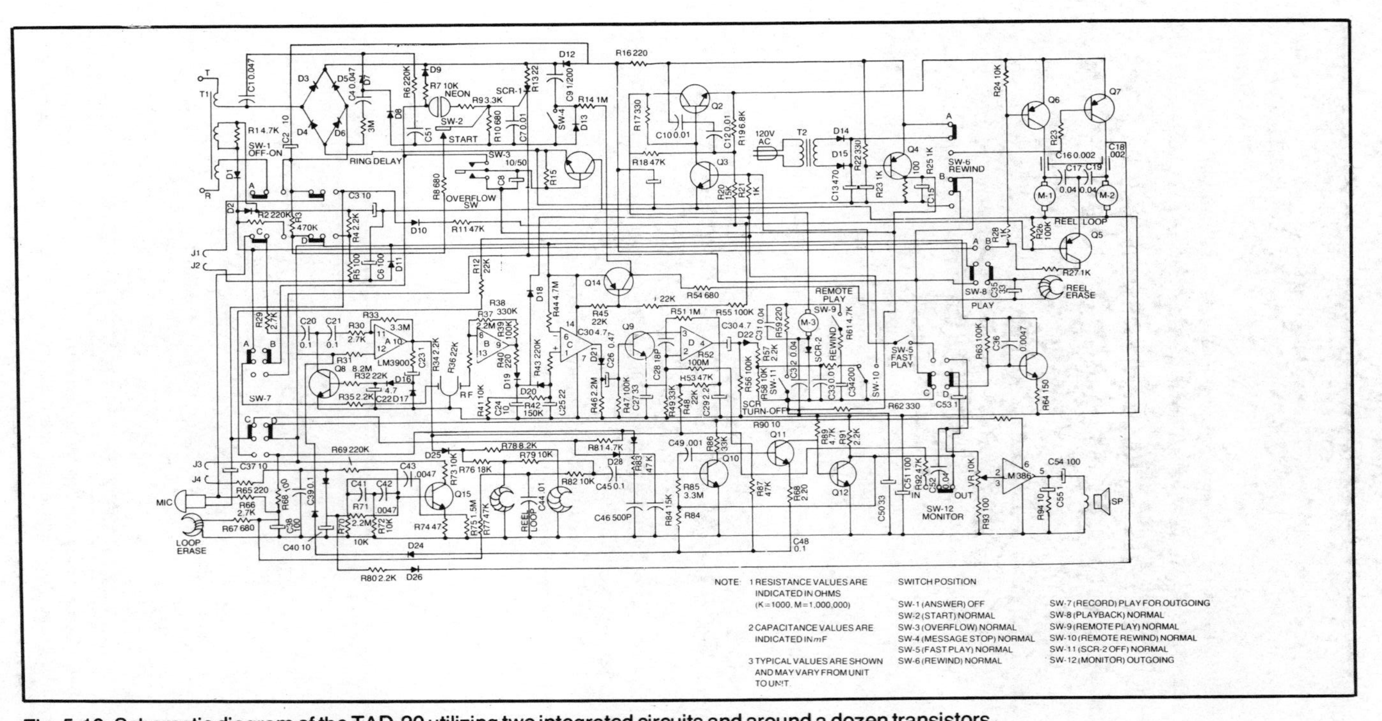

Fig. 5-10. Schematic diagram of the TAD-20 utilizing two integrated circuits and around a dozen transistors.

other foreign matter that might work into the tape drive mechanism. Rough handling, high humidity, and extreme temperatures should be avoided. The remote control unit is powered by a 9-volt transistor battery which must be replaced periodically. Radio Shack recommends a good quality alkaline battery for this unit, and due to the intermittent use involved, the battery should provide close to shelf life.

RADIO SHACK DUOFONE TAD-25

The TAD-25 (Fig. 5-12) is one of Radio Shack's top of the line answering devices. Instead of an integral tape reel as used in the Duofones discussed previously, standard cassette tapes are used for recording the messages. While the outgoing announcement length is rather short (a maximum of 20 seconds), the incoming messages can be as long as three minutes. This unit exhibits very good characteristics and maintains a speed to within ± 1%. Figure 5-13 lists the specifications for the TAD-25.

The TAD-25 utilizes two separate cassettes - one records the outgoing messages and plays it down the line to the caller, while the other records all incoming messages for playback at a later time. A built-in condenser microphone is standard, but a remote microphone jack is provided if a separate microphone has to be used. This unit even has an earphone jack for privacy when playing the messages back.

To install the unit, simply connect the mini-modular plug into the telephone jack and plug the line cord into an AC power outlet as shown in Fig. 5-14. This is one of the few answering devices which has a ring control, enabling the user to adjust the number of rings (from 1 to 10) before the machine answers. This can be used to advantage if certain messages are expected from individuals who

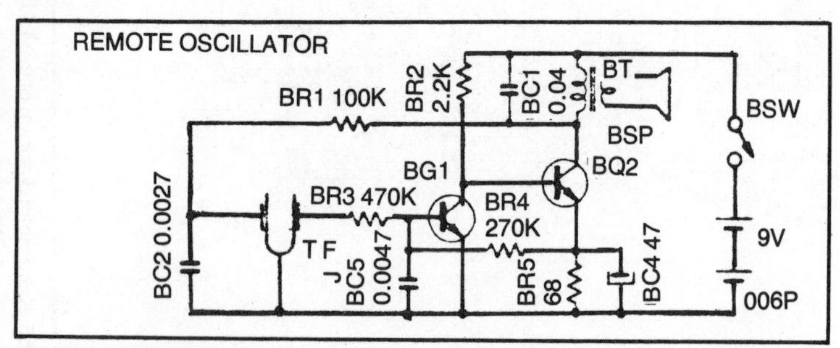

Fig. 5-11. Schematic diagram of the remote oscillator which is an audio device that injects a tone into the telephone line at some remote location to activate the TAD-20.

Fig. 5-12. Radio Shack Duofone TAD-25 is their top-of-the-line model using cassette tapes for recording answering annoucements and incoming messages.

have been informed that the answering device will only cut in after 9 or 10 rings. Most other persons not knowing this would have hung up by then.

Figure 5-15 shows the schematic diagram of the unit. Note the complexity of this diagram in comparison to other Radio Shack units. One reason for the additional circuitry is due to the dual tape head assemblies. This device, however, uses only one integrated circuit which serves as an amplifier for the speaker output. The balance of the circuitry is comprised of transistors, diodes, and other standard components.

RECORDaCALL REMOTE 90A

The RECORDaCALL® Model 90A with VOX (Fig. 5-16) is a complete telephone answering system with the capability of changing your outgoing announcement by remote control and also to backspace by the same system. To use, just dial your phone number and signal the Dual/Remote 90A to play back all your stored messages automatically. If you should miss any part of the message the first time around, signal the 90A to backspace to any point so you can hear all or part of the messages played back, over and over, if you wish.

The TAD 90A offers complete remote control command with the advantage of changing your outgoing announcement when away from your phone. For example, if you have a change of plans and want to change your announcement, simply press the buttons on the remote key and re-record a new announcement over the phone. The old announcement is replaced with the new message and is automatically played back to you for your okay. You can re-record as many times as you want to and when you're satisfied, merely hang up. The system will reset itself automatically with your new announcement.

Another feature, the VOX® with voice actuation, lets your caller speak as long as they wish. They aren't cut off before they are finished talking. Voice actuation will sense the absence of voice when the caller stops speaking and automatically hang up. When messages are long, the voice actuation will get it all. However, you still have the option of switching to a 30-second fixed time if so desired.

FEATURES

* VOX operation. The caller can leave a message as long or as short as they wish
* 20-second outgoing message length
* Call Monitor - listen to who is calling without picking up the telephone
* Selectable ring delay - answers your phone after 1 to 10 rings.
* Can be used as a standard tape recorder or dictating machine
* Two-cassette operation

For your own protection, we urge you to record the serial number in the space provided.
You will find the serial number on the bottom of the unit.

SPECIFICATIONS

Outgoing Announcement Length	20 Seconds
Maximum Incoming Message Length	3 Minutes
VOX Release Time	10 Seconds
Audio Power Output	500 mW (Max)
Frequency Response	300-3000 Hz ± 2dB
Ring Adjust	1 to 10
Cassette Speed Variation	±1%
Cassette Wow & Flutter	<0.2%
Rewind Time (C-60 Cassette)	140 Seconds
Power Requirements	120 VAC, 60 Hz
Power Consumption (Standby)	3.5 Watts

Note: The TAD-25 must not be connected to:
* Coin-operated systems
* Party-line systems

This Telephone Answering System has been approved and registered for direct connection to the telephone lines. The FCC registration number and the ringer equivalence number are noted on the label of your unit.

Fig. 5-13. Specifications for the TAD-25.

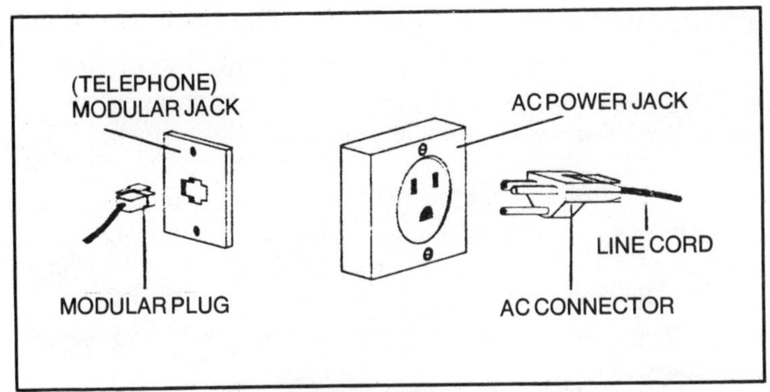

Fig. 5-14. Method of connecting the TAD-25 to the phone and power lines.

All of the RecordaCall functions are handled with only one control. By a turn of a dial, the 90A does it all; that is, turn the dial to answer your phone, turn to rewind, turn to play back your messages, turn to announce only, turn to record new outgoing messages, turn to record two-way conversations, turn to use your unit as a dictating machine if you wish, and so forth.

Incoming messages are recorded on a standard size cassette, which makes it convenient to store for future reference. The basic circuitry of the RecordaCall 90A consists of twin integrated circuit amplifier. A high quality dynamic microphone handles the sound and its power supply is conventional 120-volt AC. Some of the features of the 90A include:

—One control operation.

—Two cassette system.

—VOX® with voice actuation.

—Accepts up to 120 messages.

—Ring control from 1 to approximately 10 rings.

—Exclusive silent monitor: only the caller's voice is heard.

—Announce only, gives your message to caller and then hangs up.

—Fast forward to speed up message review.

—Rapid rewind saves time in rewinding message tape.

—Rapid erase, allows you to erase old messages when rewinding.

—Two-way conversation record with built-in lawful beep.

—Message light, flashes when message received, glows to indicate power is on.

—Limited warranty, one year parts, excluding accessories, and 90 days on labor.

RECORDaCALL 70A

The RecordaCall VOX® 70A offers many features found only in the top of the line machines and reflects many years of developmental research and technological advances by TAD Avanti Inc.

Like the 90A, the 70A has VOX with voice actuation that lets your callers speak as long as they wish. The system senses the absence of voice when the caller stops speaking and automatically hangs up. When the messages are long, the 70A has the capability of recording it all, unless you desire to switch to the 30-second fixed time mode.

A real convenience built into the 70A is ring control. It allows the selection of the number of rings desired for the machine to answer your calls. Such a device is extremely useful when you're busy at home. Then you can set your control for more rings to give you time to get to the phone or screen your calls, such as annoying nuisance calls. If you recognize your caller's voice, you can answer yourself, giving you complete control of your phone.

The control of the 70A is almost identical to the 90A discussed previously; that is, you handle all the functions with only one control - a turn of a dial. Personalizing or changing your outgoing calls is also convenient. Using the endless loop cassette, you can change your outgoing message as many times as you wish by re-recording a new message or by popping in different cassettes as the need or occasion dictates. Incoming messages are recorded on a standard size cassette, which makes it convenient to store for future reference. The 70A also has twin integrated circuit amplifiers with a high quality dynamic microphone.

RECORDaCALL 60A

The RecordaCall 60A is the junior TAD of the TAD Avanti, Inc. line, and as one would imagine, it is also their least expensive model. Still, it has a few features of its own. One of them allows you to control the length of incoming calls to either 20 or 40 seconds. With a flip of a switch, you can limit incoming calls to short messages or longer messages, letting your caller speak longer.

Like the other Avanti products, the Model 60A allows you to select the ring you want your machine to answer your phone on and also, like the other models, all functions are completed with one control - the turn of a dial. This unit also uses the trouble-free twin cassette system as described for the other RecordaCall units.

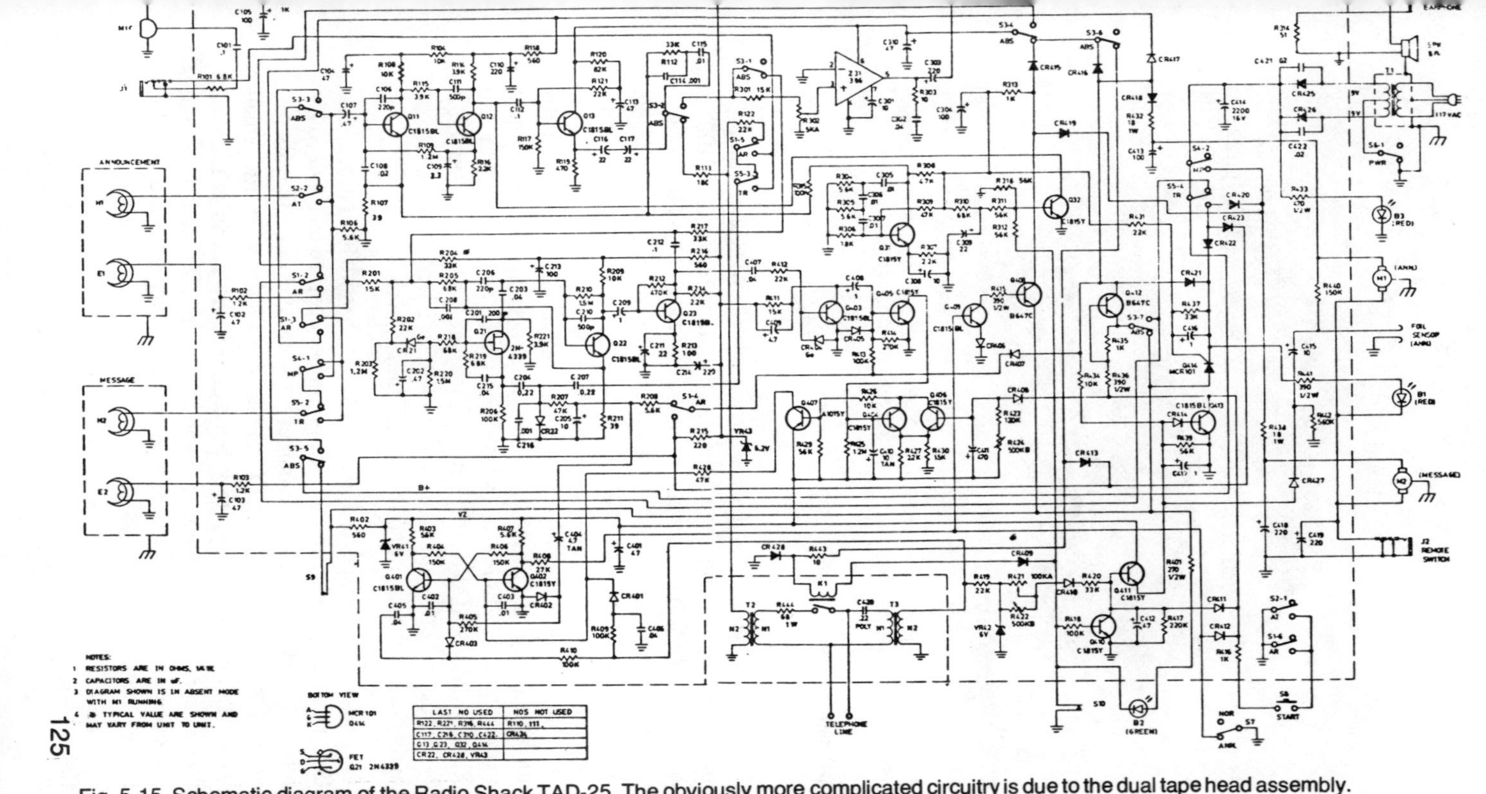

LAST NO USED	NOS NOT USED
R122, R221, R316, R444	R110, 111,
C117, C216, C310, C422,	CR424
C13, Q23, Q32, Q414	
CR22, CR428, VR43	

Fig. 5-15. Schematic diagram of the Radio Shack TAD-25. The obviously more complicated circuitry is due to the dual tape head assembly.

Fig. 5-16. Record a Call Model 90A TAD which has the capability of changing your outgoing announcement by remote control plus many other features.

PHONE-MATE 900

Phone-Mate, Inc. offers seven models of TADs which range in price and quality from their Model 900 (selling for around $100) to their Remote 930 that retails right around $300. The Phone-Mate 900 was one of the first TADs that featured a dual tape and still sold for a relatively low price. With the dual tape system, the outgoing announcement only needs to be recorded once and the self-rewinding tape resets itself automatically. This eliminates the inconvenience of having to record your personal announcement many different times as with "one-tape" systems. With the Phone-Mate 900, your incoming messages are recorded on a separate cassette, eliminating the need to listen to the outgoing announcement before each and every message received.

As might be expected, the Model 900 has its limitations, one being that only 30 half-minute calls can be received on a side. This might not be enough for a business taking in dozens of calls per hour. On the other hand, it might be exactly what the small business needs, or definitely the person who needs a TAD for home use. Other features include Power On and Message Received Lights, Audio Scan, Call Monitor, Fast Forward, and AC Power. See Fig. 5-17.

PHONE-MATE REMOTE 905

The Remote 905 is similar to the 900 model, except that in addition to the features found on the TAD 900, it offers remote capabilities. Besides a coded pocket tone key that retrieves messages from any telephone, the Remote 905 includes a Dual Tape system with a self-rewinding announcement tape, separate from the message cassette. This dual tape system eliminates the inconvenience of repeatedly recording announcements, as well as listening to the outgoing announcement before each message received. This model records up to 30 half-minute messages per side and includes other quality features such as Audio-Scan for instant location of messages on fast forward or rewind. See Fig. 5-18.

PHONE-MATE 910

A moderately priced TAD is shown in Fig. 5-19 and is a multi-functioning unit offering a wide selection of features, including twin cassettes, Audio-Scan, Ring Adjust, Call Monitor, and Announce Only. Individual cassettes pop in and out easily for replacement. This unit records up to 60 half-minute calls per side. With twin cassettes, you can file important messages for future reference while also establishing an "announcement library" for recurring needs.

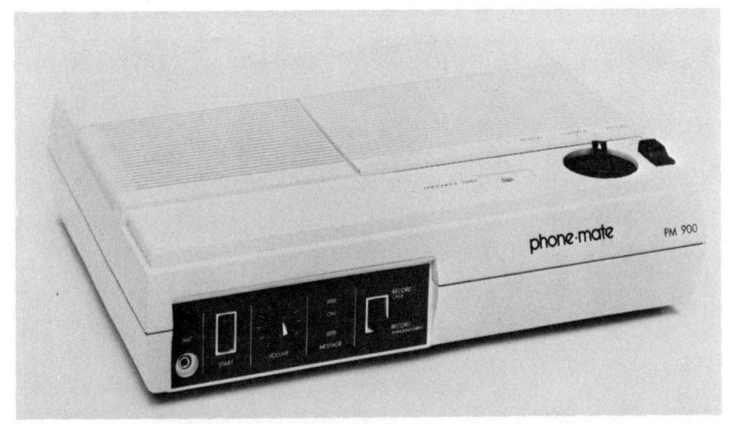

Fig. 5-17. Phone-mate 900 is the least expensive model of three offered by this firm and therefore has less features, but it may be just the thing for home owners with minimal needs.

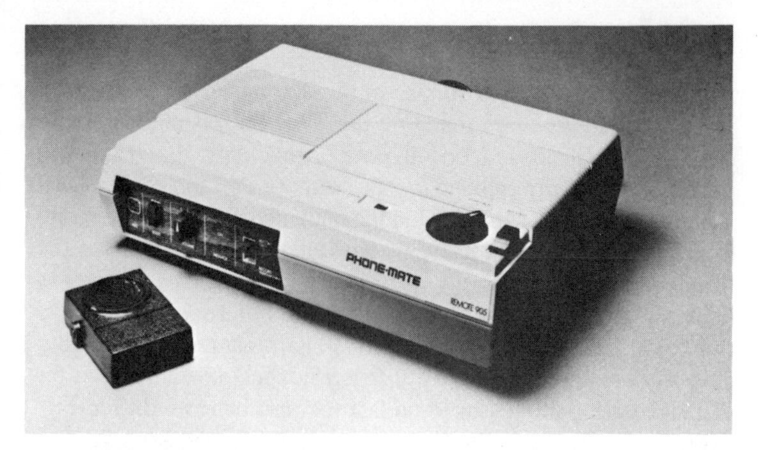

Fig. 5-18. The Phone-Mate 905 is similar to the one shown in Fig. 5-17 except for the remote capabilities on the 905.

PHONE-MATE C-VOX 920

This microprocessor engineered unit is another model featuring all sorts of "extras" on a TAD, again at a moderate price. The Controlled Voice Activation (C-VOX) allows for lengthy or involved messages. As long as this system recognizes the sound of a voice, it will continue to record. It automatically stops recording when the caller finishes speaking, leaving no blank space before the next call.

This unit records approximately 120 calls per side, and you can adjust the length of the outgoing announcements. See Fig. 5-20.

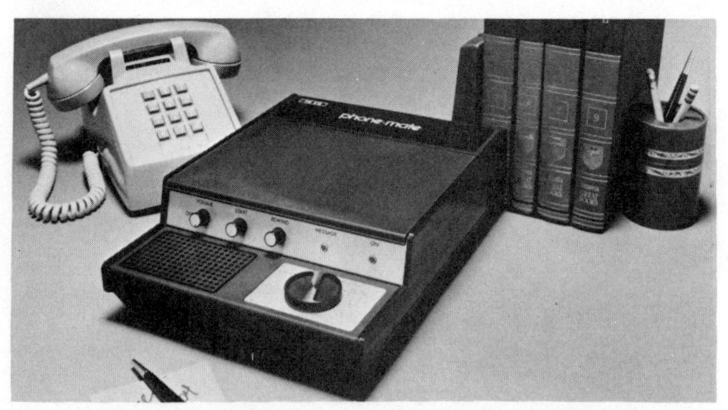

Fig. 5-19. The Phone-Mate 910 is a multi-functioning unit offering a wide selection of features including twin cassettes, Audio Scan, Ring Adjust, Call Monitor and Announce Only.

PHONE-MATE REMOTE 925

With the introduction of this model, Phone-Mate established a new low-price point for a twin cassette, remote TAD. Remote control, as on other devices, enables the user to retrieve messages from any telephone anywhere and at any time. All you need is the coded pocket tone key that comes with the unit and is shown in Fig. 5-21.

Phone-Mate's easy-to-replace cassettes eliminate the expensive repair charges necessary to change a tape on reel-to-reel machines. Furthermore, it records up to 60 half-minute calls per side - enough for the average business that doesn't have an answering service or person in the office to take the calls. Other features are as described for the 900 through 920 TADs.

PHONE-MATE REMOTE 9000XL

This multi-featured microprocessor controlled TAD (not shown) seems to be one of the most popular models in the Phone-Mate line. The state-of-the-art Fail-Safe design automatically corrects common user mistakes with maximum reliability and quality operation. With a turn of the function selector, this unit can be used to record two-way conversations, tape record/dictation or provide for announcement only capability. Other features include remote control, twin cassettes, C-VOX, fully variable outgoing announcement, and other features discussed previously.

PHONE-MATE REMOTE 930

This model represents the top-of-the-line for Phone-Mate. Such features include Fail-Safe Design, LED Digital Call Counter, Broadcast Timer, and Remote Backspace, which allows you to

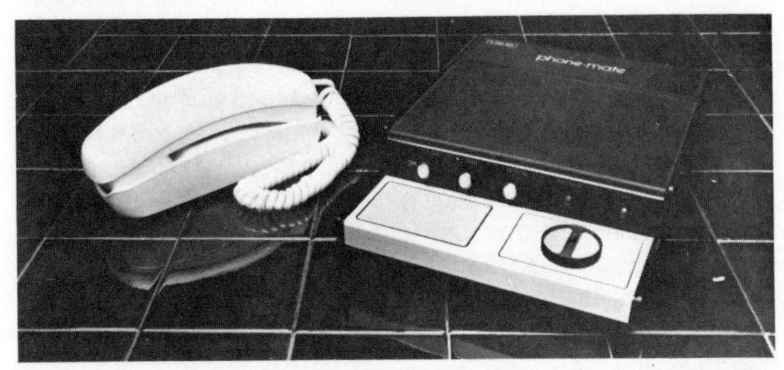

Fig. 5-20. The controlled voice activation (C-VOX) on the Phone-Mate 920 allows for lengthy or involved messages from callers.

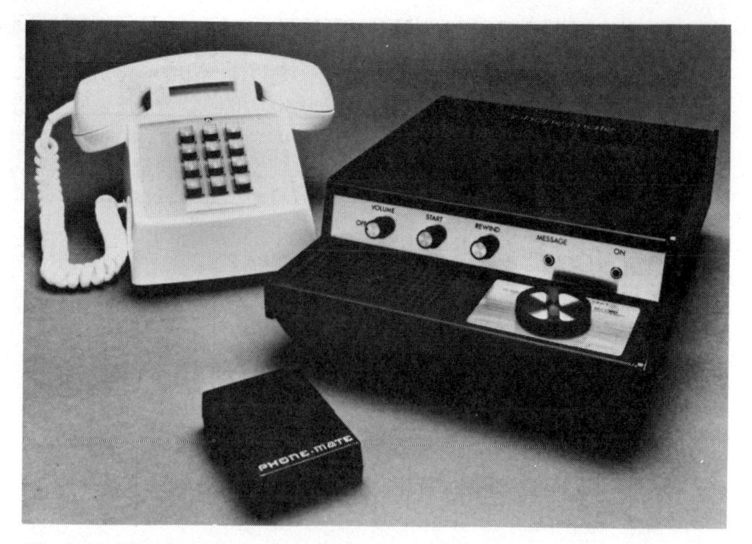

Fig. 5-21. With the introduction of this Phone-Mate Remote 925, the manufacturer established a new low price unit for a twin cassette, remote TAD.

remotely replay individual messages instantly without waiting for the entire tape to rewind and replay. Other features are the same as described for the 9000 XL. See Fig. 5-22.

ITT OWN-A-PHONE PC 6000

The Perfect Answer TAD, as shown in Fig. 5-23, is ITT's top of the line answering device designed for the discriminating user. This device has the capabilities of setting the circuits to your own individual code through the use of five microswitches. Then, from any telephone in the world, you can play back your messages by using your voice to activate the machine.

A built-in telephone incorporated into the unit enables its owner to use this as a microphone for recording any outgoing messages. Dual cassette system features a tape for variable length outgoing messages and a voice activated incoming message tape which allows callers to leave any length messages up to 3 minutes.

The LED alpha/numeric display is a message counter and can be used to time your outgoing messages as you record. It also indicates and confirms the unit's operating functions. Other features include push-button operation with fast forward, fast rewind, and volume controls.

ITT PC4000

The Easy Answer™ unit by ITT is designed for the entire family to use, since its single-knob selector allows the user to dial

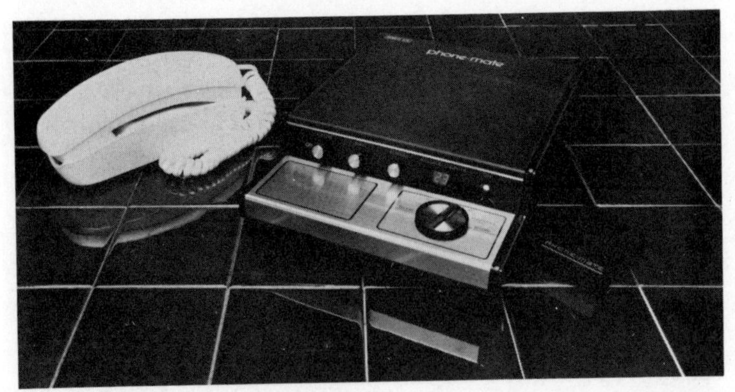

Fig. 5-22. The multi-featured Phone-Mate Remote 930 is one of the most popular TADs in the Phone-Mate line.

recording, answering, playback and dictation functions. This unit also has dual cassettes, a message counter, fast forward and fast rewind controls, a built-in microphone plus indicator lights for power and the answering mode.

Outgoing messages on this model are recorded on a replaceable 20-second endless loop cassette, while incoming messages are taped on a 30-second per call basis on a standard 60-minute two-sided cassette. This device also has an on/off volume control and tape ejectors. See Fig. 5-24.

QUASAR TADs

Quasar Microsystems, Inc. currently offers three different models of TADS, one at under $100! The least expensive model is ideal for home use, and it will insure that you will never miss another call while you're out or wonder who it was that tried to reach you. With this model, you can politely answer every call with

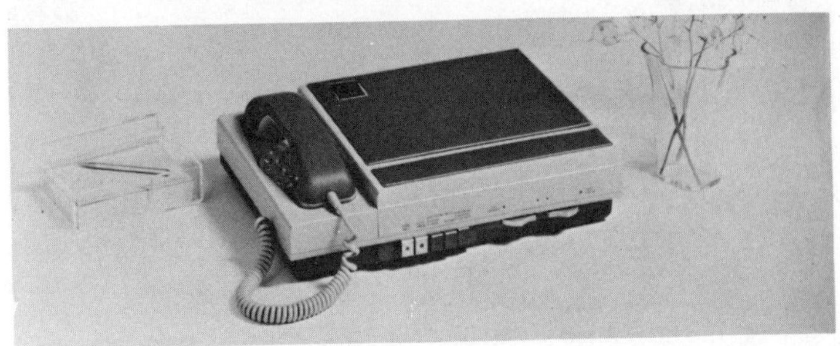

Fig. 5-23. The Perfect Answer, as it is called by ITT, is their top-of-the-line answering device.

a pre-recorded greeting, inviting callers to leave a message if they wish.

Quasar's Call Jotter comes equipped with a cassette tape on which you can record the message you wish to use for greeting callers, or just use the prerecorded cassette tape which is furnished with the Call Jotter.

You can screen incoming calls with this TAD by turning up the volume control to monitor incoming calls. The Call Jotter will then automatically answer incoming calls on the first ring with a pre-recorded message, and you can hear the caller's name when he replies (without his knowing that you are there). You then have the option of ignoring the call or coming in on the call if you wish to speak with him or her.

On this model, callers have 36 seconds to leave their message, which allows more than enough time in most instances for then to leave their message. Research reports have determined that most messages left are between 25 and 30 seconds in length. The cassette tape on which these messages are recorded has a 30-minute capacity per side.

This model TAD has the added advantage of a portable cassette player to be used by family members for instructional tapes or to simply listen to pre-recorded music. The electronic

Fig. 5-24. The Easy Answer TAD by ITT is designed for the entire family as it features a single knob selector to dial recording, answering, playback and dictation functions.

circuitry utilizes all microprocessor techniques designed for trouble-free operation. The smaller model is shown in Fig. 5-25.

Next up the ladder is Quasar's Call Jotter with a remote telephone answering system. The specifications are similar to the model previously discussed, except that you can receive your messages remotely from anywhere in the world! This system allows the user to call from any telephone and have his messages updated and the unit reset to accept new messages all in one call. This model uses one C-60 cassette tape for its pre-recorded announcements and incoming messages. These tapes are readily available in any retail store where audio goods are sold. See Fig. 5-26.

The top of the line model from Quasar features a dual-cassette remote telephone answering system. This model automatically answers your phone on the first ring or can be set to answer on a number of rings, up to four. See Fig. 5-27. The greeting announcements is automatically variable in length, which means there is no need to manually reset the announcement time. Callers are not forced to wait for the announcement tape to "cycle out", since the caller may start recording immediately following the announcement.

The length of the caller's message is voice controlled to provide each caller with as much time as necessary (within

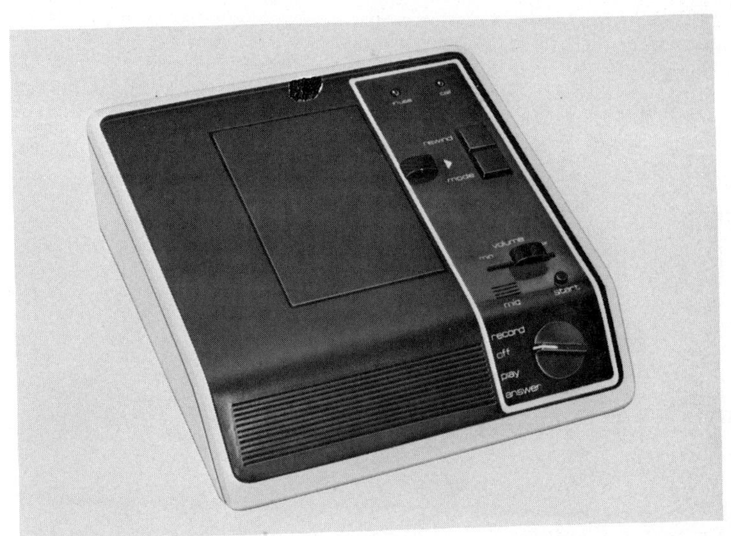

Fig. 5-25. Call Jotter by Quasar is a very low-priced TAD that is excellent for home use.

reason), up to one minute, and will be repeated word for word from any telephone by means of its remotability features.

This model has all of the features mentioned for the other Quasar models, plus some of its own. The specifications for this model follow:

Model Description: Automatic answer, announce, remote control system, Model 10A.

Power Requirements: AC/DC power pack U/L approved.

Recording Capacity: Outgoing message variable time, incoming message voice-activated, up to one minute.

Message Capacity: One outgoing message, 60 incoming messages of 30 seconds each.

Microphone: Built-in-dynamic.

Shipping Weight: 9 Pounds.

Dimensions: 2″ × 2½″ ×¾″.

Remote Capability: Receive or replay messages from any telephone.

FCC Number: AAR99N 62592 AN-N O.0B.

SUMMARY

Telephone answering devices are most abundant on today's consumer market. The models described and pictured in this chapter are but a random sampling of all that is available. They do,

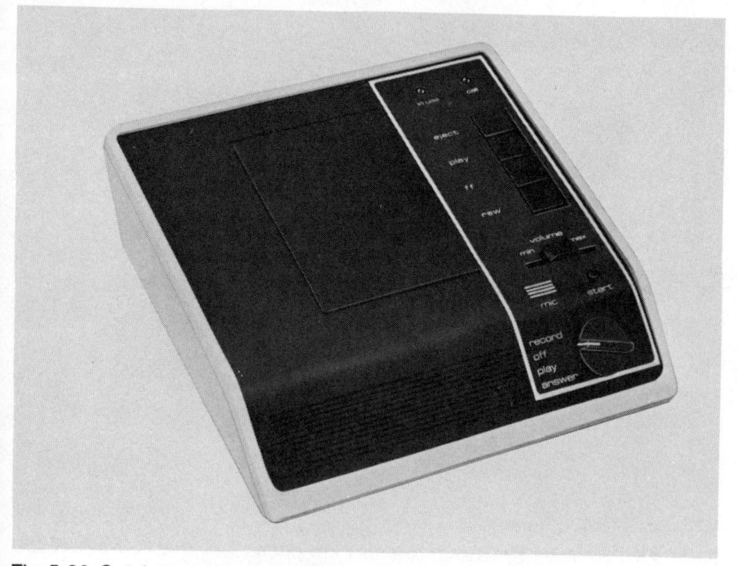

Fig. 5-26. Call Jotter with remote telephone answering system.

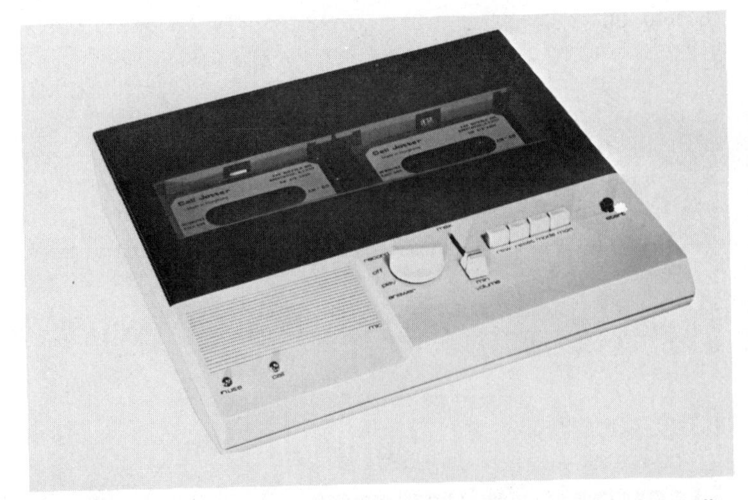

Fig. 5-27. This top-of-the-line model from Quasar features a dual-cassette remote telephone answering system as well as many others.

however, provide the reader with a cross-sectional example of all there is to choose from.

The TAD can take the place of a secretary or other office employee when it comes to answering the phone while no one is present to do so. Some models will take a recorded message from the caller, while some will not. The remote type will even play your messages down the phone line when you call from across the street or from the other side of the world when keyed by the proper tone signal. Due to the wide variation of conveniences offered by telephone answering devices, it is an excellent idea to shop around and consider all the conveniences offered as compared to the price and your individual requirements before actually making a purchase.

Once the telephone answering device is installed in your home or office, persons calling your number will have a means of keeping in touch with or, at least, getting a message through to you on a 24-hour-a-day basis. While TADs are a supreme convenience when used in the home, many businesses consider them to be vital necessities to their operations, saving perhaps millions of dollars each year.

Chapter 6

Electronic Telephones

The discussion of telephone systems, operations, and instrumentation presented so far in this text has covered a large category of services and devices available to the average consumer. This chapter will deal exclusively with electronic telephone instruments. True, all of the instruments and devices discussed to this point have used electronic circuits to varying degrees. This chapter's discussion is different in that the electronic circuits in the devices featured form a major part of the instrument and provide a spectrum of conveniences and services that are not common to most telephones. It should be remembered that communication is still the main purpose of these instruments, but they represent modifications and improvements upon of some of the phones mentioned previously. They use electronic circuitry to perform many of the manual functions of less sophisticated devices.

Many electronic telephones feature self-dialing capabilities through programming and memory circuits. Many offer built-in intercom features along with control of various desk units from a central location. Some even offer intercom conversation with callers. This means that the user does not need to take the handset off the cradle to talk to the calling party. Other devices may offer built-in calculators, automatic music playdown circuits, and even a combination of these.

Many of the instruments in this chapter may be used universally for a myriad of communication functions. Others may be applicable only for a moderate to large-sized office. Regardless of their uses, all of the models discussed in this chapter depend heavily upon electronic circuits to provide their design conveniences. Because of this, they differ from most of the previously discussed instruments in this book.

AUTOMATIC TELEPHONE MEMORY DIALER

Only a few years ago, making a phone call involved looking up the number in your phone book, writing it down, and carefully dialing each digit in order to avoid a wrong number. Some of the more adventuresome telephone users even attempted to memorize a number viewed only seconds before in the phone book and then hoped that their memories were adequate to dial the correct number. All too often, memory failed and many wrong numbers were dialed in the course of the day.

This bother is no longer a necessity of telephone use with the invention of automatic telephone dialers which take all of the human memory work out of the dialing process and commit selected numbers to an electronic memory which is far more accurate and reliable than its human counterpart. The dialers on today's market can store from about 10 to 32 telephone numbers which are programmed ahead of time. The dialing process is accomplished by simply pressing a button on the automatic dialer. The internal memory and switching circuits do the rest.

Fig. 6-1. Radio Shack Duofone-32 memory dialer.

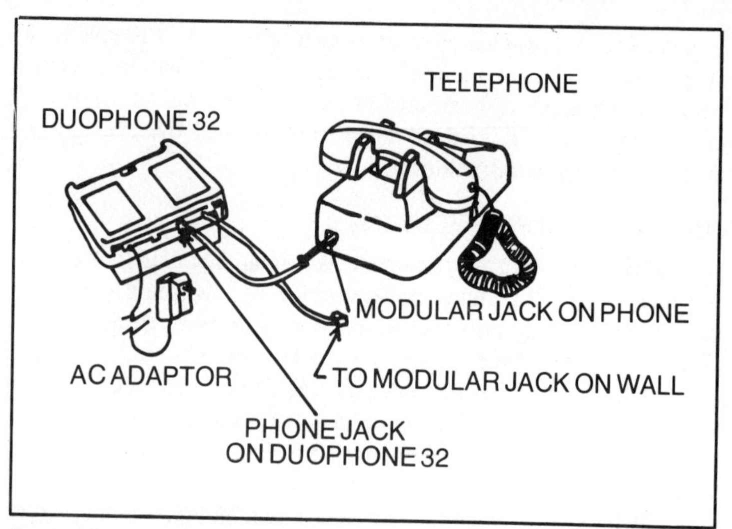

Fig. 6-2. Connection of the Duofone-32 to telephone instrument and phone line.

These devices connect to your standard telephone in a few seconds using modular plugs. Some even incorporate the memory dialing circuitry in the body of a complete telephone instrument. All of these devices are convenient time-savers when programmed with numbers most often dialed from your home or business.

Figure 6-1 shows the Radio Shack Duofone-32, which can be pre-programmed with as many as 32 phone numbers. A single button is pressed which automatically sends the dial information down the phone line for you. It features quick-connect installation using modular connectors or telephone adapters. It is compatible with both rotary dial and touch-tone telephones and has a switch selectable output pulsing rate compatible with all telephone exchange systems. The self-contained microcomputer circuitry allows the programming of the 32 phone numbers with up to 14 digits per numbers. This means that even long distance phone calls may be keyed up with the punch of a button. Other convenient features include a clock which displays time continuously or converts to an elapsed timer allowing the timing of phone calls up to one hour in length. An internal 9-volt battery is installed within the unit, but the main power source is from an AC adapter. Should the AC power fail, the internal battery still supplies power to the memory circuit, preventing loss of programming. If this feature were not built in, a power failure would erase the memory and reprogramming would be necessary.

Figure 6-2 shows the connection of the Duofone-32 to a telephone instrument and to the modular jack from the wall. This installation takes only a few minutes to perform, provided a power outlet is located near your present telephone. If not, it will be necessary to run an extension cord to this location.

Installation of the 9-volt battery for auxiliary power in the event of a power failure is simple and can also be completed in a few minutes. The battery compartment is located in the base of the unit and is accessed by snapping off the bottom plate with a small screwdriver. Figure 6-3 shows this installation. Radio Shack cautions that the battery should be replaced only while the AC adapter is plugged in to prevent loss of programming. Shelf life can be expected from this battery as it is not necessary for normal operation and provides no power except when the power line fails.

Programming the Duofone-32 appears a bit complicated upon reading the instruction manual and, indeed, eight pages of a 16-page instruction manual are devoted to programming the various numbers and sequences into the circuitry. A re-dial feature is also available which will automatically re-dial the last number selected as many times as you push the re-dial button. Once programming has been established, the unit need not be adjusted again unless it becomes necessary to change a programmed number.

On the rear of the unit is found the dial rate switch. This is a two-position switch, one side of which is labelled 10 and the other side 20. The 10 position is used for most telephone systems, especially the rotary dial type. The 20 position provides much faster dialing speed and some phone systems may not be able to process this rate, which will result in erratic or inconsistent

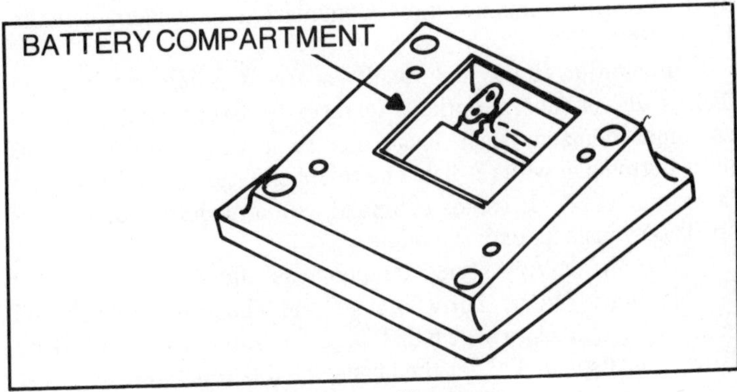

Fig. 6-3. Installation of 9-volt battery for auxiliary power.

dialing. The two positions are accurately named, as the lower number gives a dialing rate of 10 pulses per second (pps) and the 20 position provides 20 pps.

One note of caution: This device and many others like it are especially susceptible to static electricity. Under the right conditions, the human body can build up a significant amount of static electricity which, if discharged into the keyboard of the Duofone-32 from your hand, may completely wipe out the memory. Radio Shack cautions through its user's manual that you discharge your body by touching a metal object before applying your hand to the surface of the Duofone-32.

Figure 6-4 shows the specifications for the Duofone-32. A schematic is not provided in the owner's manual. As with all other such devices, it is necessary to notify the phone company of your intentions to install it in their lines. Once installed, the automatic dialer is very convenient and efficient in saving time which might normally be required to look up the number and then manually dial it on your telephone. Obviously, only the more important numbers and those which are used most will be entered into the memory and then recorded on the faceplate index.

MODULUS I

The Modulus, a new design in telephone sets, is manufactured by Tele-Devices Ltd., Montreal, Canada and Tele-Devices Corp. of Plattsburgh, New York. Shown in Fig. 6-5, this basic model is available with different electronic circuits to provide the features desired by the customer.

This new telephone desk and wall set is fully comparable in quality of performance to the sets now in use on the American continent and is compatible with existing systems. It is approved by the Federal Communications Commission for connection in the United States.

In addition to the high quality of transmission, ringing, and dialing, it has several optional features to meet recent trends in customer demand, such as an electronic ringer and universal push-botton dial with re-dial. The telephone can be used as a main or extension set. It comes equipped with a modular plug or with spade terminals for export models.

The attractively styled telephone is offered in three color combinations (ivory, burnt orange and chocolate brown) and contrasting chocolate brown dial face. The subscriber's telephone number is displayed under the handset and is protected by a clear cover.

The housing has a positive cradle for the handset. The housing, handset, and base are of high impact ABS material. Unitized assembly of the set facilitates easy installation and servicing. Printed circuit boards and standard electronic components are used throughout for compactness and dependability of operation.

The dial is designed for reliable operation in excess of one million operations and the two-tone electronic ringer is highly sensitive over a range of frequencies. Its volume can be externally adjusted by means of a small knob located in the side of the set. The handset is light and shaped for optimum acoustic performance and

Specifications

Capacity of phone numbers:
32 phone numbers (with up to 14 digits per phone numbers)

Power Requirements:
AC: AC adapter supplied (provides 15VAC 300mA at 60 Hz) for operation from 120VAC 60 Hz.
DC: 9V battery for Memory protection during power failures.

Clock and Timer:
Accuracy—same as the 60 Hz AC line accuracy.
Operation—same as an AC wall clock. That is, if power goes off, the clock and timer stop and will resume operation when the AC power returns.

Telephone Connections:
Uses modular connectors (USOC type RJ-11C)

Telephone Specifications	10 PPS Dialing Rate	20 PPS Dialing Rate
Dialing Rate	10.2 Hz	19.6 Hz
Make/Break Ratio	62.7%	62.7%
Interdigit Period	700 mSec	390 mSec
Ringer Equivalence	0.0B	0.0B

Fig. 6-4. Duofone-32 table of technical specifications.

comfort to the user. The four composition feet and additional weight ensure that the set will not slide when the cord is extended.

As was previously mentioned, there are several different circuits which may be found in the same case style. One model provides hand-free operation by providing a built-in speaker and sensitive electronic microphone that lets you talk without a handset. Another allows the user to connect a stereo system or some other source of music to the instrument which will, in turn, play this music down the line when a caller is placed on hold.

The Modulus Ampliphone provides a call monitor which means you never have to pick up the phone to dial a call. The user simply presses the speaker button, listens for dial tone and then dials the number. When the call is answered, you simply pick up the handset and speak in a normal manner. The user may also dial in a normal manner.

All of the models feature last number re-dail and allow the last number dialed to be redialed simply by pressing the numbers (#) button. The Ampliphone's universal push-button dial works anywhere with either touch-tone or rotary dial systems. A conference speaker and hold feature provides group listening conference calls at the touch of a button. The speaker also permits hands-free holding by monitoring the line whenever the caller is placed on hold. It also has a two-tone adjustable ringer, is completely solid-state, and is lightweight in design.

The Modulus Serenader offers many of the features of the rest of the Modulus line, plus the advantage of playing music down the phone line to the caller who has placed on hold. This feature is ideal for business offices which must take a large number of phone calls simultaneously. Under these conditions, callers are often placed on hold for moderate to long periods of time. An input to the Serenader will accept the output from a stereo system, tape recorder, broadcast receiver, or many other types of audio devices. Many users connect their Serenader to FM radio receivers turned to special sub-channels which supply continuous mood music.

Technical Specifications

The Modulus telephone set is designed to work with battery-operated manual or automatic switchboards.

☐ **Transmitter.** Operates over a range of 300 to 3400 Hz within the band as shown in Fig. 6-6. Dynamic resistance of the transmitter is 130±50 Ohms as measured in the handset, with

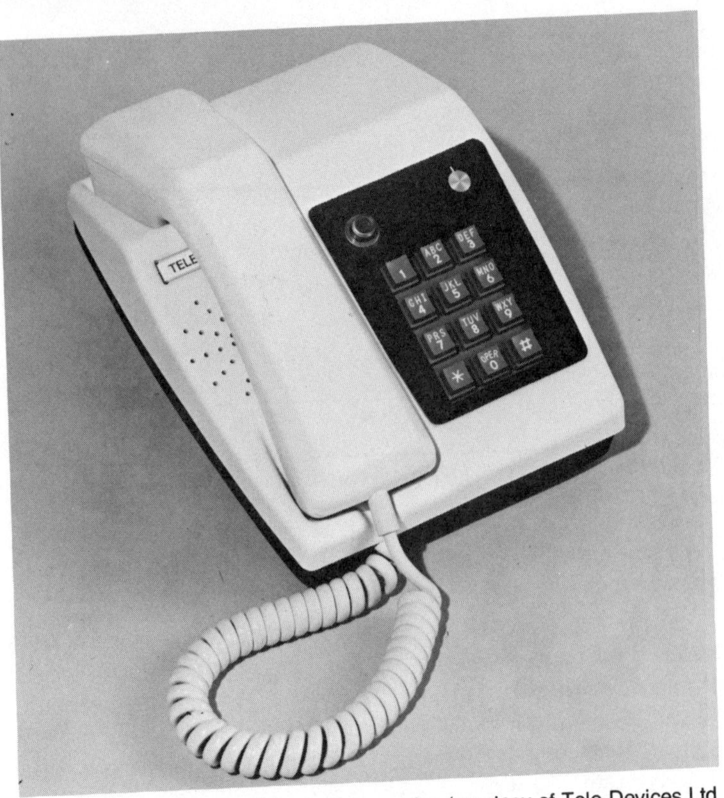

Fig. 6-5. Modulus electronics telephone series (courtesy of Tele-Devices Ltd and Tele-Devices Corp.).

diaphragm in a vertical position. Reference equivalent in transmission as measured by objective method for subscriber line of 4.3 dB (0,5N) is ±.43 to ±6.5 dB.

□ **Receiver.** Frequency response is in the range of 300 to 3400 Hz as shown in Fig. 6-7. Reference equivalent in receiving as measured by objective method for subscriber line of 4.3dB (0.5N) is ±2.6 to 7.8 dB.

□ **Ringer.** Two-tone electronic ringer is designed to operate at frequencies above 10 Hz. Ringer loudness at maximum setting as measured .5 m (1.6 ft.) from the telephone set ... ... min. 70dB. Regulation of ringer loudness ... ... 0 - 70dB

□ **Dial.**

Pulsing frequence ... ... 10± Hz
Percentage break ... ... 61±2 (Export 67±2)
Minimum number of operations ... 1,000,000
Interdigital pause ... ... 420 ms

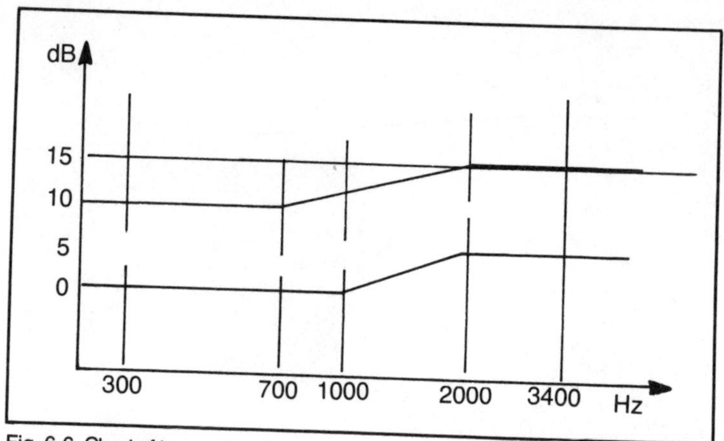

Fig. 6-6. Chart of transmitter frequency response (courtesy of Tele-Devices Ltd and Tele-Devices Corp.).

☐ **Sidetone.** Effective attenuation of sidetone in the frequency band of 300 - 3400 Hz is . . . . min. 22 dB.

All of the Modulus models are available in units which are designed for touch-tone areas only and for systems which accept only rotary dial instruments. These latter Modulus instruments are still push-button models but they are pulse triggering rather than conventional tones and are completely operational with the rotary system.

RADIO SHACK DUOFONE-16

A combination telephone instrument and automatic dialer is found in the Realistic Duofone-16. Its custom designed features provide one touch automatic dialing and hands-free conversations from the home or office. This device differs from a previous automatic dialer discussed in that the Duofone-16 is a complete telephone instrument. It is not an adjunct to your present telephone. Because of the sophisticated electronic circuitry incorporated in this device, it is not intended to be used while another phone on the same line is off the hook or engaged. Should a second phone be picked up, the Duofone-16 level may drop drastically or may even go dead and cause static and other noises to be present on the phone line.

The device is shown in Fig. 6-8 and features self-contained microprocessor circuitry, automatic memory dialing of up to 16 phone numbers, each of which may contain as many as 15 digits, and a built-in amplifier for hands-free talking or listening. A memory protection circuit stores the programmed numbers for up to a period of one year. As is customary with most electronic telephones, the last number dialed is automatically stored in memory and may be re-dialed by pressing the numbers button.

While most automatic dialers have certain differences, all of them do operate in similar manners. A brief explanation of the various controls should help the reader to understand the use of this entire category of devices, although different models may require slightly different operational procedures. The keyboard is used to dial phone numbers or to enter them in memory. The automatic keys on the right of the instrument are used for automatic dialing as well as for entering complete numbers into the memory. A red priority key is available to store a frequently used or emergency telephone number, such as Fire, Police, or Rescue Squad. Another control called the Store key prepares the unit for entering a specific phone number into memory.

As was previously explained, the numbers key is used to automatically re-dial the last phone number which was manually dialed. This is important to remember, because a number which has been played from memory will not be registered in the last number memory section. Only manually dialed numbers are stored here. To the left of the 0 key is the Pause key which is indicated by an asterisk (*). This is used when a pause is necessary between numbers such as when accessing an outside exchange from an office. There are many indicators on the Duofone-16. Just above the push-button keyboard are the dialing indicator and the memory

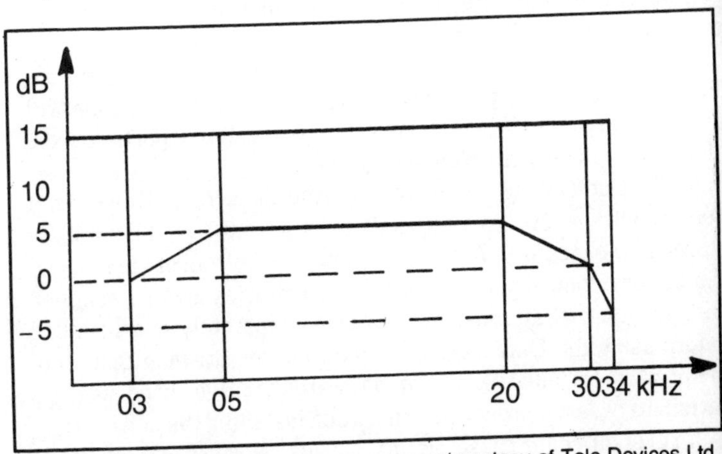

Fig. 6-7. Chart of receiver frequency response (courtesy of Tele-Devices Ltd and Tele-Devices Corp.).

store indicator. The latter shows that the unit is ready to store a memory, while the other simply indicates that the unit is activated by either removing the handset from the cradle or by depressing the amplifier button. This latter control is used to activate the built-in telephone amplifier for hands-free operation.

Since everyone talks at a different audio level, the Duofone-16 has a sensitivity control which adjusts the volume of the microphone. The minimum position is used in areas where high background noise may be a problem. This control is set to maximum for normal areas.

Almost everyone has suffered through the inconvenience of being interrupted with private business matters during the middle of a telephone conversation. In this situation it is usually necessary to cup your hand over the transmitter portion of the handset to prevent the party on the other end from hearing your conversation. The Duofone-16 offers a privacy button which turns off the built-in microphone when using the amplifier, so the person on the other end cannot hear you.

The handset operates in pretty much the same manner as the handset on a standard telephone. There is a catch which locks it to the main body of the unit to prevent it from falling when mounted vertically on the wall. To disengage the handset, it must be pushed toward the back of the unit and then lifted away at a 90-degree angle.

One additional control is connected directly to the speaker which is mounted within the unit. This is the volume control and can be set so that the party on the other end is heard at a comfortable level. The microphone and speaker are mounted next to each other behind the grill on the front panel. When using the amplifier, you must speak toward the grill so the microphone will pick up your voice. When in the mode, the incoming portion of the phone call will be heard from the speaker.

The Duofone -16 is powered from the phone line. However, it does require 4 "AA" penlite cells to protect the phone numbers stored in the memory. Figure 6-9 shows the bottom of the unit and the battery compartment. Since these batteries are not supplied with the unit, they must be purchased separately and installed before using the Duofone-16. Notice from the drawing that all of the positive terminals are located towards the right-hand side. Be certain to observe proper polarity when installing these batteries, as a reversal could cause circuit damage. Normal battery life is about one year, as there is a very low current drain from the

Fig. 6-8. Radio Shack Duofone-16.

memory circuit. When you change batteries, be sure that the mini-modular plug is connected to the wall jack so as not to lose the stored numbers. The memory circuit is actually powered by the telephone line. The batteries form a back-up circuit which supplies current only when the phone line is disconnected or when the Duofone-16 is moved to another location. It is conceivable that no current drain would be present for the entire year should this unit be constantly connected to the phone line at all times. Replacement of the batteries at yearly intervals assures that a fresh battery supply is always available in case of a line failure. The expected shelf life of most batteries is about one year, so after this period of time even batteries which have never been used tend to fail.

Finding a proper location for mounting the Duofone-16 and similar units which offer hands-free operation may take a little experimentation. Remember that room acoustics have a great effect on proper operation. You may want to try different placements of the unit until you find a location that provides good microphone pickup as well as good sound projection from the

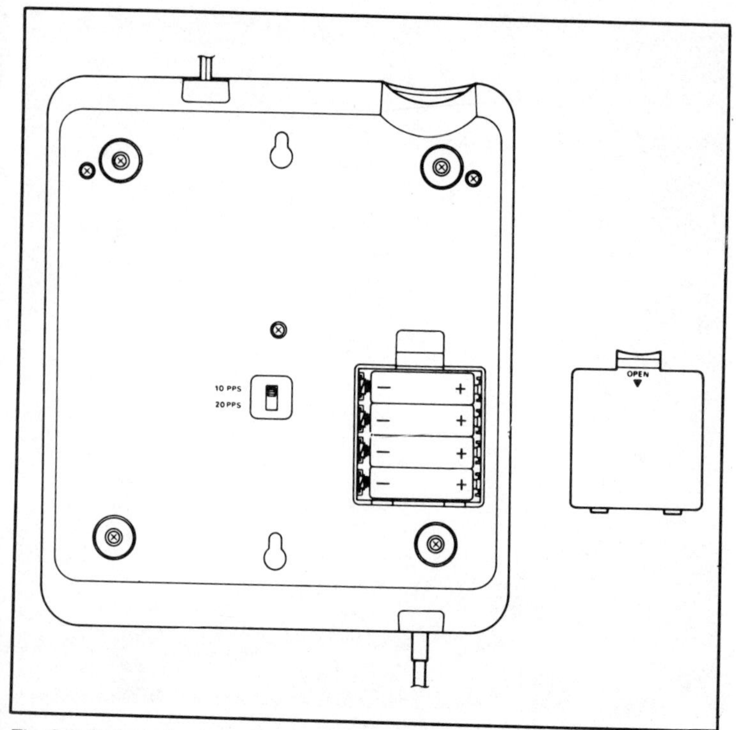

Fig. 6-9. Bottom of unit showing battery compartment and dialing rate switch.

speaker. Radio Shack points out that any vibration or noise on or near the mounting surface will tend to be picked up by the microphone and cause erratic operation. Areas which exhibit high levels of background noise should be avoided.

This unit connects to the phone line in the normal manner through modular-type connector plugs and jacks. It can be used with both rotary dial and touch-tone systems. If your present telephone installation is more than a few years old, you may have different termination blocks which means that you will have to call your local phone company and have them install a proper receptacle. If your present installation has one of the older four-prong receptacles, Radio Shack sells an adapter which will conform with the modular plug on the Duofone-16.

Mounting the Duofone-16 on a table top or other smooth surface is done in an obvious manner with the base of the unit resting on the four rubber feet. This same unit may also be hung on a wall. This mounting requires only that two screws be placed in the wall 7 ⅛″ apart. Figure 6-10 shows how the Duofone-16 is mounted to these screws. You may want to change the slant angle by exchanging the front and rear rubber feet. If this is done, you must also change the extension of the lower screw head which must be further from the wall than the top screw.

Care and maintenance is minimal with the Duofone-16. The manufacturer points out that the batteries must be changed every year or when weak. Weak or dead batteries often leak damaging chemical which can destroy delicate electronic circuits. Even the batteries which are advertised as being "leak-proof" can still leak under certain conditions.

The Duofone-16 is much more complex than a standard telephone instrument. For this reason, it must be handled with the care you would exercise when handling a stereo, radio, or other electronic device. While it is very rugged, it cannot withstand many of the abuses which are regular occurrences with many standard telephones. Avoid rough handling, areas with high levels of dust and humidity, and temperature extremes. For example, a basement or garage location might not be the best for this sensitive device if these sites are not adequately insulated and/or de-humidifed as is necessary.

The Duofone-16 has two switchable dialing rates. In most instances, the unit will be set for 20 pps, which provides 20 pulses per second. The selection switch is at the bottom center of the base. Some of the older exchanges may not be able to adequately

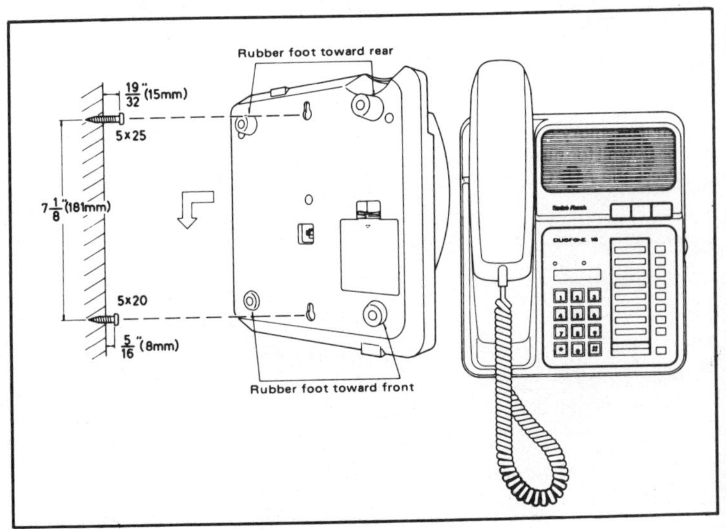

Fig. 6-10. Mounting of Duofone-16 to wall.

handle this dialing rate and will require that the rate switch be set to the slower 10 pps. position. Figure 6-11 provides the telephone specifications for the two dialing rates.

While the internal circuitry of the various devices discussed in this book may not be a prime concern, Fig. 6-12 indicates the complexity of the circuitry which provides so many advantages for the user. In addition to the discrete components found in the schematic, seven integrated circuits are also used which may contain hundreds of transistors, diodes and other solid-state components. Ten or fifteen years ago, if it had been possible to build a device with this many conveniences, it probably would have filled several large rooms using the components and technology of the time. Today, these and many more user conveniences are built into small packages which can be held in the palm of the hand.

SOFT-TOUCH TONE DIAL CONVERTER

Certain service areas which have switched from rotary dial service to a full tone dial system often charge extra rent for telephone instruments which offer the newer touch-tone keyboards. The old rotary dial phones will still work with this new system, but they are much slower and time-consuming when it comes to dialing the desired number. Tone dialing is much quicker and highly efficient, but many customers do not feel that this convenience is worth the extra monthly rental price.

Fortunately, a device is available which will instantly convert your rotary dial telephone to a true tone-dialed operation. Called the Soft-Touch, it is available from many telephone outlet and hobby stores. This device is shown in Fig. 6-13 and consists of an electronic tone generator coupled with a circular keying arrangement. All of this is mounted along with a sensitive microphone in a plastic holder which replaces the transmitter element in the handset of your present rotary dial phone.

Telephone Specifications	10 PPS Dialing Rate	20 PPS Dialing Rate
Dialing Rate	10 Hz	20 Hz
Make/Break Ratio	61.3%	64.5%
Inter-digit Period	920 mSec	530 mSec
Ringer Equivalence	0.9B	0.9B

Fig. 6-11. Telephone specifications for the two dialing rates of the Duofone-16.

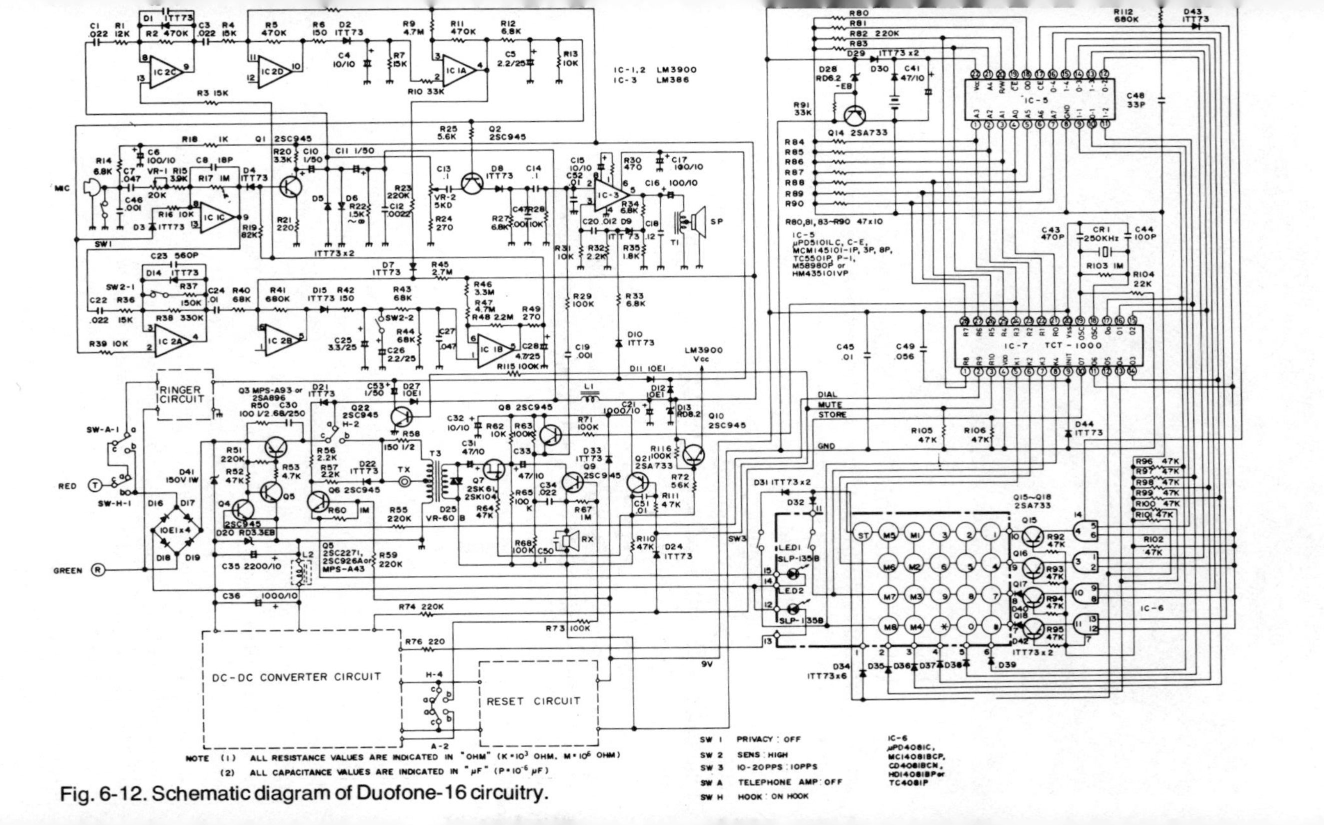

Fig. 6-12. Schematic diagram of Duofone-16 circuitry.

Fig. 6-13. Tone dial converter from Radio Shack.

To install this device, which sells for about thirty-five dollars, you simply unscrew the mouthpiece and remove the transmitter element. The threads on this converter match the threads on the handset and replace the original part. Once the converter is attached to the telephone, the built-in microphone will provide much more sensitivity than the original carbon element, and the electronic pad will send tones down the phone line when the keyboard elements are depressed.

Again, this tiny component shown installed in a conventional telephone handset as shown in Fig. 6-14 completely replaces the carbon element and screw-on mouthpiece. The numbers may be lined up on the circular keyboard simply by rotating them around their base before tightening the threaded element.

After the initial purchase price of about thirty-five dollars is paid, there are no monthly payments for the convenience of tone dialing. Another advantage is found in the portability of this little device. It may be attached to a telephone in about fifteen seconds, which includes the removal of the original mouthpiece and element. Likewise, the handset may be converted to its original state in the same period of time. Many persons carry the soft touch converter with them when on business. Salesmen and businessmen may carry them in order to access their computers, which are set up in tone dial areas. Of course, your area must have central office equipment designed to respond to the pulses. When businessmen access their computers when calling from a rotary dial only system, they must dial the computer number in the conventional manner and then use the converter to generate the tones. The rotary dial tone system does not respond to these tones; only the computer does.

The Soft-Touch is manufactured by Telephone Electronic Corporation/Buscom Systems and uses a calculator type of integrated circuit, which contains over 4,000 transistors in a miniature package. The tones are crystal controlled and have an accuracy of plus or minus one quarter of one percent. Again, persons living in areas which offer rotary service only will not be able to tone dial numbers with this device.

CALL DIVERTER

Some homes and many business locations may have need of a way to answer phone calls when no one is in. Often, a telephone answering device or a private answering service is called upon, but sometimes these are not desirable. A Call Diverter is an alternative which may solve the particular problem. These devices are not answering machines but electronic circuits which divert all the

Fig. 6-14. Converter mounted in handset of Telephone.

incoming calls to any phone number the owner selects. In other words, party A calls your office phone number at 555-1234. You are not at this number, but your Call Diverter has been set up to automatically divert this incoming call to your home phone number of 555-4321. Your home phone will ring and you may talk to party A even though he or she originally called your office number. These devices are ideal for persons who expect important calls and want to answer the call themselves rather than have this done by a secretary or answering machine.

Telephone Electronics Corporation/Buscom Systems offers a popular Call Diverter know as the Pulsar. The front panel of this low-profile unit is shown in Fig. 6-15. The desired number to which all incoming calls are to be diverted is dialed on this panel. The installation of the Call Diverter is not a simple plug-in procedure, as has been the case with most of the devices previously discussed. It requires close cooperation with your local central office, because the diverter incorporates dial delay, an adjustment of the dial tone speed return that is available from the central office. This amounts to a time of zero to ten seconds. Another element, speed, is also incorporated and is an adjustment of ten or twenty pps to match the speed of the central office. The third factor is patch delay, an adjustment for the speed of central officeswitching from zero to twenty seconds.

Operation is really quite simple and requires that two telephone lines be installed at the diverter location. One is used as the incoming call line, while the other handles the call to the diverted location. Using the above phone numbers as an example, the diverter front panel would be set up to ring 555-4321. If this were a long distance exchange, the area code would be used in front of this number, along with a 1 in front of the area code if this is the digit used to access long distance lines. When the on switch is activated, the unit will automatically dial the front panel number when an incoming call is received and switch the output of the incoming line to the input of the outgoing line. In other words, the incoming line is patched through to the remote telephone. The caller doesn't even realize what is taking place.

Call diverters have not received a great deal of interest from average homeowners and small businesses. However, some of the medium to large businesses and a few of the smaller ones which are involved in critical applications are using them quite regularly and with a great deal of satisfaction. While this has not actually been tried by the author, it is highly conceivable that several call

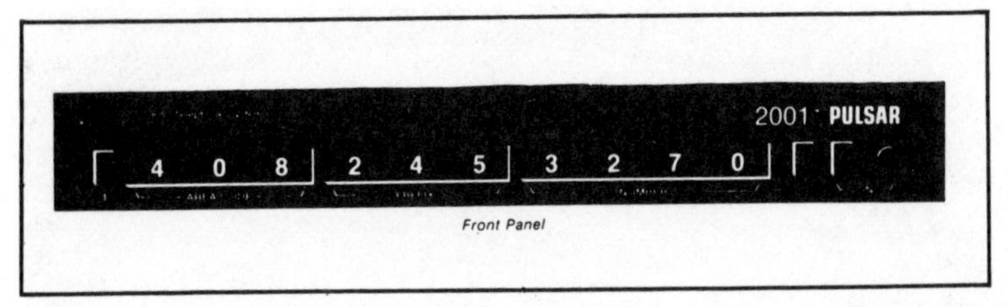

Fig. 6-15. Front panel of 2,001 Pulsar Call Diverter (courtesy of Telephone Electronics Corp./Buscom Systems).

diverters could be used in several different locations, all of which would be tied together. This would be for the executive who might be in several different places. When the incoming call is transferred by the first diverter, a second diverter at the called location might switch it to yet another number, and another diverter here might call yet another number. In this manner, one phone call might ring four or five different telephones at different locations until an answer is finally received. This would not be a very practical system. It would certainly be expensive, and a businessman in the situation might prefer to forego the complex equipment and install a radiotelephone in his automobile or carry a briefcase radiotelephone with him at all times.

RADIO SHACK ET-350 CORDLESS HANDSET

Cordless telephones are dealt with in another chapter of this text. However, the Radio Shack ET-350 Cordless Handset is more like an electronic telephone falling into the categories of this chapter than it is a true cordless unit. All of the true cordless telephones normally work in conjunction with a standard telephone instrument, although they may be used as discrete devices. The ET-350 is designed as a discrete device and is a complete answer and dial telephone with a cordless handset rather than a true cordless telephone. Shown in Fig. 6-16 and Fig. 6-17, the unit may be either wall or desk mounted and consists of a base/cradle which stays in a fixed location and a removable handset which is cordless. The universal dial system, using a push-button keyboard, is mounted in the cradle. When dialing out, the desired number is punched in from this fixed location. The user may then carry the cordless handset to any location up to about 50 feet away from the base unit.

The ET-350 is more like having a standard telephone instrument with a 50-foot cord than it is a true cordless telephone, which may have a range of 300 feet or more. This Radio Shack

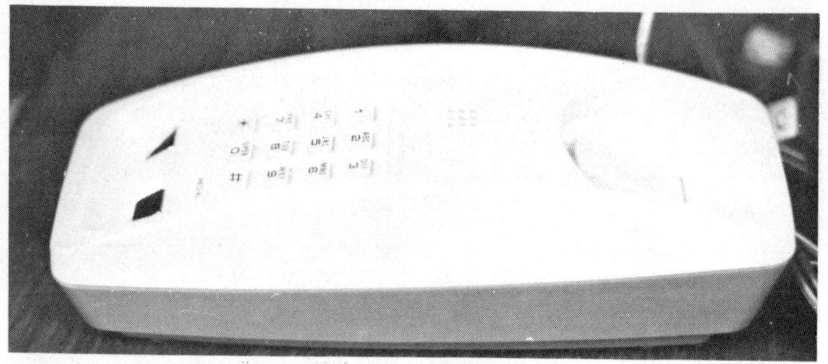

Fig. 6-16. Radio Shack ET-350 cordless handset.

device sells for about one hundred dollars, which is one-third to one-half of what most good cordless systems cost.

This device works by containing a transmitter and receiver in both the base and the handset unit. Once the call is established, the base unit patches the phone line into a transmitter which transmits to the receiver contained in the handset. The person with the handset transmits back to the base receiver, which routes this audio information into the hard-wired telephone line. The base unit is powered by the AC line and it offers a built-in battery charger which automatically supplies charging power to the nickel-cadmium batteries installed in the handset. Two electrode prongs exit the bottom of the handset case just below the microphone and contact the charger output points in the base just below the keyboard when the handset is resting in the cradle. Normal use of this device should never completely drain the internal batteries in the handset; however, should the user talk for long periods of time and often, some charging problems may develop. For this reason, the ET-350 is designed for normal telephone uses which are required by about 99% of all private telephone subscribers.

This device could be put to good use in a large office environment. In these situations, secretaries must often put callers on hold while going to a file cabinet to check certain records. Using this cordless handset, the secretary could remain in touch with the caller the entire time the filing cabinet was being checked. This is especially important if the searching process takes quite a long time. Callers on hold often become frustrated and hang up, or office personnel must constantly go back to the fixed instrument and inform the caller that they are still searching. At a price of only one hundred dollars per unit, the ET-350 closely matches the cost of many types of decorator phones and some fixed, electronic models.

In the home, the cordless handset might be especially appreciated by a housewife when she makes or receives a call near suppertime. The cordless handset allows her to answer the phone and then tend to the evening meal while carrying on a conversation. True, many modern homes have telephones installed in the kitchen and dining areas, but a coiled cord can present many difficulties or outright hazards when used in close proximity to a hot oven burner.

While the cordless handset from Radio Shack offers many conveniences which cannot be found in fixed telephone instruments, the author does not recommend that this device be used in areas subject to a high amount of radio frequency transmission. These and all similar devices are subject to interference from citizen's band, amateur, and commercial broadcast transmissions of high intensity. If your home is located extremely close to a commercial broadcasting station, the interference may make practical use of this cordless device impossible. This is not to indicate that the ET-350 is defective in any way. It does contain bypass circuitry to prevent undue interference, but when radio frequency transmission levels reach very high potentials, adequate bypassing is nearly impossible. Since a great majority of the population does not live in areas exhibiting intense RF fields, more complex bypass circuitry would not be an advantage, and it would certainly increase the overall price of this relatively inexpensive telephone instrument.

The Radio Shack ET-350 is compatible with nearly every phone system in the United States. It offers push-button dialing which will work with either tone or rotary dial systems. Like so many of the electronic telephones made today, it offers last number re-dialing by depressing a single button on the push-button keyboard. This device is easily installed to the present phone company block inputs through mini-modular plugs. It is necessary to have a standard AC outlet within reach of the power cord

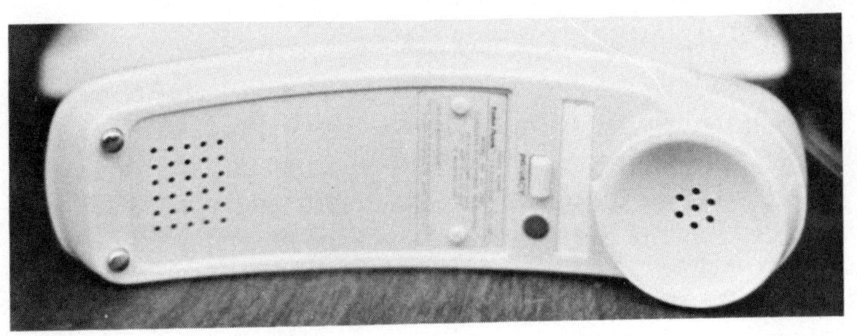

Fig. 6-17. ET-350 remote unit.

attached to the bottom of the base unit. While extension cords may be used safely, the owner may wish to locate the base unit at a point on the wall which offers a short run to the power outlet. As with all other privately owned devices which are to be installed to the phone lines, you must contact your local company, giving them the licensing data contained with this unit, which has been approved by the FCC. No separate license is needed by the user, as is the case with citizen's band radio equipment.

CONFERENCE UNITS

A conference line is actually a series of telephone lines all tied together so that many different parties may speak in round table fashion from widely separated locations. A conference device ties together a number of phone lines to enable these conference calls to be made.

Special balancing circuitry is required to prevent signal drop each time an additional line is switched into the conference circuit. If enough lines are tied together without this balancing circuitry, eventually all that is heard is line noise. The balancing circuits may often take the form of amplifiers that boost the gain of each line so that normal listening levels are achieved. Clipping amplifiers, which also use compression, may even be used in the more expensive conference devices to bring up gain variations in a line and to suppress inputs which are too high. This automatically keeps the gain at a more or less consistent level.

These devices are offered by many different companies and are not terribly expensive, as they are designed to be used with a multi-line, business phone system. It is necessary to have all of the persons who are to be engaged in the conference call to phone the available lines at the office where the conferencing device is installed. Once all parties are on the various lines, the device combines their calls so that each person may talk with any of the others.

AUTOMATIC DIALING INTRUSION ALARMS

Another interesting adjunct to the telephone instrument is found in the burglar alarm systems which may be connected to a standard phone line. When the alarm is triggered, the electronic circuitry within the master unit automatically dials a pre-set number. Some of the least expensive models simply play a tone down the line when answered, while others play a recorded message stating the fact that an alarm has been triggered at a specific location, the address of which is also given. Often, the pre-set number dialed by the alarm is the local police department. The Heath Company, better known as Heathkit by many hobbyists and experimenters, offers one of these devices, as do most professional burglar alarm retailers and manufacturers.

The operation of these alarm systems is very similar to the call diverters discussed earlier in this chapter. Whereas the diverter was triggered by an incoming phone call on another line, the alarm system dialer is activated by a break in the alarm system. Both the diverter and the alarm system automatically dial a pre-set number when triggered by their respective sources.

SUMMARY

It can be seen that electronic telephone instruments offer the consuming public a great variety of conveniences which cannot be obtained from a standard telephone. Many of the devices discussed incorporate microelectronic circuitry to memorize phone numbers, dial these numbers upon manual command, and to allow the user to communicate without the necessity of lifting the handset.

Still other devices are almost completely automated systems, which respond automatically to inputs from another phone line. Only a few years ago, most of these devices would have been beyond the technological capabilities of the electronics industry, or at least so expensive as to be completely impractical for even the larger businesses, not to mention the average consumer.

The telephone has become an integral component in the home and office environment. The average home in the United States has several of these devices. Every day new models with many new convenience features are being introduced on the market. Most of these are available to average consumers at reasonable prices. Some contain electronic clocks. One even has its own built-in electronic calculator. A certain model may not ring but may alert the household resident that a call is coming in by buzzing, chiming, or even by playing a recognized musical piece.

If your present telephone does not fully suit your requirements, you would do well to consider many of the attractive alternatives presented by the manufacturers, many of which have been discussed and pictured in this text. With the dependence factor of the modern telephone being as high as it is, it seems only reasonable that the average consumer would want to get the most for his or her money. Depending upon individual requirements, a rental instrument from the phone company may be adequate, but others may want to fully explore the world of modern electronic telephone conveniences that were only dreamed of a decade ago.

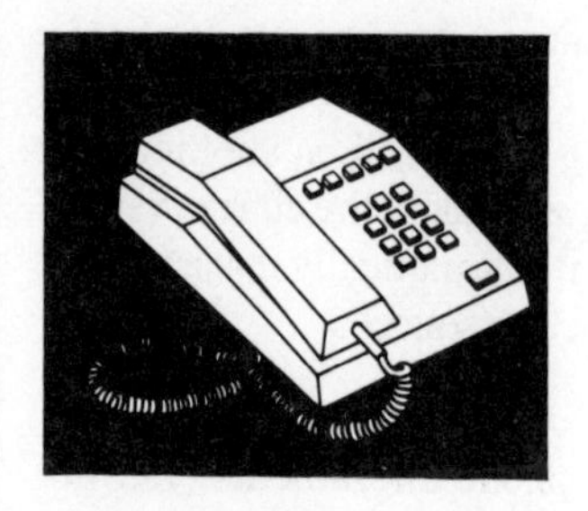

Chapter 7
Accessories

After having discussed the types of different telephones and telephone systems available to the average consumer on today's market, we now approach a very large field which is made up of devices which are designed to be attached to and used with the standard telephone instrument or phone line. Many of these devices extend the versatility of the standard instrument and may even convert it to a total package delivering much the same convenience as an expensive electronic telephone.

Most of the products mentioned in this chapter are relatively inexpensive. They may be purchased for less than five dollars on the lower end of the scale, but may range to a hundred dollars or more for some of the more exotic circuits. Some of these items will require notification of your local phone company before installation, as they will attach directly to the telephone line. Others will be externally connected and will not require that this notification be given. In each case, the manufacturer will provide the necessary information regarding legal installation. In cases where notification is required, the manufacturer must provide certain technical information in order to comply with FCC regulations.

ARCHER TELEPHONE LISTENER

The Archer Telephone Listener pictured in Fig. 7-1 is a small amplifier which is designed to pick up incoming calls from a telephone instrument and amplify them to provide drive for a

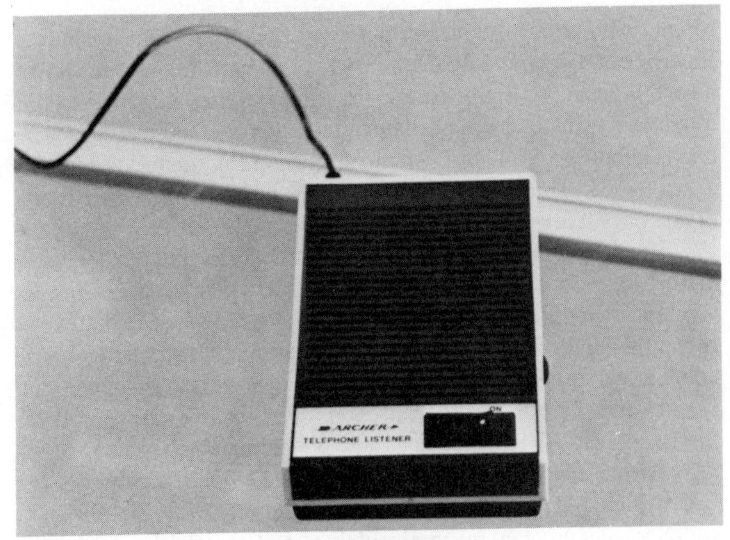

Fig. 7-1. Archer telephone listener from Radio Shack.

built-in speaker. This simple device is powered by a 9-volt transistor radio battery and does not require a separate power supply. No direct wiring connections are made to the telephone. All that is necessary is to attach the suction cup onto the phone and then turn the switch to the *on* position.

For most telephones, the best pickup location will be near the left rear of the phone base. However, it may be necessary to experiment with various placement positions of the suction cup in order to find the spot where most audio transfer takes place. On some phones, the best positioning may be directly to the back of the earpiece on the handset.

Once the suction cup is in place, its cable is connected to the telephone pickup jack on the back of the main unit. A volume control is provided and is usually adjusted to the maximum position when initiating the call and backed down as is necessary for a comfortable listening level. When not in use, the Telephone Listener is turned off to conserve battery life.

The Archer Telephone Listener only amplifies the voice of the party at the other end of the line. It does not amplify your voice going into the phone system, and the phone must be active at the time of its use. In other words, someone must be talking to you.

The main unit can be mounted on a wall or shelf, using the mounting holes which have been located in the back of the case. Radio Shack advises the removal of the case back which serves as a template by which the proper placement positions for the holding screws can be marked on the mounting location. Figure 7-2 shows how this is done. Small screws or nails are then attached to the mounting surface at the predetermined points. The nail or screw heads should extend out from the surface of the mounting area for about 3/16″. This enables the case back to be slipped over these points once it is re-attached to the Telephone Listener.

When volume begins to drop off or the sound becomes distorted, this is an indication that the battery is beginning to weaken. For this type of application, a long-life alkaline battery is recommended. If a great deal of use is intended, it might be more economical in the long run to purchase a rechargeable 9-volt battery and charger. Radio Shack does offer a 9-volt adapter jack on the rear of the Telephone Listener, which may be used to access the internal circuitry with a power line driven 9-volt supply.

Figure 7-3 shows the schematic diagram of this simple unit, which consists of a magnetic pickup coil which drives a pre-amplifier circuit and finally a balanced amplifier. The amplifier is capacitively coupled to the small output speaker. The entire circuit, with the exception of the magnetic pickup and speaker, is located on a printed circuit board.

This simple device comes in very handy when it is desirable for a group of persons to listen to the remarks of someone on the opposite end of the telephone conversation. Voice quality from the output of the amplifier is reasonably good, considering the small size of the speaker, and is effective in conference table situations. Due to the self-contained nature of this circuit and to the fact that no hard wiring connections are necessary to the phone line, this unit may be carried in a coat pocket and attached to any telephone system in a matter of seconds. It will also fit easily into a briefcase for executives who must travel constantly.

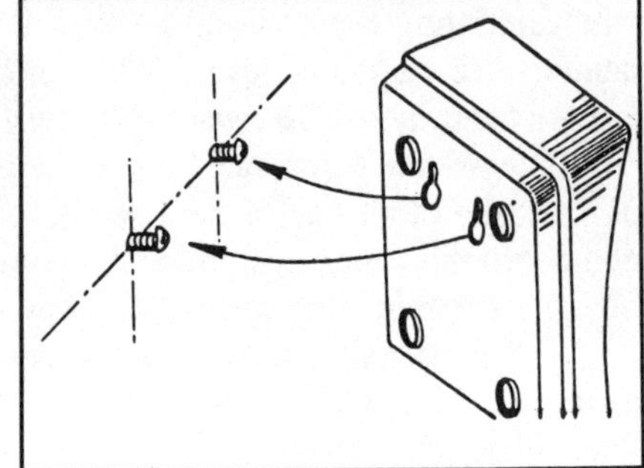

Fig. 7-2. Mounting of telephone listener to wall.

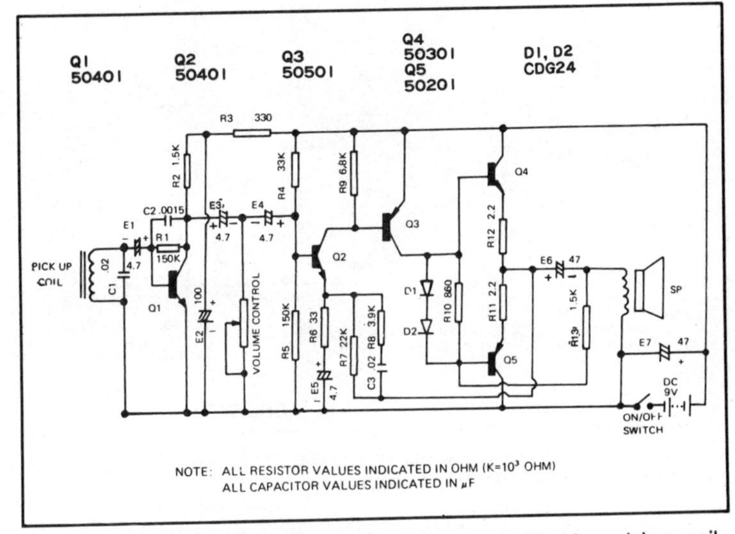

Fig. 7-3. Schematic diagram of telephone listener showing pickup coil, pre-amplifier, and balanced amplifier.

REALISTIC TELEPHONE AMPLIFIER SYSTEM

Another product from Radio Shack is shown in Fig. 7-4 and is called the Realistic Telephone Amplifying System. This system permits hand-free telephone conversation and can be used for round table conference calls when the caller needs to be heard by all members of the conference.

Referring to the last figure, it can be seen that the Realistic TAS provides a cradle for a standard telephone handset. Radio Shack points out that this cradle will not fit GTE telephones, as it is made for the conventional type of handset only. Only the incoming caller's voice is amplified using this device. This is not designed for two-way amplified conversations.

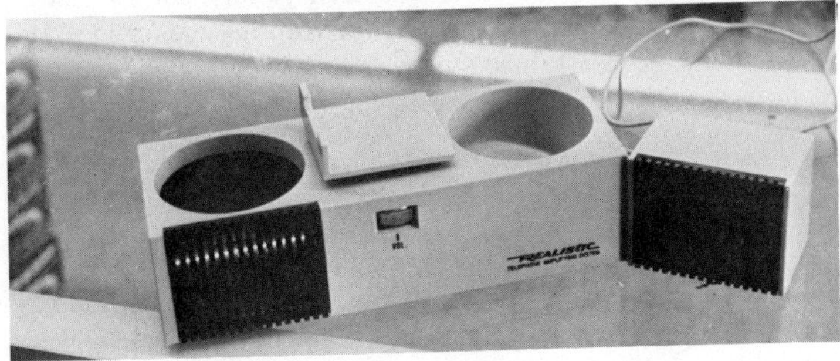

Fig. 7-4. Realistic telephone amplifying system.

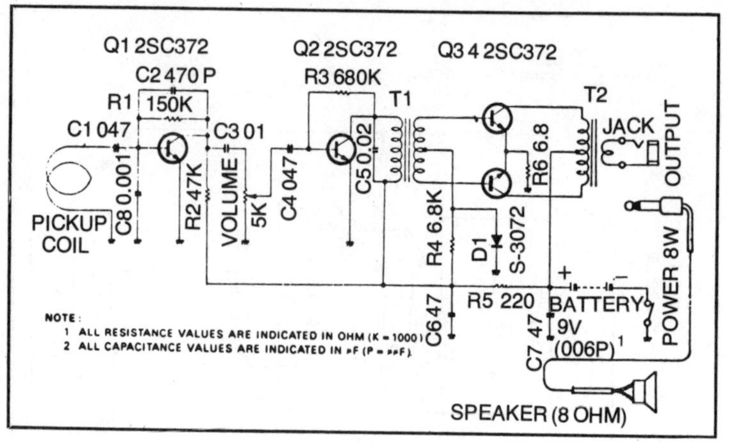

Fig. 7-5. Electronic circuitry of telephone amplifier consisting of four transistors and a pickup coil (courtesy of Radio Shack).

Figure 7-5 shows the electronic circuitry of this device. It consists of four transistors and a pickup coil. The audio from the telephone line is sampled by this pickup and then fed to a pre-amplifier stage. From here the amplified signal is passed on to an amplifier driver circuit and finally to a balanced amplifier which is transformer coupled to an 8-ohm speaker. This speaker is housed in its own separate enclosure and is connected to the electronic circuit by means of an audio cable and phone plug. This allows for placement in areas where audio feedback is minimized or avoided. A master volume control is used to set the desired output level. This is also used to avoid feedback.

The power source for this telephone amplifying system is a 9-volt transistor radio battery. If this device is to be used often, a rechargeable 9-volt battery might be opted for to avoid the expense of constant battery replacement.

This simple device costs less than twenty dollars and is perfectly adequate for many types of conferencing applications. Its audio output is clearly understandable, and the volume control features provides enough versatility for it to be used in very small to medium-sized conference rooms.

SPEAKERPHONE 1000

Another type of conferencing amplifier is shown in Fig. 7-6. The Speakerphone 1000 from TT Systems Corporation allows hands-free telephone conversations without using your telephone handset. It features a built-in electronic ringer, allowing this instrument to be used in a room even if a telephone is not present. The SP-1000 is attached to the phone line through any existing mini-modular jack and is then immediately ready to receive incoming calls. It offers Answer, Hold, Sensitivity, and Volume controls for ease of operation. The internal circuitry is composed of integrated electronics design, assuring a very stable device. Another convenience feature is found in the fact that operating power is obtained directly and completely from the telephone line. No separate batteries are required, nor is a power plug attached to your wall outlet. The unit is FCC approved and requires notification of your phone company for installation.

A built-in condenser microphone is mounted in the front panel and is very sensitive to the human speech pattern. A speaker is mounted in the top portion of the case which broadcasts the voice of the caller. An LED indicator lights up when the unit is transmitting. There is also a two-way recroding jack which may be used for

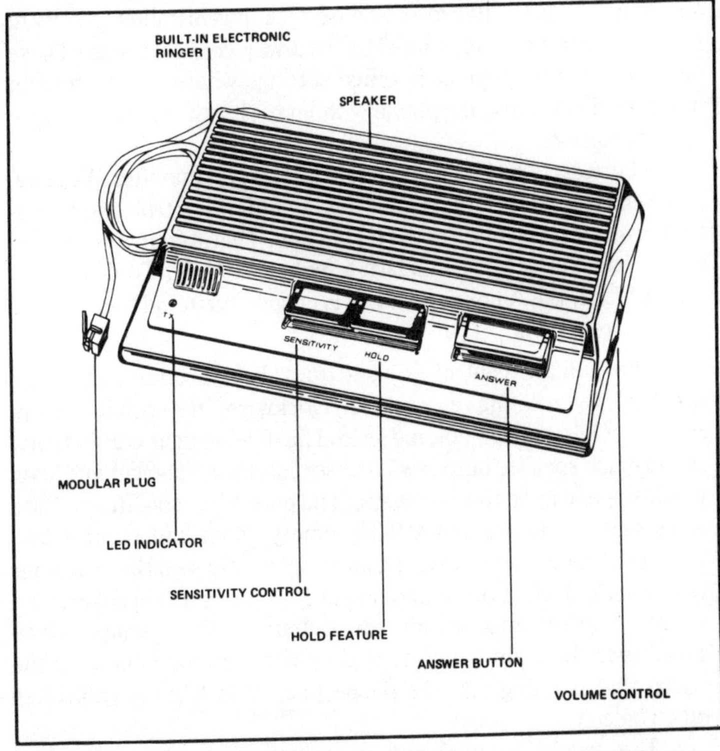

Fig. 7-6. Speakerphone 1000 conferencing amplifier (courtesy of TT Systems Corp.).

accessing a tape recorder. Through this output, both sides of the conversation may be committed to tape.

Operation is very simple. When the electronic ringer is heard, the Answer button is depressed and the conversation carried on in a normal manner. Should you wish to place the caller on Hold, the Hold button is depressed. This ties up the line until it is depressed again. The Volume Control on the right-hand side is used to vary the output level of the caller's voice being heard through the speaker.

This is a compact telephone accessory which weighs only two pounds. It measures 5″ × 7″ × 2″ and is contained in a molded plastic case with a simulated wood-grain finish. This gives it an attractive businesslike appearance, and the Speakerphone 1000 blends well with other desktop accessories.

DUOFONE ELECTRONIC TAS

Another type of hands-free electronic amplifying system is shown in Fig. 7-7. Offered by Radio Shack, it is installed directly to the telephone line and is used for two-way conversations. These devices are most popular in office settings where a stenographer may take dictation on the phone without having to leave her desk or typing position.

There are four controls, labeled Privacy, Amplifier, Volume, and Sensitivity. By depressing the Amplifier control, the unit is turned on and connected to the phone line. This is similar to picking up the handset of your standard telephone to talk. To disconnect the Duofone, the Amplifier control is pushed again and reverts back to the off position.

The volume control is used to vary the output from the speaker. By rotating the control clockwise, the volume is increased. Only enough volume should be used to hear comfortably, as too much speaker output will cause distortion. The volume level would depend upon room acoustics and phone line conditions. Bare walls and tile floors and a fairly empty room will require less volume than one with carpeting and drapes. These latter trappings will absorb much of the sound output from the small speaker. The volume control affects only the output of the speaker which broadcasts the caller's voice. It does not vary the volume of the signal that is going into the phone line. This level is controlled electronically.

The sensitivity push-button control affects the pickup sensitivity of the built-in microphone. To reduce the sensitivity, the

Fig. 7-7. Duofone electronic TAS from Radio Shack.

button is pushed downward so that it locks. Depressing it again will return it to its normal sensitivity position. The reason for needing a decrease in sensitivity comes into play when the TAS is used in an office environment where there may be a high level of room noise. This noise may be picked up by the condenser microphone, causing hard to understand phone conversations. In the decreased sensitivity position, the microphone will be responsibe only to the voice of the user from a short distance away. This improves on the effectiveness of the overall communications.

The privacy push-button control is pushed down in the same manner to lock it into place. This turns the microphone completely off and allows you to carry on a side conversation with an employee in the same office without being heard by your caller. This is the electronic equivalent of covering the mouthpiece of a handset with your palm. Depressing the button again returns it to the normal position.

When choosing a location for this device, you must be aware of the room acoustics as they apply to the TAS. This factor will have a large effect on the efficient operation of this device. It may be necessary to experiment a little with placement position until you achieve a location that provides good microphone pickup while giving good sound projection from the speaker. Bear in mind that any vibration or noise in or near the surface on which you have placed the unit will probably be picked up by the microphone and cause unusual sound effects and amplifier action. Avoid these locations, as even the sensitivity control will not do much to get rid of them. In general, avoid all locations with high background noise levels when possible.

The Duofone TAS is designed to be installed separately or with your standard telephone instrument by means of mini-modular plugs and jacks. Figure 7-8 shows the proper installation connections between the telephone, the TAS, and the phone line.

Install your Duofone near the standard instrument in such a way as to avoid the possibility of it being knocked to the floor. Position the wire connection free of normal telephone activities. Since more wires are involved with the installation of this device, make certain that none of these conductors are in paths of pedestrian traffic. An entanglement could pull the cord completely out of the Duofone or, at worse, cause a serious accident.

If the device does not work properly, check wire connections to the phone line. Since power is derived from the line, there is no possibility of inoperation stemming from weak batteries as could be the case with some of the other types of accessories discussed. Radio Shack points out that telephone calls are automatically switched and routed through a complex group of equipment. You will occasionally encounter a "weak" or "noisy" line. In some cases you may have to ask the other party to speak up or do so yourself. The connection may be so poor that you will actually have to redial the call. Remember, your telephone amp can only amplify the signal it receives from the phone line.

Key or business telephone systems are not an intended use for this amplifier system. However, the Duofone can be connected to

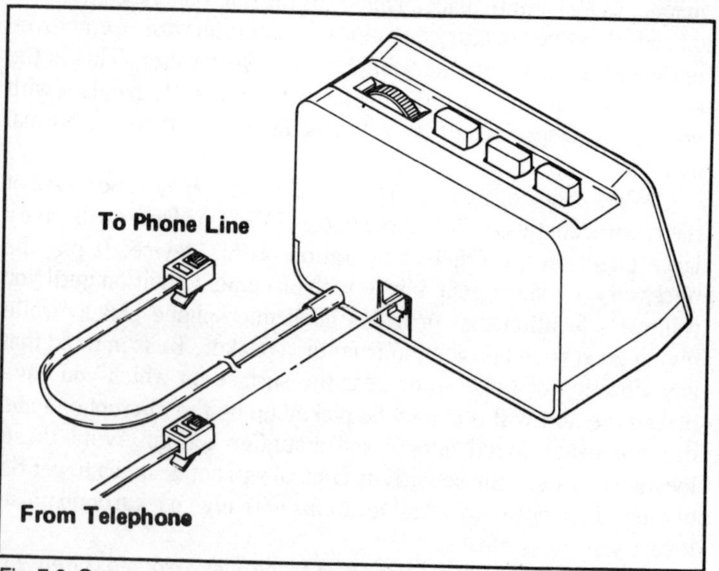

Fig. 7-8. Connections between telephone, TAS, and the phone line (courtesy of Radio Shack).

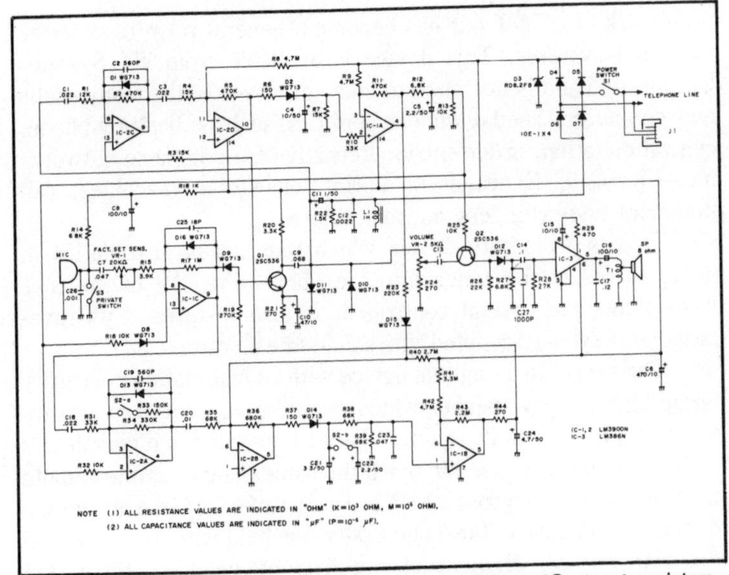

Fig. 7-9. Schematic drawing of amplifier circuit using three ICs, two transistors, and other solid-state components (courtesy of Radio Shack).

provide proper operation on many such systems. Generally, the telephone amp will not provide adequate switching capability to cause the telephone company "hold" circuit to stay disengaged when the telephone handset is "on-hook" during use of this device.

Figure 7-9 shows a schematic drawing of the amplifier circuitry which is composed of three integrated circuits, two transistors, and a number of diodes. The Duofone utilizes a signal-sensing circuit which detects the first signal applied to it. This will be either the outgoing signal from the microphone or the incoming signal from the phone line. Upon sensing the appropriate signal, the circuit turns on the proper internal amplifier controlling that signal. This circuit remains activated until the signal ceases. At this time, the signal-sensing circuit returns to normal and waits for the next signal. In other words, the person that speaks first has priority, and the second person cannot gain control until the first stops talking. This can be likened to a "shared" system with both the caller and the called party using the Duofone circuitry on a sharing basis.

TONE CODER-1

The device shown in Fig. 7-10 provides touch-tone signalling with a standard telephone instrument. Touch-tone is a registered trademark of AT&T but has become a general term to describe tone dial systems. This device is available from TT Systems Corporation and is designed to take advantage of the many useful new computer-based telephone services, such as: bank-by-phone, central dictation, order entry, alternative long distance networks like City-Call, Execunet or Sprint, answering machines, call diverters-line extenders, and many more.

This is a discrete device which requries no installation in order to insert the tones into the phone line. All of the tone frequencies are crystal controlled. It also features a positive-response keyboard coupled with solid-state electronics.

The secret to using this device with no installation necessary is the high-fidelity speaker which produces superior tone quality and volume. This unit, again, is battery operated and completely self-contained. It is used in much the same manner as the remote message-keying devices which are a part of telephone answering devices. In this case, the Tone Coder-1 is held with its speaker to the transmitting element of a telephone handset. When the push-button keyboard elements are depressed, the tones are broadcast through the speaker and into the telephone handset. Owing to the fact that there is no direct connection to the telephone instrument or line, this device requires no FCC or phone company approval. It is reasonably priced at about $65.00 and is popular with computer operators and ham radio enthusiasts who both use these tones for accessing their respective equipment.

TELE-RECORDER 150

Obtaining a good tape recording of a two-way telephone conversation is often difficult when using external pickups. The Tele-Recorder 150 provides direct access to the telephone line and gives its owner control of a cassette tape recorder through switches installed in the device case. This is an FCC approved circuit which will require advising your telephone company of its installation. It works with most cassette recorders and automatically activates the motor drive when the handset is lifted. The recorder stops when the receiver is replaced. No batteries are required, as all operating power is obtained from the phone line. This is an ideal device when automatic taping of all phone conversations is desired. It may be installed, tested for operation, and then forgotten about. Again, any cassette tape recorder which offers a microphone-switched pause control can be used with the Tele-Recorder 150 for automatic taping.

Fig. 7-10. Tone coder-1 device for tone dialing with a standard telephone instrument (courtesy of TT Systems Corp.).

To properly set up the tape recorder, one of the Tele-Recorder's plugs is inserted in the microphone input jack, while the other plug is attached to the remote control input. The recorder is then placed in the Record mode of operation. It will do nothing as long as the handset is in the cradle, because the remote control cable is in the open position. When the handset is lifted, this cable is switched closed, current flows to the tape recorder motor, and taping begins. There is a control on the Tele-Recorder 150 which will allow the tape recorder to play back its recorded information to the user pushing the switch to the listen position.

The Tele-Recorder 150 is available from TT Systems Corporation for less than $40.00 and may be installed in a few minutes. It comes equipped with a mini-modular plug which will mate with most modern telephone installations. Older installations may necessitate a call to the phone company to have them install the proper receptacle. They are authorized to charge an installation fee for this service.

Figure 7-11 shows the device under discussion, which is very tiny and may be attached to the back of a desk-mounted telephone

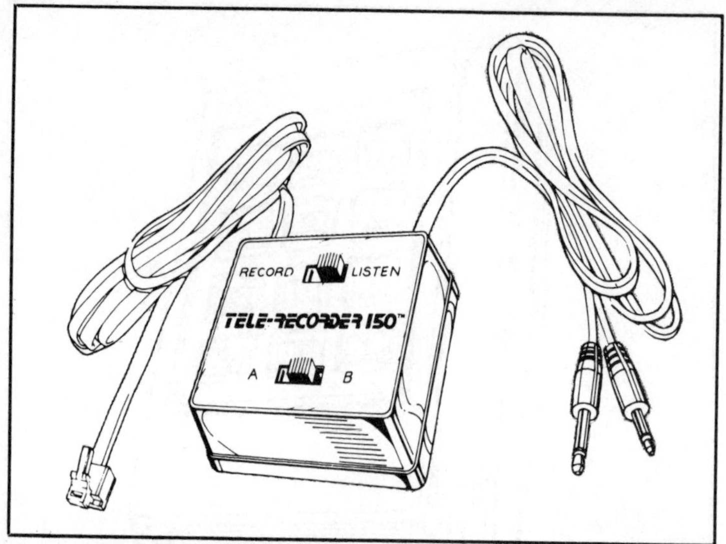

Fig. 7-11. Tele-Recorder 150 two-way telephone conversation taping device (courtesy of TT Systems Corp.).

for an unobtrusive placement. Adequate cord length is provided between the mini-modular plug and the device case for most installations. Should additional cord length be needed between the tape recorder and the Tele-Recorder 150, extension cables are available from most hobby stores.

TELE-RECORDER 250/350

TT Systems Corporation also offers the Tele-Recorder 250/350 line of one-piece telephone recording systems. Basically, what these units consist of is the Tele-Recorder 150 circuit installed within a cassette tape recorder. Since one-piece construction is used, both of these models are simple to install and activate as soon as the handset is lifted from its cradle. The recorders automatically stop when the receiver is replaced.

The Tele-Recorder 250 is a one-piece unit which records all telephone conversations at a standard speed. The Tele-Recorder 350 is basically the same unit but records at one-half the normal speed. This means that normal taping time is doubled when using a typical cassette tape. The slower speed, technically, cannot provide as good a reproduction as the faster speed, but voice communications do not require high-fidelity, so the slower speed is certainly a worthwhile feature. The Model TR 250 sells for just under $100.00, while the 350 sells for about $140.00. Both require direct connection to the phone line, are FCC licensed, and contain the TR 150 circuitry mounted internally.

Figure 7-12 shows the TR 250/350 which contains the usual assortment of cassette tape recorder controls. It can be used just like a cassette recorder when not connected to the telephone line. This makes it a very versatile device.

DUOFONE TELEPHONE HANDSET AMPLIFIER

Persons who make a lot of long distance phone calls or receive a large number often experience difficulty with noisy lines or lines which are weak. In both cases, it is often quite difficult to clearly understand the distant caller. The Telephone Handset Amplifier from Radio Shack shown in Fig. 7-13 is ideal for use in these situations and also for persons who are hard of hearing.

It installs to the phone instrument in seconds, connecting between the handset and the phone base. It features a volume control which is important when line levels are low. A tone control

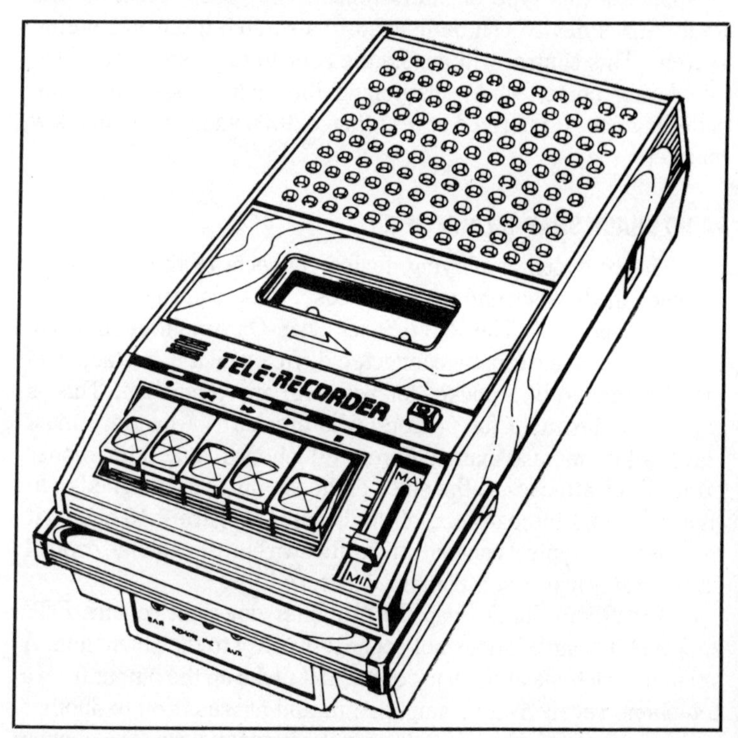

Fig. 7-12. Tele-Recorder 250-350 complete telephone conversation recording device (courtesy of TT Systems Corp.).

Fig. 7-13. Telephone handset amplifier from Radio Shack is ideal for persons with hearing impairments.

adds additional versatility when line noise is a problem. Since this noise is similar to the noise which is passed through a stereo system when playing a scratchy record, increasing the bass response while eliminating much of the treble will often get rid of or suppress this type of interference. The tone control on this Radio Shack device can be likened to a scratch filter on a stereo system. This control will enable the user to tune out much of the interfering noise, while sharpening the audio response of the caller's voice. It is priced at less than $20.00 and installs in a few minutes.

RADIO SHACK SNAP-ON AMPLIFIER

The previous amplifying device connects to the phone line between the handset and telephone base, thus requiring telephone company approval. The Radio Shack Snap-On Amplifier shown in Fig. 7-14 is an externally connected device which is portable and may be carried in a pocket for use with any telephone. This is especially desirable for persons who are hard of hearing, must travel a lot, and use many different telephones. The thick rubber strap which attaches to the sides of the amplifier is designed to fit tightly around the earpiece of the handset. Figure 7-15 shows it installed on a typical handset. It has its own built-in volume control and operates from a self-contained battery.

The Radio Shack Snap-On Amplifier does not require FCC approval, because it does not connect directly to the phone line. A small microphone at the back of the unit picks up the output from a telephone receiver unit, amplifies it, and passes it on to another receiver element which is held near the listener's ear. This device is very similar to an accessory which your local telephone company

Fig. 7-14. Radio Shack snap-on amplifier may be easily carried from phone to phone.

will install for hard of hearing subscribers who need the additional amplification. The phone company version is built into a standard handset and the volume control is contained in the handgrip. The nice thing about the Radio Shack unit is that it does not require a monthly rental fee and can be purchased for about $12.00. This includes the battery.

AUTOMATIC DIRECTORIES

While not exactly a phone instrument attachment, automatic directories are becoming more and more popular as adjuncts to the modern telephone. These directories make telephone number

Fig. 7-15. Snap-on amplifier attached to telephone handset.

information available at your fingertips and are improved, electronic versions of the old snap-open telephone number directories which have been used in homes and business offices for many years.

Figure 7-16 shows the Radio Shack Top Automatic Directory. Its motor is operated from two "C" batteries and allows the user to recall as many as 300 numbers with a touch of the keys mounted on the front panel. The numbers appear through a clear, plastic window as the data file rotates. This is an attractive directory which mounts in a small space and can be accessed instantly. It sells for about $25.00 and measures 4⅝ × 5¾ × 7⅜ inches. It is the top of the line Radio Shack directory model.

Figure 7-17 shows the Radio Shack Portable Hand Directory. Two "AA" batteries power its internal motor which rapidly locates up to 390 names and numbers. The data file may be rotated either forward or in a reverse mode, so if a particular number is suddenly passed over, the rotation may be stopped and the data file backed up for faster access. This device may be held in the palm of the hand and sells for less than $7.00.

Figure 7-18 shows the Miniature Radio Shack Pocket Directory, which holds up to 200 names and numbers and may be kept in a shirt pocket. Two "AAA" batteries are required for operation. This model is often chosen by salesmen and busy executives who may make a lot of phone calls in distant towns and territories that they cover. Since it is highly portable, it can always be on hand and

Fig. 7-16. Radio Shack desk top automatic directory.

The GTE Flip-Phone is a compact instrument which is unfolded for use as shown.

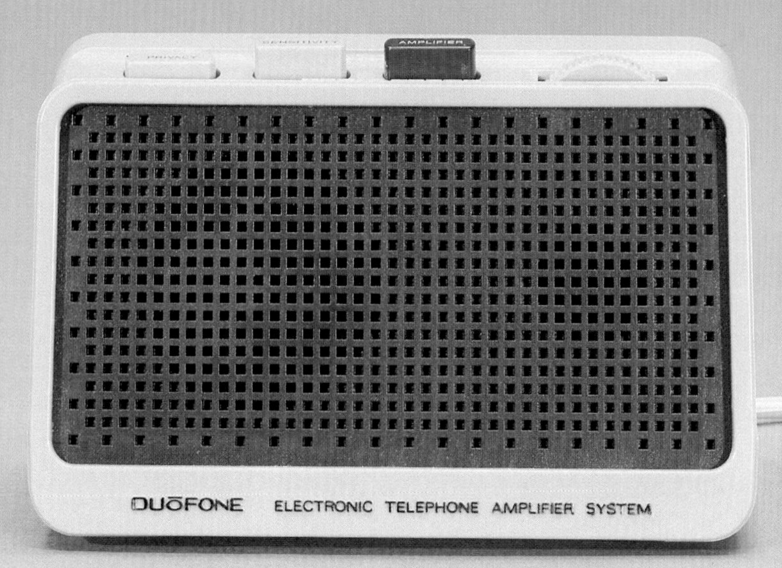

This Duofone Telephone Amplifier allows you to make conference calls at home or in the office.

This convenient snap-on amplifier attaches to the telephone earpiece and boosts the callers voice.

The SNOOPY & WOODSTOCK Phone manufactured by the Bell System is one of the most popular new designs. The housing is manufactured by American Telecommunications.

A Digital PABX Console may be combined with memory storage systems to allow an operator to have immediate access to important information.

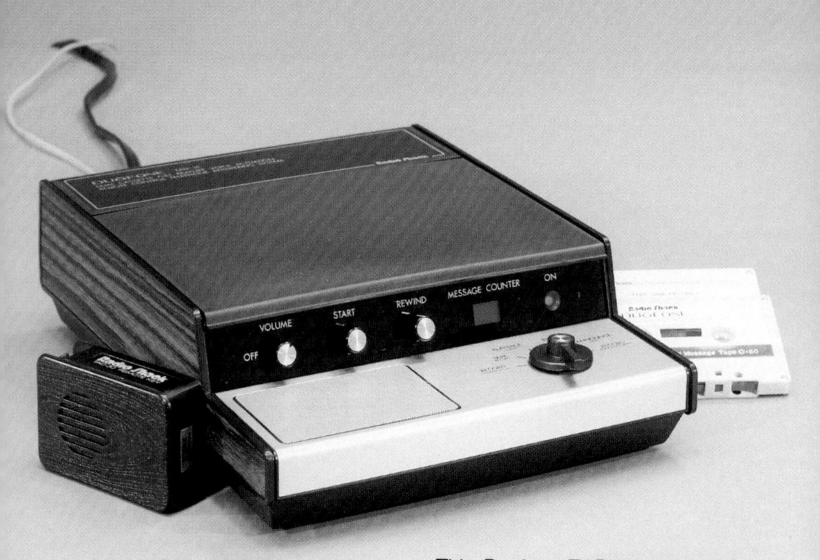

This Duofone TAD-31 not only records messages up to three minutes long but also has a remote control to allow you to hear your messages from any phone.

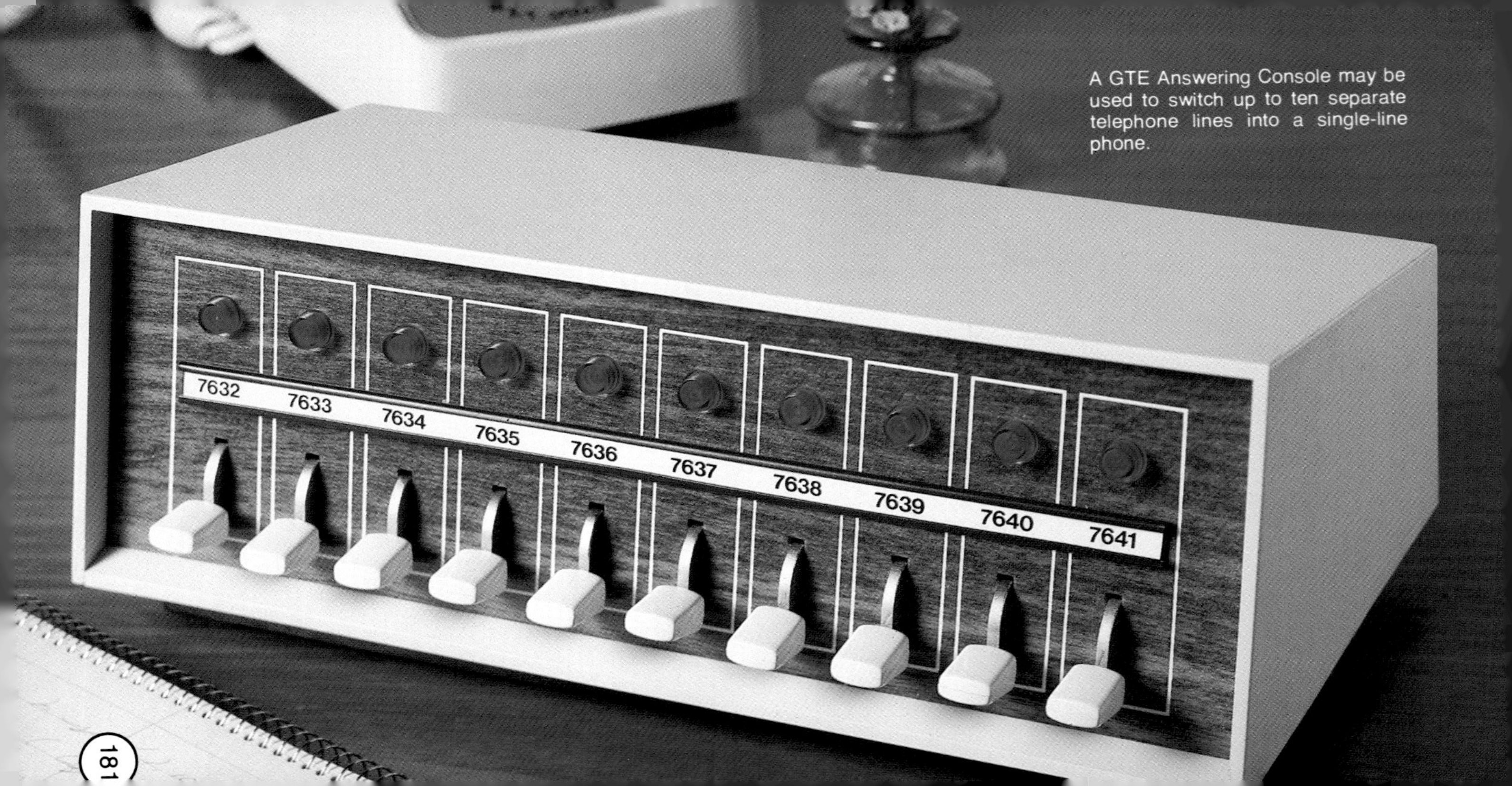

A GTE Answering Console may be used to switch up to ten separate telephone lines into a single-line phone.

The Duofone-16 features single button dialing of any one of 16 different numbers which are stored in its memory. It also has a two-way amplifier, automatic redialing of busy numbers and a battery powered backup to maintain the memory in the event of a power failure.

This elegant French Continental telephone features 14-carat gold filigree and ivory-colored base.

The GTE Panel Telephone features modern design and styling and is available in a number of different colors.

Using the GTE-60 Electronic Switchboard, a single operator may channel calls to many different locations.

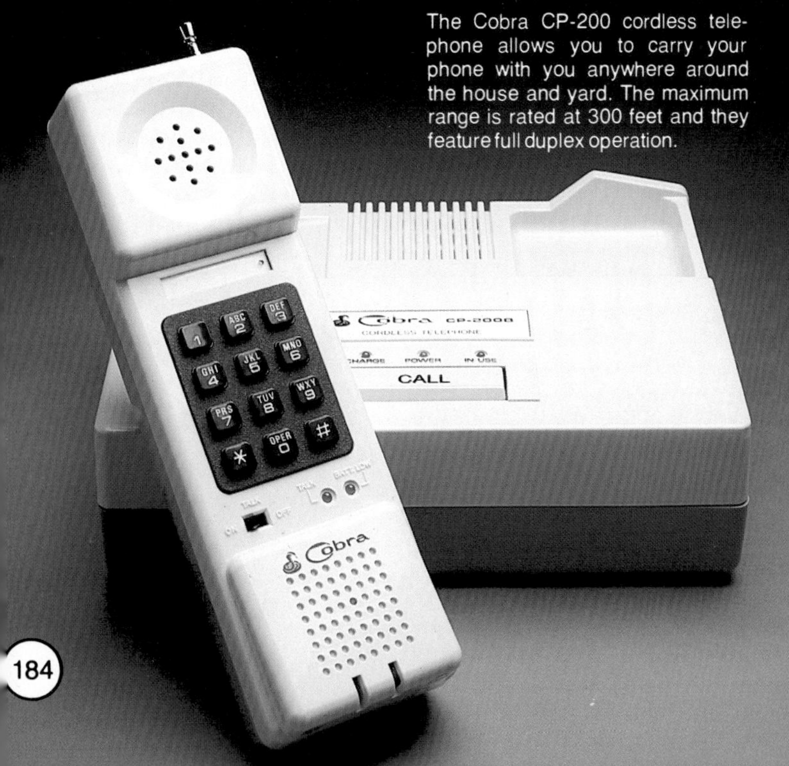

The Cobra CP-200 cordless telephone allows you to carry your phone with you anywhere around the house and yard. The maximum range is rated at 300 feet and they feature full duplex operation.

Fig. 7-17. Radio Shack portable hand directory.

serves as an instant reference source which is immediately available.

PHONE SILENCER

Another device that requires connection to the internal wiring of a telephone instrument is the Telephone Silencer available from many different manufacturers and hobby stores. It is designed to attach to the bottom of your telephone, as is shown in Fig. 7-19, and prevents the phone from ringing when you don't feel like answering. This is a convenience feature which precludes the user from having to take the phone off the hook or disconnect it at the wall plug. The Telephone Silencer eliminates the ringing sound of your telephone, but the caller still hears the ringing indication in his or

Fig. 7-18. Radio Shack miniature pocket directory.

her earpiece. What it really does is give the caller the impression that no one is home.

This device may only be attached to a telephone which you own and cannot be used with a rental unit from your local phone company. Its connection requires that the case be removed from the telephone base and that certain inner connections be made near the bell ringer. Installation takes less than a half hour, and complete instructions are provided by the various manufacturers. Most of these devices sell for about seven to eight dollars, depending on the manufacturer or outlet chosen.

RADIO SHACK MULTI-LINE SELECTOR

The Radio Shack Multi-Line Selector shown in Fig. 7-20 allows a connection of single-line amplifiers, telephone, and

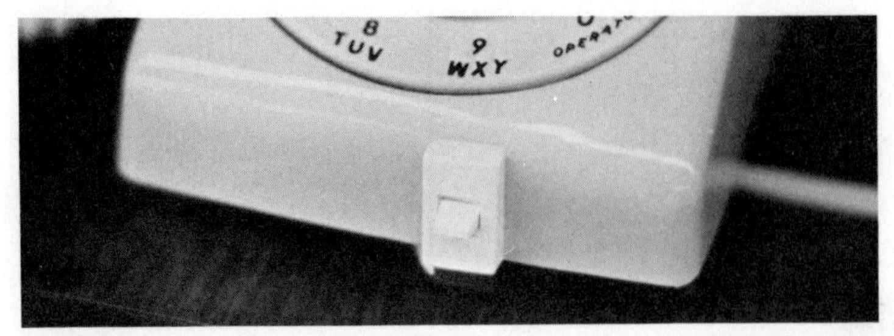

Fig. 7-19. Snap-on telephone silencer prevents telephone from ringing when activated.

answering machine to multi-line systems. This is an FCC approved device which installs in minutes to a standard multi-line system terminal. The device used with this selector is attached by means of a mini-modular plug. The rotary switch on the front of the selector allows the user to choose any one of up to four telephone lines in the multi-line system. It comes with a three-foot cord and a mating Amphenol connector. Since it is attached directly to the phone line, telephone company approval is necessary. This device is approved by the FCC and is priced at around $40.00.

HARDWARE

There are many types of telephone receptacles, adapters, and plugs which are used to connect the instrumentation cable to the phone line. The mini-modular plugs that we have been discussing throughout this text are the most common type of connectors used, but they come in many different styles and mounting configurations. The older four-prong mounting jacks are still seen from time

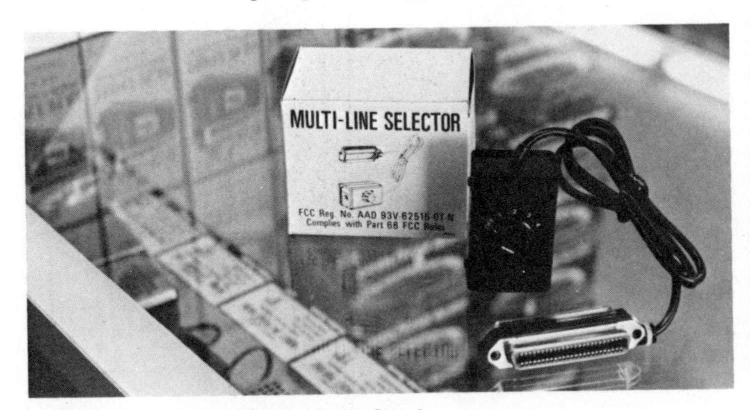

Fig. 7-20. Multi-line selector from Radio Shack.

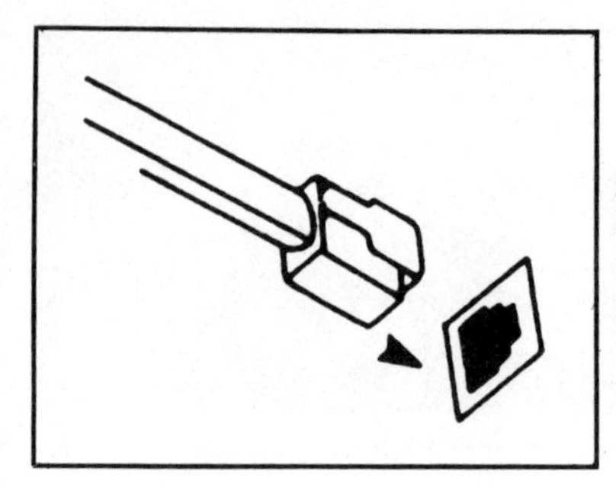

Fig. 7-21. Mini-modular plug and receptacle found between the handset cord and the base of the telephone unit.

to time, although they are not usually installed by phone companies today.

Figure 7-21 shows a standard mini-modular plug and receptacle which is found between the handset cord and the base of a standard telephone instrument and between this base and the wall outlet. Figure 7-22 shows the older four-prong style.

Since many home and business phone installations were made years ago, it is often necessary to make a conversion before the newer types of telephone instruments and accessories can be easily installed. This necessitates calling the phone company and having a serviceman come to your home or office to make the modification. Fortunately, there is an alternative which does not include a service fee.

Those individuals with the older style four-prong receptacle may purchase a converter or adaptor such as the one shown in Fig. 7-23. This particular model is supplied by TT Systems Corporation and plugs directly into the four-prong receptacle and provides an

Fig. 7-22. Old-style four-prong receptacle used to connect instrument to line.

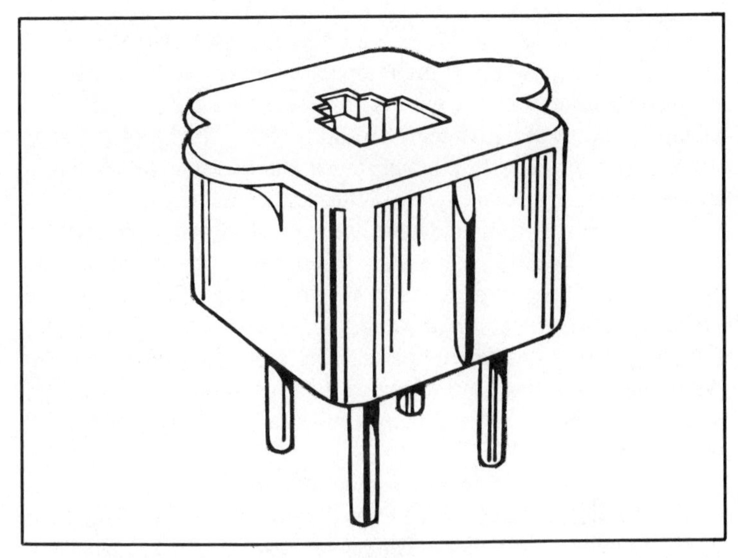

Fig. 7-23. Mini-modular to four-prong adaptor.

input for a mini-modular plug. This adaptor is inexpensive, selling for about $5.00, and the complete conversion can be made in less than a minute.

Some phone instruments may be equipped with a four-prong plug which will not directly fit a mini-modular system. The adaptor shown in Fig. 7-24 will immediately convert from the four-prong style to a mini-modular plug. This, too, sells for around $5.00.

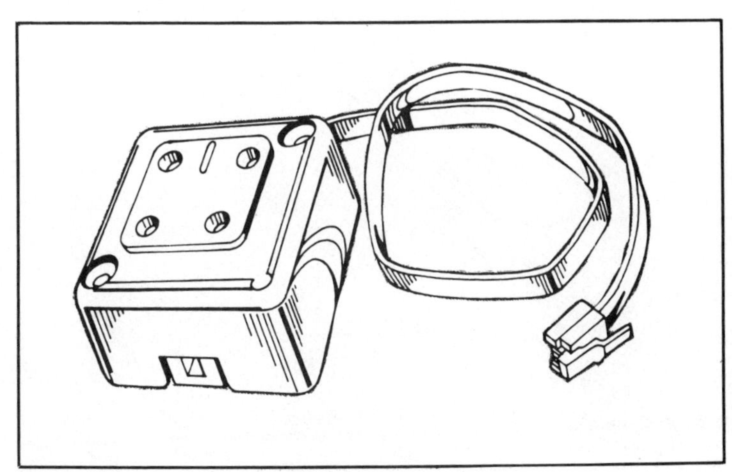

Fig. 7-24. Four-prong to mini-modular adaptor.

If you would like to extend the distance between your telephone instrument and the telephone line connection, this can be accomplished with the extension cord shown in Fig. 7-25. This adaptor, selling for less than $10.00, creates a 20-foot modular extension cord. The connecting block has two modular receptacles. For longer distances, two of these adaptors may be combined in series for an instrument/phone line separation of over 40 feet.

When it becomes necessary to connect two devices to the same phone line, the adaptor shown in Fig. 7-26 might be necessary. Retailing for about $7.00, it converts one modular female jack into two female jacks. This type of adaptor might be desirable when attaching a device such as an automatic dialer, which must be used in conjunction with a standard telephone instrument.

The FCC has adopted a group of standard plugs and jacks to be used with all modern phone systems. The hardware discussed here complies with this ruling. Telephone instrument and ancillary devices are to be connected through six or eight conductor modular plugs and jacks, or through fifty point microribbon connectors. These latter connectors are used for multi-line purposes. If the telephone company makes any changes in facilities, equipment, operations, or procedures which may affect the operation of customer provided equipment, it must notify the customer in writing.

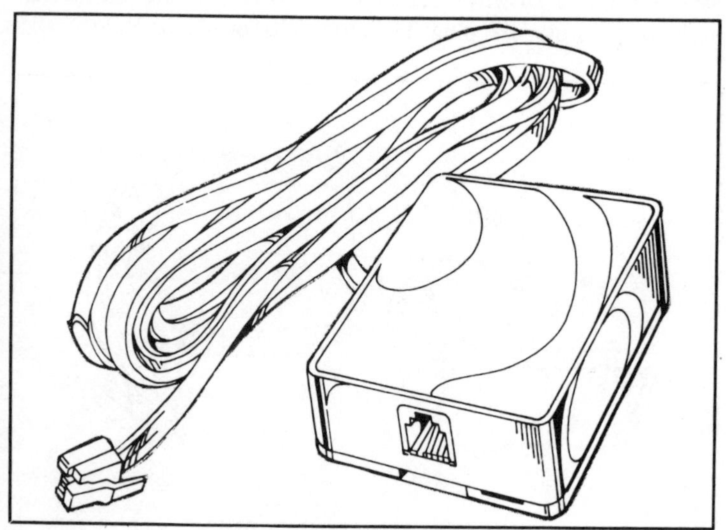

Fig. 7-25. Mini-modular extension cord .

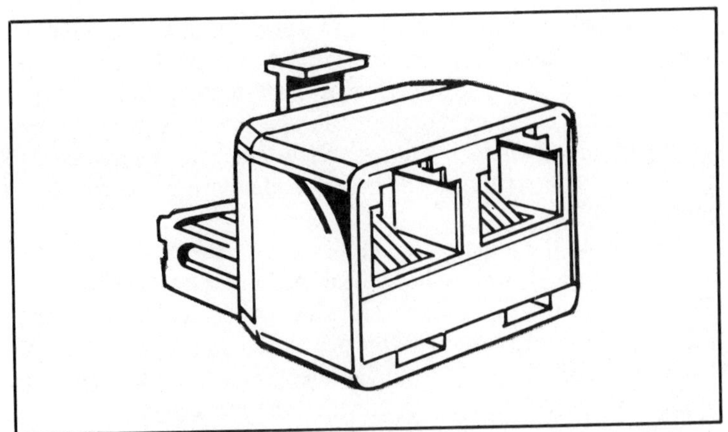

Fig. 7-26. Dual adapter for converting one modular female jack into two jacks.

What all of this means to the private telephone owner is that all of your hard work regarding purchase of adaptors, installation of telephone instruments, etc., will not suddenly be outdated by a surprising telephone company policy change. Should your phone company decide to change to a different type of plug or receptacle, it would be required to notify all customers in writing of this intent. This is a good regulation which helps to standardize all telephone installations and connections. It would be terribly complicated and expensive if every manufacturer of telephone instruments used a special type of plug or connector. It would be equally as confusing for the various phone companies to do likewise. With the FCC ruling in this matter, the entire industry is more or less confined to a set pattern of plugs and connectors.

Figure 7-27 shows the various types of modular hardware available from GTE Automatic Electric. You will notice that the adaptor plugs and jacks closely resemble some of the previous types discussed, although there are certain appearance differences. Notice, too, the wall receptacles which are similar to some of the standard telephone company installations. Figure 7-28 shows some of the modular hardware from Tel Products, Inc. which provide a similar range of convenience and adaptability features.

MECHANICAL ACCESSORIES

Two accessories which have been with us for many, many years are shown in Fig. 7-29. These are add-ons to the present telephone base or handset which provide, in a most non-complex manner, real creature comforts and conveniences. The Rest-A-Phone is a shoulder attachment which fits around the grip of the handset and allows busy secretaries and stenographers to type while holding the handset to the ear by applying pressure between the head and shoulder. Fatigue can be partially overcome by moving the device from shoulder to shoulder during long dictation periods.

Probably the simplest accessory is a stick-on ballpoint pen which is attached to its adhesive-backed holder by a coiled cord. The holder is attached to the front, side, or back of the telephone base, and the cord assures that the pen will always be within reach. Paper may still disappear from the telephone site, but you can be assured of always having a pen with this $.89 accessory.

FONE-A-LERT EXTENSION RINGER

Most of the amplifying devices discussed to this point have sampled the audio information from the telephone line, passed it through several amplifying stages until it was finally emitted from a loudspeaker. The Fone-A-Lert shown in Fig. 7-30 is a device which allows you to hear your phone ring regardless of where you may be on your premises. It is a telephone bell amplifier of sorts, although it does not actually increase the signal strength of the bell. Rather, when the phone rings, its circuit is triggered and it emits a piercing audible signal which is synchronized to the telephone ringer.

This device is very simple to install. Just apply the suction cup to the end of any telephone, as is shown in Fig. 7-30. Unreel the wire between the pickup and the amplifier, which is forty feet long, and carry the device with you whenever you know you'll be out of hearing range of the phone or where environmental noises might make it difficult to hear. While the pocket-sized device can be carried, the cord will be a bit of a nuisance, so most users simply place the device on a table or stand near where they will be located.

The Fone-A-Lert operates from its own power source, which is a standard 9-volt transistor radio battery and lasts for months in normal usage. Its circuit is a durable, solid-state design, and no mechanical parts are included. This means the device should be much more dependable than a standard telephone ringer. It measures 4¼″ wide by 3¾″ high and 2½″ in diameter. An integral wire reel takes up the slack when the full forty feet of extension cord is not used. Other cord sections may be added to increase this length.

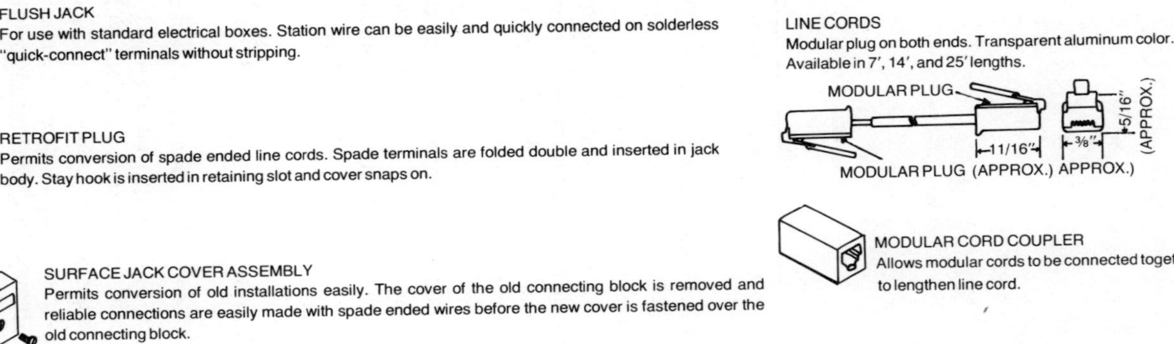

FLUSH JACK
For use with standard electrical boxes. Station wire can be easily and quickly connected on solderless "quick-connect" terminals without stripping.

RETROFIT PLUG
Permits conversion of spade ended line cords. Spade terminals are folded double and inserted in jack body. Stay hook is inserted in retaining slot and cover snaps on.

SURFACE JACK COVER ASSEMBLY
Permits conversion of old installations easily. The cover of the old connecting block is removed and reliable connections are easily made with spade ended wires before the new cover is fastened over the old connecting block.

LINE CORDS
Modular plug on both ends. Transparent aluminum color. Available in 7′, 14′, and 25′ lengths.

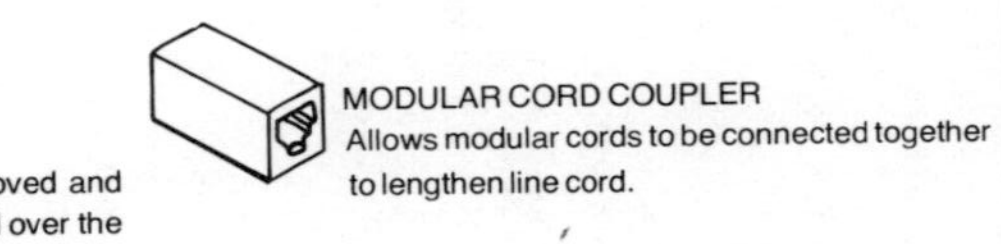

MODULAR CORD COUPLER
Allows modular cords to be connected together to lengthen line cord.

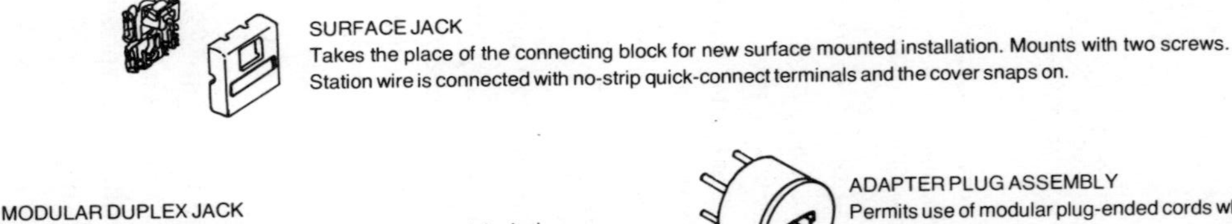

SURFACE JACK
Takes the place of the connecting block for new surface mounted installation. Mounts with two screws. Station wire is connected with no-strip quick-connect terminals and the cover snaps on.

MODULAR DUPLEX JACK
Allows two telephones to be plugged into a single modular jack

ADAPTER PLUG ASSEMBLY
Permits use of modular plug-ended cords when older four-point jacks are already in place.

Fig. 7-27. Various types of modular hardware for phone connection (courtesy of GTE Automatic Electric).

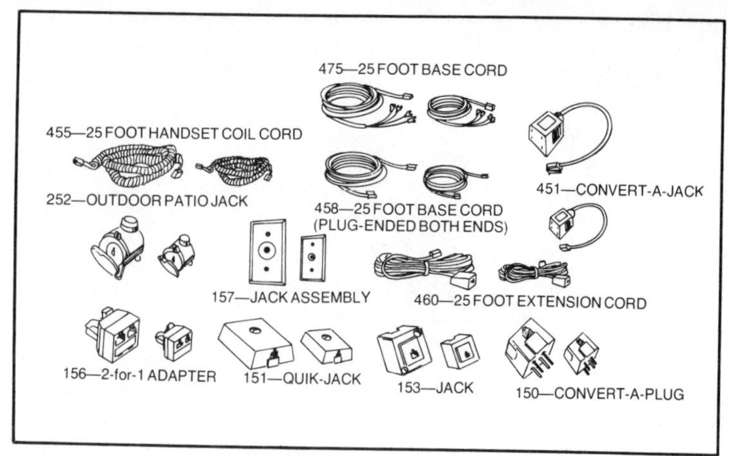

Fig. 7-28. Modular hardware from various manufacturers all conform to certain required specifications (courtesy of Tel Products, Inc.).

Operation is extremely simple and merely involves the throwing of the switch on the unit face to the on position. The Fone-A-Lert will then signal every time your telephone rings. When you don't need this penetrating signal any longer, the switch is turned to the off position. Changing this device from one phone to another involves merely the removal of the suction cup from one instrument and attaching it to another. This is accomplished in a few seconds.

An additional advantage of this device is the fact that it will work with almost any type of chime or ringer. Its manufacturer states that it is excellent as a doorbell extension. Here, the suction cup is attached to the door chime which is then extended by the master unit. The extension cord is flexible and made up of forty feet of 24 AWG, two-conductor speaker wire. The additional lengths, if desired, are easily spliced into the wire which is mounted to the takeup reel.

Selling for about $22.00, the Fone-A-Lert is a useful device which will be especially appreciated in the larger homes where a limited number of telephone instruments have been installed. With this unit, it is possible to get an indication that a phone is ringing while on the patio, at a backyard swimming pool, or from the noisy confines of a garage or repair shop. It may also come in handy only a short distance from the telephone when appliances such as vacuum cleaners and electric mixers are in operation. The noise created by these devices may make it nearly impossible to hear the standard ringer on your telephone.

BEL-RINGER

Another type of electronic ringer installs in virtually any kind of telephone, extension phone, or phone accessory. It has lower current requirements than conventional ringers and allows larger numbers of ringers on a single telephone line. The device is known as the Bel-Ringer and is much lighter than conventional ringers, with heavy bell-tapping suppression. With its unique micro-power electronic design, the Bel-Ringer can be an aid to those with impaired hearing or in noisy surroundings.

Shown in Fig. 7-31, the Bel-Ringer emits a pleasant warbling sound when the telephone is activated. It also features a unique acoustic control system that allows for the use of wide-range volume settings and replaces the harsh, raucous sound associated with conventional telephone bells. It's ideal for the quiet home or office, and its piercing warble will also often cut through the noisy environment of production areas. As a replacement, it affords a more pleasant sound without compromising excellent attention-getting quality. Specification drawings of the Bel-Ringer are shown in Fig. 7-32. This miniature device installs in a privately-owned telephone instrument in a short period of time and requires only a fraction of the current drawn by conventional electromechanical

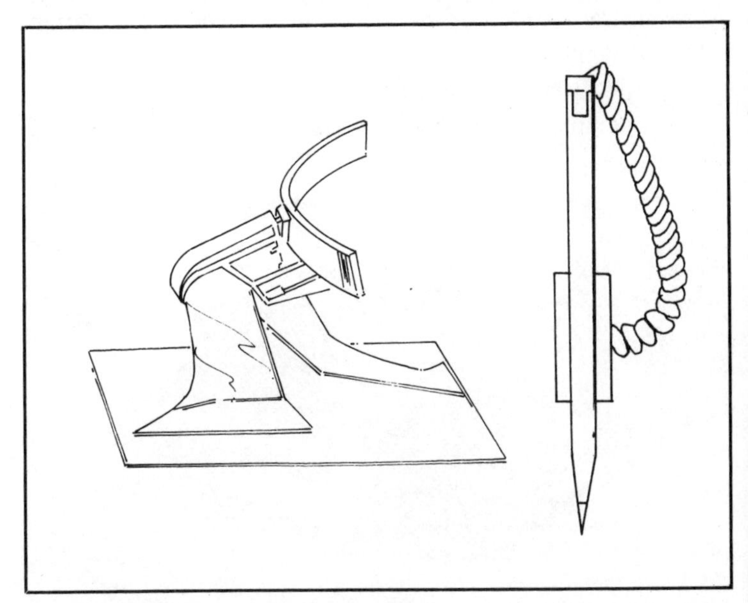

Fig. 7-29. Handset shoulder rest and attachable telephone base pen are some of the more common telephone accessories (courtesy of Radio Shack).

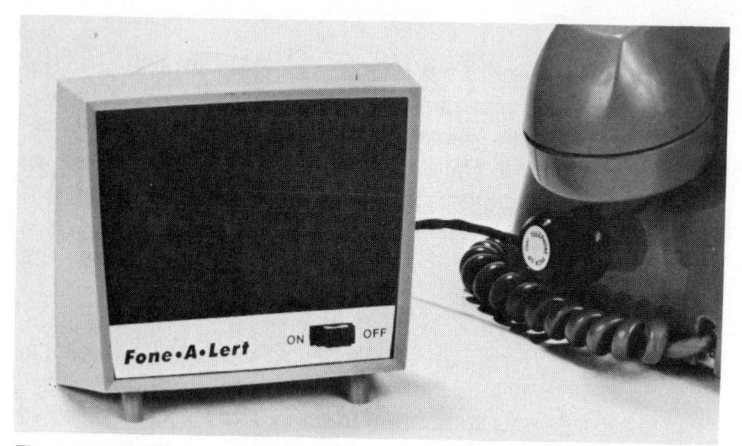

Fig. 7-30. Fone-A-Lert ringing amplifier (courtesy of Floyd Bell Associates, Inc.).

ringers. Many additional telephones employing the Bel-Ringer may be attached to any ordinary telephone line before ringer overload results. The device is much smaller than an ordinary telephone bell and installs easily in the standard instrument models, as well as a variety of special models which usually will not accept a conventional bell. Convenient mounting brackets, telephone standard spade lug connectors, and color-coded leads permit rapid conventional assembly.

Its specifications are as follows:

—Operating voltage: 90 volts RMS AC ± 25% 16 Hz to 100 Hz.

Fig. 7-31. Bel-Ringer electronic ringer (courtesy of Floyd Bell Associates, Inc.).

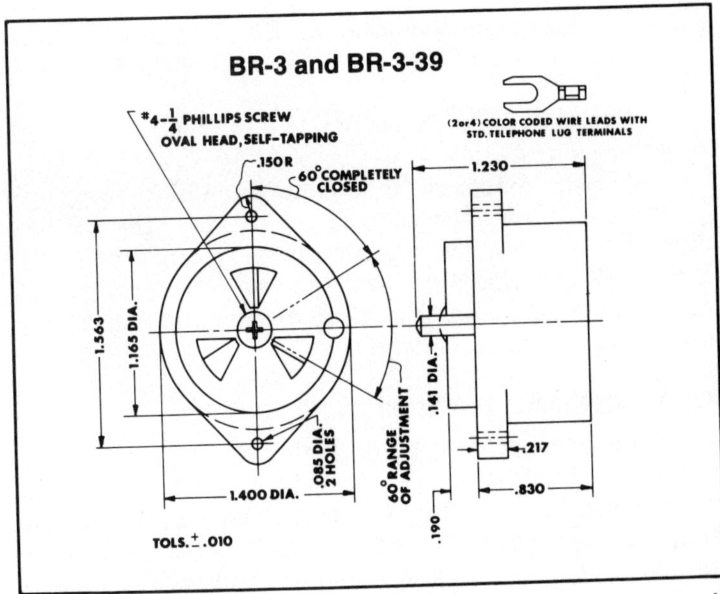

Fig. 7-32. Specifications of Bel-Ringer models BR-3 and BR-3-39 (courtesy of Floyd Bell Associates, Inc.).

—Transients: Withstands voltage transients in excess of 2000 Volts and 100 microseconds (no sound output).

—Operating current: 5 milliamperes maximum ringer equivalency = 0.4B at 16 Hz 90 V RMS.

—Bell-tapping suppression: Very heavy, suppresses all common dialing hookswitch and random line impulses. Typically 800V 2 milliseconds.

—Loudness (typical): 85 db at one meter on axis in free field with an A weighted ANSI type 2 sound level meter. (Acoustic volume control open.)

—Turn-on voltage: 65V or less 16 Hz with .47 db blocking capacitor.

—Range of volume control: Typically 15 db.

—Operating frequency: Models BR-3-39 and BR-3-39EV: typically, lower frequency 1700 Hz, upper frequency 2100 Hz, modulated at approximately 11 Hz. Model BR-3: typically, lower frequency 3000 Hz, upper frequency 3600 Hz, modulated at approximately 11 Hz.

—Operating temperature: −20 degrees C to +65 degrees C.

—Storage temperature: −40 degrees C to +85 degrees C.

—Humidity range: 0 to 85%.

—Exposed construction materials: KJB plastic, brass.

—Standard color: Beige on BR-3-39 and BR-3-39EV, black on BR-3.

—Size: 1.39 Cubic inches (vs. 15.5 cubic inches, standard ringer).

—Weight: .060 pounds (vs. .903 pounds, a 15:1 advantage).

There are several models to choose from. The BR-3-39 discussed here has the acoustic volume control, while the BR-3-39EV is equipped with an electronic volume control that clips to the telephone housing for added convenience. All of these devices have similar specifications.

TEL-A-TONE REMOTE EXTENSION RINGER

Electronic telephone ringers have become very popular today and provide flexibility to a standard phone system. The Tel-A-Tone Remote Extension Ringer shown in Fig. 7-33 plugs into any telephone extension outlet and produces a pleasant, distinctive remote signal whenever your telephone is called. The signal is high volume in nature, typically 85 decibels, and can be adjusted by means of a wide-range acoustic volume control.

Once this device is emitting a signal, you simply return to your telephone, wherever it may be located, and take your call. It features rugged, weatherproof construction which easily permits usage out-of-doors and is ideal for those who want to enjoy their

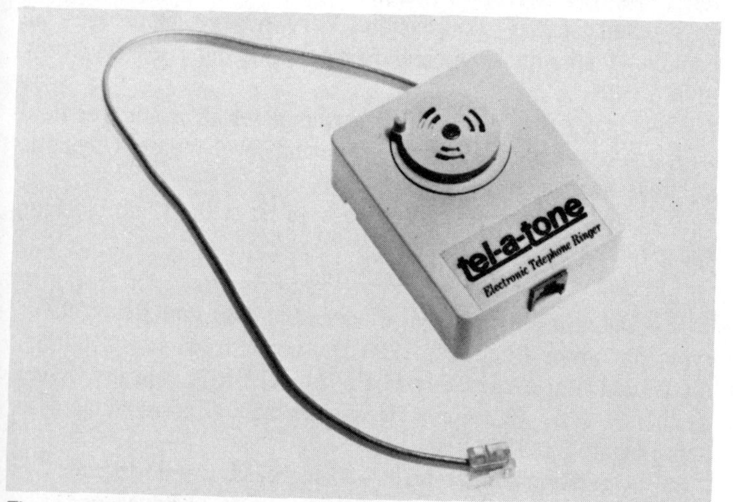

Fig. 7-33. Tel-A-Tone remote extension ringer plugs into any telephone extension outlet (courtesy of Floyd Bell Associates, Inc.).

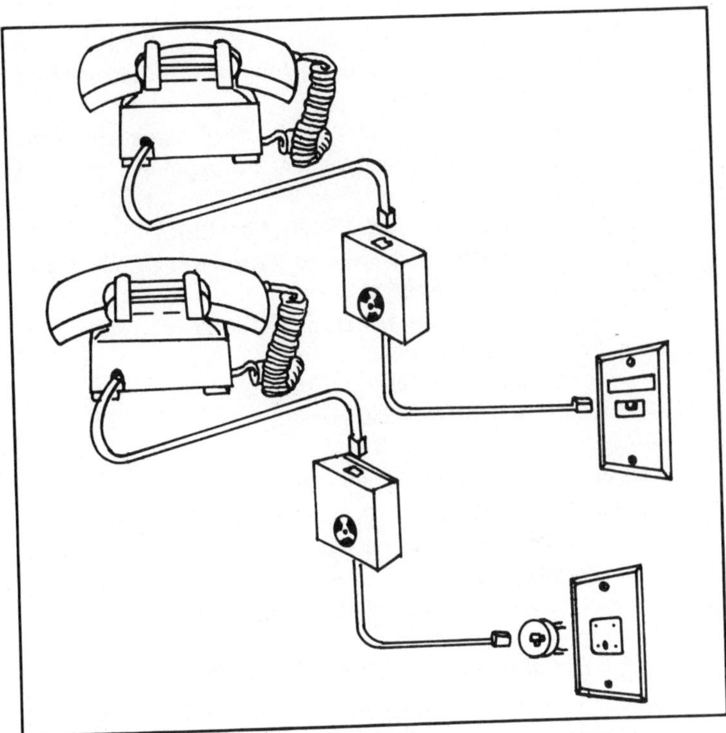

Fig. 7-34. Installation of Tel-A-Tone electronic ringer which requires no external power source (courtesy of Floyd Bell Associates, Inc.).

pools, yards, or patios without worry of missing a phone call. Like the former models, its solid-state design permits the use of more individual ringers on a single line without creating an overload condition.

The Tel-A-Tone electronic ringer requires no external power source, as it is powered directly from the phone line. It is easily installed as is shown in Fig. 7-34. Double-sided tape is applied to permit permanent wall or baseboard mounting. It should be completely installed within ten minutes. Once installed, the device becomes part of the overall system and may be all but forgotten about.

The Tel-A-Tone specifications are as follows:

—Operating voltage: 65 to 96 Volts RMs. 16 to 100 Hz.
—Operating current: 5 milliamperes max.
—Operating frequency: 1700-2000 Hz. Modulated at 11 Hz.
—Operating temperature: –20 degrees C to +65 degrees C.
—Storage temperature: –40 degrees C to +85 degrees C.

—Humidity range: 0 to 85%.

—Loudness: 85 dB at 1 meter with volume control open. Range of volume control 15 dB.

—Bell-tapping suppression: Very heavy, suppresses all common dialing, hookswitch, and random line impulses. Typically 800 V, 2 milliseconds.

—Ringer equivalency: 0.4B at 16 Hz. 90 V RMS.

—Transients: Withstands voltage transients of 2000 V and 100 milliseconds (no sound output).

—Exposed construction materials: KJB plastic, brass, FCC approved cord with modular plug.

—Color: Cream with black lettering.

—Size: 2.35 × 2.98 × 1.45 inches.

—Weight: 75 grams.

This device is licensed only for private, single-line telephone installation. It is not intended to be used on party line systems nor with coin actuated line and pay phones. Use of this device in these prohibited applications may cause severe line interference problems and is in violation of phone company regulations.

EXTENDER II WIRED-IN-PLACE ELECTRONIC RINGER

Very similar to the device just discussed, the Extender II electronic ringer from Floyd Bell Associates, Inc. is the wired-in-place version. Shown in Fig. 7-35, this device offers wide-range acoustic volume control, a standard modular connector, and a cover which is the same size and shape as a standard modular receptacle. This device is designed to take the place of the modular receptacle to which your present phone line and instrument are attached. Replacing the standard wall or baseboard mounted connector, the permanent connection of the electronic tone ringer to the line may eliminate service calls, as unterminated lines are also eliminated. It fastens to the wall or baseboard in a conventional manner and the cover snaps into place. The hookup is the same as with a standard modular jack and all fasteners and connections are hidden by the snap-on cover. In most areas it will be necessary to have the telephone company make this installation for you.

This device is FCC licensed and passes all tests for connection to telephone lines. A very effective bell-tapping suppression avoids many problems and its completely solid-state design assures dependability and long device life.

It will be found that this arrangement is very economical when compared to separately installed ringer components. The elec-

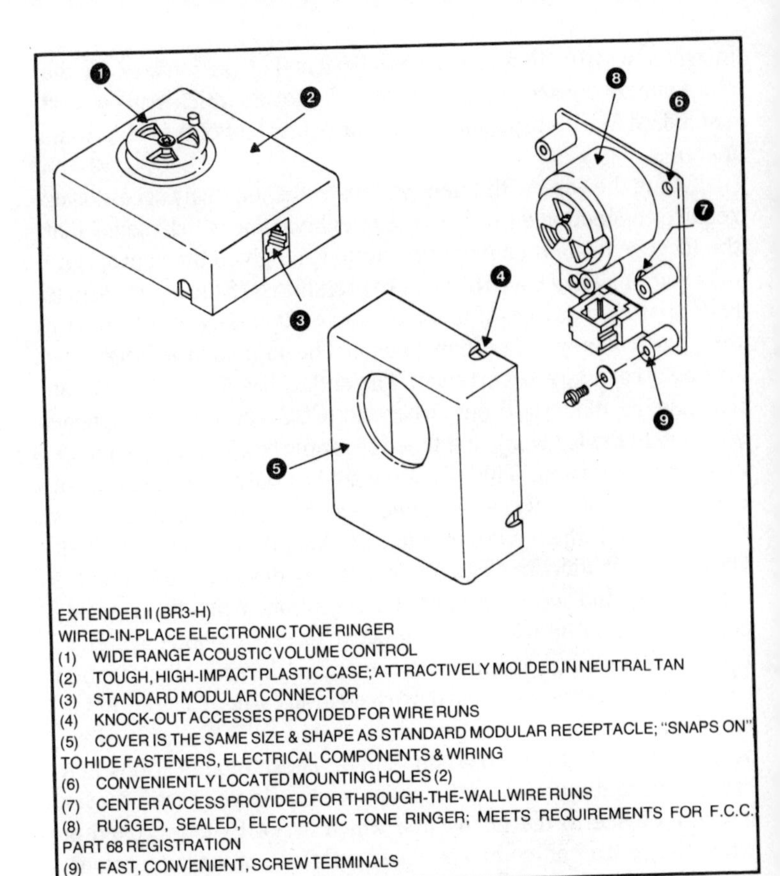

Fig. 7-35. Extender II electronic ringer is a wired-in-place version of the previous electronic device (courtesy of Floyd Bell Associates, Inc.).

tronic ringer is part of the modular connection, making for a simpler installation which is both inexpensive and more reliable. A piezoelectric sound transducer is used instead of a conventional loudspeaker. This device should last indefinitely.

SUMMARY

Never has there been such a large assortment to choose from in the telephone accessory line. This selection is certainly appropriate, owing to a similar choice in telephone instruments for the home and office. Much of the technological advancement which has developed many of the electronic telephone instruments has also been applied to the accessory line. The manufacturers have wisely chosen to offer the consumer the completed telephone instrument with all of its conveniences and attractions, or for the more budget conscious, the add-on accessories which may convert a standard Ma Bell special into a near equivalent of that electronic marvel.

As is the case with telephone instruments, many accessories require connection directly to the telephone line which means that the local central office must be notified and given the appropriate information. They have the right to prohibit you from using various FCC approved devices if it can be proven that they interfere with their particular system. Then, you must be notified in writing. This will be a very rare occurrence. The author has never heard of an instance in modern times when an FCC approved telephone attachment device would not function properly with existing phone company equipment. This of course applies only to those types of equipment which were installed according to manufacturer's directions and which were in sound working order when installed. Of course, malfunctions can occur with any device, and telephone instruments and accessories are no exception. When these occur, certain types of interference can occur to entire phone systems. For this reason, it is important to remember the rule which applied to telephone instruments and discussed in an earlier chapter: never modify through circuit changes a device which is FCC approved and attaches directly to the telephone line. These modifications may cause severe disruption of the phone service in your area. Never install a device to the phone line which is not FCC approved or when the central office has not been informed of your intentions. Never install a device to the telephone line which is knowingly defective, as this defect may change the entire circuitry and create interference, even though the design is FCC approved and the phone company knows and approves its interface with their lines. Remember, only the design and general manufacture has been approved by the FCC. Individual units can be changed or damaged. When this occurs, the approval may be null and void.

The reader who is in the market for some of the many accessories on the market is urged to shop around. Many of the products described in this chapter are available from many sources and by a multitude of manufacturers. It's a competitive market, so don't spend extra for something you might find at half the price if you shop around. The devices presented in these pages are all from reputable and proven manufacturers, but don't be lulled into thinking that these are the only devices of their kind in the world. There may be many other similar accessories which will perform the same functions and provide similar conveniences.

On the other hand, be cautious. If a reputable manufacturer sells one device for three times the price of another, you might be wise to see what other good manufacturers charge for similar devices. The one, unknown dealer who is way below the manufacturers you know to be established may not be telling you all the facts. A good buy is not always the least expensive.

With the many different conveniences which can be had by installing some of the accessory equipment discussed here to your present telephone, it may not be necessary nor even desirable to replace what you already have. Examine your current and future needs in telephone services. If you feel that your present system would be wholly adequate with the addition of a self-dialer, then maybe your best bet would be to purchase an auxiliary unit rather than to replace the entire telephone instrument with a self-contained telephone, self-dialer, intercom, etc. You may not really need the intercom and some of the other features. If you cannot conceive of ever needing them, then why go to the extra expense and bother of installation. Don't slight yourself either by spending money for something which will be obsolete in regard to your needs within a few months.

For some, the large selection available in the accessory line may be a bit confusing, but sensible browsing coupled with asking questions of the manufacturers and dealers should quickly clear up any doubts which cannot be answered through the discussion in this text. Most dealers will even give you the opportunity of testing out certain accessory devices in your home after purchase. If you are not satisfied with the performance, you may return the merchandise (in new condition) for a complete refund or exchange on a more appropriate accessory or telephone instrument.

Accessories have closely shadowed telephone instruments in development and popularity. This may be due to the fact that most of the modern electronic instruments are later developments and conglomerations of add-on accessories which have been in use for some time. The reader is warned to keep an eye out for further innovations and modifications of existing circuits. They are occurring on a daily basis.

Chapter 8

Telephone Scramblers And Security Devices

In this day of publicity about illegal phone taps, espionage, and industrial spies, voice scramblers and communications security devices which guarantee the privacy of telephone communications have become popular. This popularity, however, is mainly enjoyed in the industrial world. Since this text is devoted to telephone systems, many of which are practical for the general consumer, it may not seem appropriate to get into a full discussion of security devices. But it is the purpose of this book to inform the reader of the many devices which he or she may have a present need for *and* to inform this same reader as to what may be available for future needs and wants. Voice scramblers and security devices may fall into this latter category.

VOICE PRIVACY DEVICES

Technical Communications Corporation manufactures different types of communications security devices, commonly called Voice Privacy Devices (VPDs), designed to provide different levels of communications security. Selection of the appropriate VPD is dependent upon security requirements, communications system parameters, the specific type of terminal to which the VPD is to be connected, and budget considerations.

The following is intended to assist the reader in selection criteria regarding optimum scrambler systems:

☐ Degree of Security Required: What are the resources of the unauthorized listeners (eavesdroppers) and what is their technical

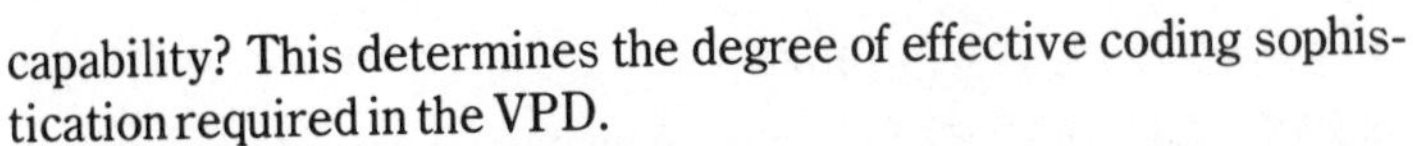
capability? This determines the degree of effective coding sophistication required in the VPD.

□ Communication Channel Quality and Characteristics: What is the complexity of the user's communication system. What type of modulation is used? Is long distance communications required, and what is the minimum acceptable reconstituted (decoded and unscrambled) voice quality? These factors may restrict the use of some scrambling/coding techniques, such as digital transmission, over certain types of channels.

□ Type of Installation: What are the types of terminals to which the VPD will be attached and what considerations should be made with regard to the entire system? This consideration indicates whether a telephone or radio model is required; how it will be installed; and the degree of operational complexity that is acceptable.

□ Budget: What budget is available as related to level of security needed? There are times when the budget determines the type of VPD.

Considering this criteria, it is convenient to look at Technical Communications Corporation's different series of communication equipment, each designed to provide a different level of security at different prices for specific systems. These are representative of some of the finer units found on today's market.

105 Series - Basic Privacy

The 105 Series of VPD equipment provides more privacy than a single inversion frequency or single code inverter at a reasonable price. They may be used when a high level of security is not necessary, but selective scrambling of certain radio transmissions is desired. Representative applications are: emergency policy action, crowd control, fire department dispatching, or private business communications. If an eavesdropper is successful in re-inverting the speech in real time, he must have two devices to unscramble both sides of the conversation since the 105's front panel switches permit selection of different inversion frequencies for transmit and receive. The 105 Series is intended to deny information to those who will not attempt to break the scrambler electronically. A modest amount of code separation from users of the same or similar types of scramblers is provided. The 105 is not intended to be used over Single Sideband channels or with telephones.

107 Series - General/Industrial Privacy

The 107 VPD Series is a low-medium level security scrambler which provides as much security as a five-band bandsplitter. One cannot learn to understand it by repeated listening as one can with a conventional inverter, and it provides a better reconstituted (decoded) quality. The 107's scrambling technique is a combination of frequency inversion, tone masking, and synchronization coding. Twenty-five (25) codes are available through the use of plug-in modules. The 107 series can be used over most normal communications systems, requires few adjustments and less maintenance than most scramblers of comparable security. The 107 is versatile, allowing it to be used in a wide variety of applications, although it is not recommended for single sideband channels. The 107 provides adequate security against the average radio technician. It is intended for use under conditions where electronic techniques will not be used to "break" the scramblers.

XL-280/PXL-280 Series - Tactical Privacy

The 280 Series of VPDs provides a tactical level of security for use over almost any type of channel. The units are small, lightweight, and particularly easy to install, adjust, and maintain. The PXL-280 has fewer codes than the XL-280 and is intended for police use and other applications where the reduction in unit cost as compared to the XL-280 is important. Millions of codes are incorporated in the 280 equipment.

Specific applications for the 280 series are: Special Police and Public Safety groups, tactical field use, provincial, state, and county use. In short, XL-280 and PXL-280 are for applications where communications security against electronic breaking, measured in days or weeks, is desired.

The 280 series is available with the dual synchronization feature permitting proper operation and increased security over long distance channels such as satellites and transoceanic phone calls.

207/307 Series - Strategic Privacy

The 207/307 series provides a high level of communications security and features TCC's "M"ary dynamic frequency and time encryption technique with billions of codes. In communications situations where the eavesdropper is considered to have good technical knowledge and equipment, the 207/307 series is strongly recommended. The 207/307 series is more secure than a bandsplitter, rolling code bandsplitter, and many narrow band analog types of scramblers. It can be used over almost any channel type including single sideband and is highly resistant to electronic deciphering methods.

The 207/307 series is available with the dual synchronization feature permitting proper operation and increased security over long distance channels, such as satellites and transoceanic phone calls.

807 Series - Strategic Security Plus

This series provides strategic security and is intended for use at the governmental and diplomatic levels. As in the case of the XL-280 or 207/307, the 807 may be used on any type of channel and provides very good recovered voice quality. Over 137 billion codes are incorporated in the 807 equipment.

Processing in the 807 is faster than that in the 307, changes take place at a rate of 250 times per second or 15,000 times per minute. In addition, additional time delay domain processing is used (in addition to frequency domain processing) to prevent identification of number counts by picking out the cadence (intersyllabic rate). The 807 is resistant to all standard breaking techniques, and computer analysis is extremely difficult.

The 807 series is available with the dual synchronization feature permitting proper operation and increased security over long distance channels such as satellites and transoceanic phone calls.

Security Summary

Each of the TCC's series of VPDs offer a different level of security. Each level and scrambling technique has been evaluated with respect to the Law Enforcement Standards Laboratory, National Bureau of Standards: Report No. 409-058, which may be referred to for detailed definitions of different levels of communications security.

The 105 Series provides a Class 2 level of security. Class 2 is adequate for situations where one does not expect the eavesdropper to tape record the message or for repeated playback and where electronic means of decoding (breaking) will not be used. The 107 Series provides a Class 3 level of security, meaning that the eavesdropper who will record the message and replay it will not decode it. Electronic assistance is needed to decode the 107.

The XL-280 and PXL-280 are resistant to breaking by the generally used inversion and parallel listening methods of electronic breaking. They provide a Class 5 level of security. The 207/307 Series provides a Class 6 level of security. This class should be used where one expects the eavesdropper to have a variety of sophisticated electronic breaking equipment. The 807 Series provides a strategic level of security which is quite high (6+) and is intended for use by high government officials and special groups, such as ministries, upper level military, and intelligence organizations.

COMMUNICATION CHANNEL QUALITY AND CHARACTERISTICS

The complexity of the customer's communication system and the type of carrier modulation used can play a factor in selecting the type of scrambling system that will provide the best performance, particularly with respect to the recovered or decoded voice quality. TCC's VPDs will give reasonable intelligibility on almost any channel including some with bandwidths as narrow as 300-2100 Hz ± 3dB. Decoded voice recognition is dependent upon having good channel quality, and in most cases the user will find that voice recognition is exceptional when considering the degree of security involved with the particular TCC system.

One should note that the recommended scrambling technique is somewhat dependent upon the type of radio modulation. This is particularly true with SSB modulation. Simple inversion scramblers and bandsplitters can be decoded with a good quality SSB receiver. Further, it should be noted that any duplex VPD intended for use over long distance channels such as satellite and transoceanic phone calls should be equipped with dual synchronization for proper operation.

Type Of Installation

TCC manufactures each series of VPDs in a number of models designed for installation with different communications equipment. The series number (105, 107, etc.) is followed by a two character code which indicates the type of installation the VPD is designed to work with. Table 8-1 illustrates the types of models and related installations available from TCC. The key indicates the meaning of each letter found in the model codes. (For example, a 107TA is a *telephone* scrambler which *acoustically* connects to a telephone handset.)

Table 8-1. Model Types and Related Installation of Various Technical Communications Corporation Devices (courtesy of TCC).

Series	RB	RV	RP	RM	TA	TW	FW	FA	Dual Sync
105	X	X	X						X
107	X	X	X	X	X	X			X
PXL-280	X	X	X	X	X	X	X	X	X
XL-280	X	X	X	X	X	X	X	X	X
207/307	X	X		X	X	X	X	X	X
807	X	X		X	X	X	X	X	X

Key: A - Acoustically coupled to telephone
B - Base Station installation
F - Facsimile
M - Manpack
P - Portable (handheld radio)
R - Radio
T - Telephone
V - Vehicular (mobile)
W - Directly wired to telephone

Further, it should be noted that the 107, 280, 307, and 807 series are available with dual synchronization to permit proper operation and increased security over long distance channels such as satellite and transoceanic calls.

Budget

It is TCC's policy to provide the highest level of security and quality for the lowest price among currently available communications privacy systems. Budgeting for a VPD series is dependent upon the level of security required. Some groups do not have the need or the budget to purchase the 207/307 or 807 systems and have found the 105, 107, or XL-280 series to provide the most security and performance available within their budget and requirements. The design of each series is such that system requirements may be expanded in the future with complete compatibility and trouble-free operation.

Summary

Neither the 105, 107, and basic inverter, or a five-band bandsplitter is recommended for use with SSB equipment. These techniques will work. However, they can be decoded by a tunable SSB receiver set to the opposite sideband. For use over SSB channels and whenever high security is needed for voice or facsimile on any kind of channel, radio, telephone, or satellite, the XL-280, PXL-280, 207/307, or 807 series is recommended.

It is extremely important for the user to understand the limitations of "scramblers." This need not, however, be carried to the point of "underselling" their value. There is a place in the spectrum or requirement for each of the TCC's series of equipment. In the final analysis, the user has to decide what will satisfy his requirement based on his own need and resources available. This discussion is summarized in Table 8-2.

The next section of this chapter will deal more thoroughly with the series of devices already discussed. Pertinent specifications will be provided for each piece of equipment.

105 Series Voice Privacy Devices

The 105 Series is mainly designed for voice communications on FM radios such as the type used for radiotelephone communications. These are analog scrambling devices which are designed to foil eavesdroppers, unsophisticated thieves, and curiosity seekers. Figure 8-1 shows the Model 105RV, which is full duplex and used for vehicular installation. It is front mounted and interfaces at the radio system's control head. It has two five-position rotary switches for code selection. The 105RB, which looks similar, is for base station use and connects at the microphone and volume control potentiometer location.

Another model, the 105RP shown in Fig. 8-2, is full duplex and is used with portable or mobile radios and connects at either the radio's accessory jack or through a universal connector. Table 8-3 shows the specifications for the 105RV and 105RB models, while Table 8-4 provides specifications for the 105RP. Admittedly, this series of voice privacy devices is mainly intended for transmitting and receiving applications and not for standard telephone systems. The 105 series is the least expensive and simplest to use.

107 Series Voice Privacy Devices

The 107 series of voice privacy devices is designed to provide a medium level of security and uses a combination of frequency inversion, tone masking, and sideband generation as a means of scrambling the voice signal. The degree of security offered using this complex scrambling method is greater than through conven-

Table 8-2. Table of Security Classification and Applications (courtesy of TCC).

Series	* Security class	Telephone Coupling	Radio Installation				Channel Modulation	** Duplex Modes	*** IMTS		FAX
		TA or TW	Base RB	Mobile RV	Portable RP	Manpack RM			Mobile	Base	FA / FW
105	(2)	None	Yes	Yes	Yes	Yes	AM,FM	H/F	Yes	Yes	No
107	(3)	TA/TW	Yes	Yes	Yes	Yes	AM,FM,Wire,	H/F	Yes	Yes	No
PXL280	(5)	TA/TW	Yes	Yes	Yes	Yes	AM,FM,SSB, Wire	H/F	Yes	Yes	Yes
XL280	(5)	TA/TW	Yes	Yes	Yes	Yes	AM,FM,SSB, Wire	H/F	Yes	Yes	Yes
207	(6)	TA/TW	Yes	Yes	No	Yes	AM,FM,SSB, Wire	Half	No	No	Yes
307	(6)	TA/TW	Yes	Yes	No	Yes	AM,FM,SSB, Wire	H/F	Yes	Yes	Yes
807	(6+)	TA/TW	Yes	Yes	No	Yes	AM,FM,SSB, Wire	H/F	Yes	Yes	Yes

* (2) basic privacy level, (3) general or industrial privacy level, (5) tactical privacy level, (6) strategic privacy level

** (H) Half Duplex, (F) Full Duplex

*** (ITMS) Improved Mobile Radiotelephone System

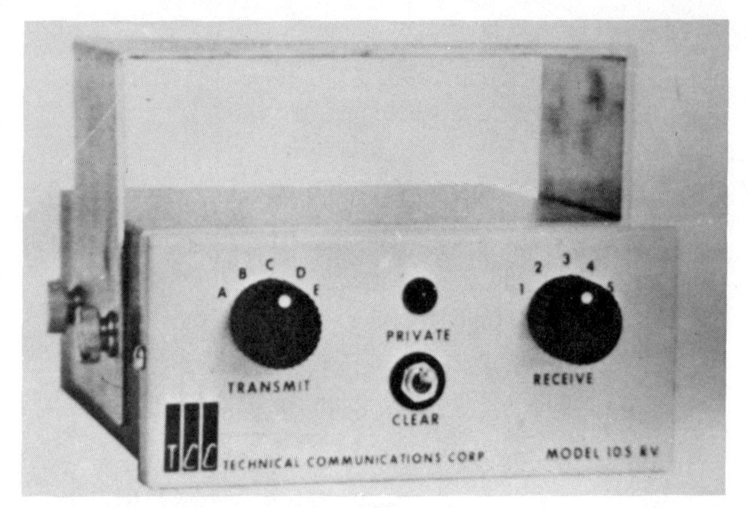

Fig. 8-1. TCC model 105RB often used for vehicular installation (courtesy of TCC).

tional inversion alone. The devices are full duplex, and there are 25 codes in each of four code families, for a total of one hundred combinations in the system. Plug-in modules are utilized to quickly change the various codes. Both telephone and radio versions are available in this series, and the devices can be supplied in such an arrangement so that one may communicate over a link consisting of a radio channel and a telephone channel in series.

This series of devices also has recoverable synchronization and clear voice override, which is the ability to receive a clear transmission when the switch is in the "private" position. A special

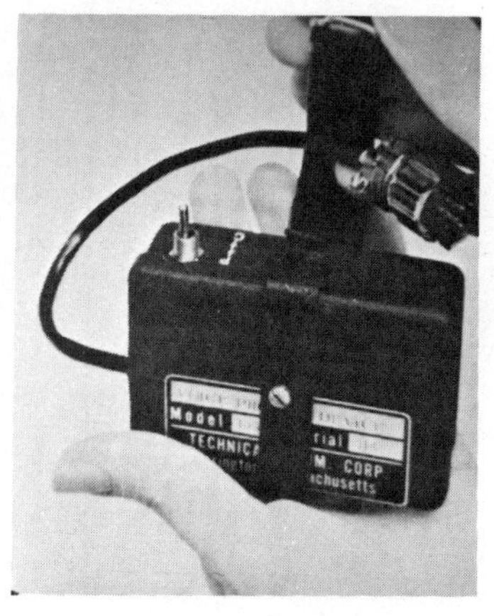

Fig. 8-2. 105RP model may be held in one hand (courtesy of TCC).

Table 8-3. Table of Specifications for the 105 Series Models RV and RB (courtesy of TCC).

Number of Codes	5 frequencies, 25 combinations
Code Selection	Rotary switches
Frequency Range	300-3000 Hz
Temperature Range	–30°C to +60°C
Controls	On/Off Switch (Clear/Private) Transmit Code Selector Switch Receive Code Selector Switch
Configuration	Vehicular or Base
Power	12V DC, plus or minus ground for mobile; base station, 120-230V AC
Size	105RV: 7″ (l) × 4-5/8″ (w) × 2″ (h) 105RB: 9-1/2″ (l) × 4-5/8″ (w) × 2″ (h)
Weight	105RV: 34 oz. 105RV: 40 oz.

version of the 107RV and 107RB called the 107-1000 series is available in special order. These units have 1000 selectable code combinations, and each is shipped with one code module. Additional modules may be ordered.

Figure 8-3 shows the 107RB units which are for base station use. They can connect at the control console or in series with various transmission lines. Table 8-5 gives the specifications for these units.

The Model 107TA is an acoustically coupled portable unit which operates from zone internal NICAD batteries and from AC power. Shown in Fig. 8-4, it can be seen that the telephone call is established using a standard instrument; then the 107TA is connected to the circuit by placing the handset in the acoustic connector. The separate handset emerging from the 107TA is used for communication. Table 8-6 gives the specifications.

Table 8-4. Table of Specifications for the 105RP (courtesy of TCC).

Number of Codes	Any 5 (inversion frequencies)
Code Selection	Rotary Switch
Frequency Range	300-3000 Hz
Temperature Range	–30°C to +60°C
Controls	Clear/Private mode switch Code Selection switch
Configuration	Handheld on portable radio
Power	From radio's batteries over a range of 8V DC to 15V DC. Power consumption 25 mA in transmit or receive
Size	3″ (l) × 2″ (w) × 3/4″ (h)
Weight	7.5 ounces

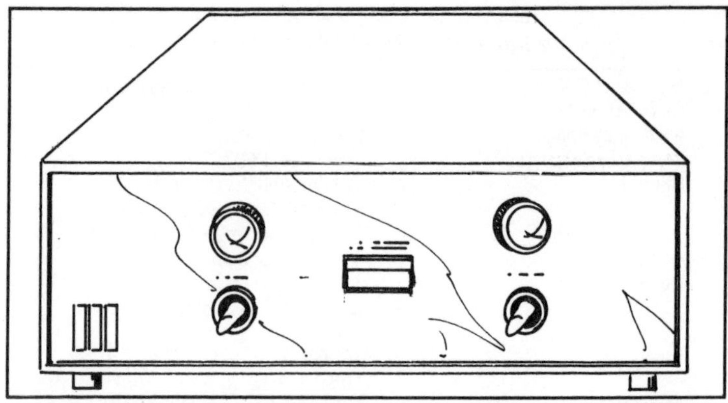

Fig. 8-3. 107RB security unit used for base station purposes (courtesy of TCC).

The Model 107TW shown in Fig. 8-5 is directly wired to the telephone line. A switch on the front panel provides the user with an option of communicating in a normal manner and then switching to the scramble mode by going to the "private" operational configuration. Table 8-7 gives the specifications for this unit, which operates from the AC line.

207/307 Series

The 207/307 series Voice Privacy Devices are designed for voice and FAX communications on all channels. This is a relatively high-cost unit and is used where information transmitted and received by telephone line is extremely valuable. This complex system is often used by corporations for protecting competitive bids on contracts and stock blocks, valuable cargo routes, proprietary processes, locations of oil and minerals, military maneuvers and special government operations. It interfaces with the telephone line and may be used in FM, AM, and SSB radio communications.

As was pointed out earlier, this series provides a high level of security using a dynamic "M"ary rolling code combination of

Table 8-5. Table of Specifications for the 107RB Security Unit (courtesy of TCC).

Controls	On/Off switch Clear/Private switch 5 position selector switch (RB-5 only) Code module socket
Configuration	Desk top or console
Power	117 volts AC 0.25 amps (230 volts available)
Size	107RB-1: 11" (l) × 15" (w) × 4-½" (h) 107RB-5: 11" (l) × 19" (w) × 4-½" (h)
Weight	107RB-1: 8 lbs. 107RB-5: 12 lbs.

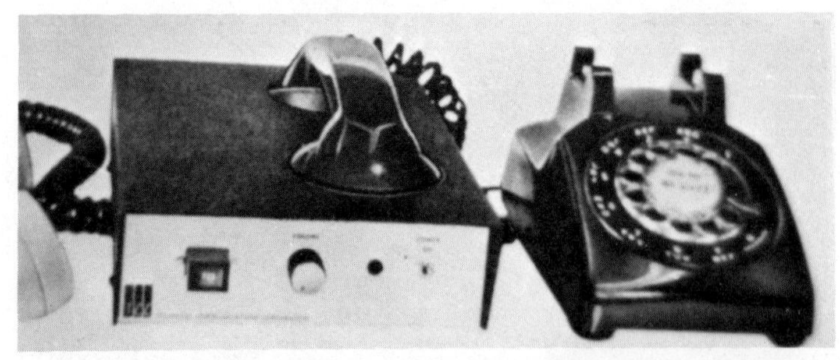

Fig. 8-4. Model 107TA security unit is an acoustically coupled portable device operating from ac power or from an internal power source (courtesy of TCC).

bandsplitting and frequency hopping in the time and frequency quadrants, effectively scrambling the audio signal. The 207 model is half duplex, using push-to-talk operation, while the 307s are full duplex.

There are 16 billion code combinations of which 262,144 may be selected at any time by means of the front panel switches. The code repetition cycle is over 1,200 hours and these units are easily installed and rather simple to use.

Figure 8-6 shows the TA model, which is acoustically coupled to the telephone and comes complete with the NICAD batteries, carrying case, and operates from AC or DC. The TW configuration is directly wired to a telephone in series with a handset and is supplied with a push-to-talk handset in the 207 configuration. These are highly sophisticated pieces of equipment offering a most complex scrambling system. Table 8-8 gives the specifications for the 207/307 series, including the various interfaces for these devices.

Table 8-6. Table of Specifications for the Model 107TA Security System (courtesy of TCC).

Controls	On/Off switch Code module socket Receive volume
Configuration	Desk or briefcase
Power	117 volts AC or internal rechargeable battery (battery supplied as standard equipment) 230V AC models available
Size	Desk: 11-⅛″ (l) × 7-¼″ (w) × 2-½″ (h) Briefcase: 12″ (l) x 18″ (w) x 5″ (h)
Weight	Desk: 5 lbs. Briefcase: 6 lbs.

Fig. 8-5. Model 107TW security system which attaches directly to the telephone line (courtesy of TCC).

XL-280 RPM-A

This particular series will not be discussed at length. These systems are designed to provide a cipher system for tactical military radios and have no direct application to the telephone line. The cipher technique involves a wide deviation in time and frequency coding and is controlled by digital codes. This ultra-sophisticated device offers a total of 18 billion code combinations and is used almost exclusively by the military for SSB, UHF, VHF, and AM radio transmitting applications.

807 Series Dynamic Multi-Code System

Designed for AM, FM, SSB, and telephone communications systems having band widths as narrow as 2kHz., the CSD 807 Communications Security System is designed for voice communi-

Table 8-7. Table of Specifications for the Model 107TW Operating From the AC Line (courtesy of TCC).

Controls	On/Off switch Receive volume Clear/Private mode switch Code module socket
Configuration	Desk top
Power	117 volts AC, 0.02 amps (230 volts model available)
Size	10-3/8″ (l) × 6-3/4″ (w) × 1-5/8″ (h)
Weight	3-1/2 lbs.

Fig. 8-6. Model 207/307TA is acoustically coupled to the telephone and features AC/DC operation (courtesy of TCC).

cations. It provides a strategic level of security and notable recovered voice quality. It employs a time and frequency division modulation technique controlled by 137 billion non-linear digital code. Maximum time delay is under 100 milliseconds.

The CSD 807 is available in telephone and radio models. The acoustically coupled telephone unit is of interest to our discussion on telephone systems. Figure 8-7 shows a drawing of the front face of the unit indicating the various control workings. The very large number of non-linear codes, coupled with the fine increment frequency processing and the unique form of time division delay permits a very high degree of security which defeats both conventional and computer analysis. The use of time delay breaks up the intersyllabic content of speech, removing the ability to extract information based on cadence, such as a number count from 1 to 10.

Operating simplicity is featured along with clear voice override and dual synchronization. Clear voice override permits intelligible reception of both clear and secure messages when in the "private" mode. Dual synchronization allows communications over long distance links, such as via satellites, without the need to resynchronize every time one speaks.

The AC power supply operates from 115/230, 50/60 Hz VAC and comes complete with NICAD batteries and charging circuit. Thus, about four hours of operation is sustained in the event of an AC power failure.

The CSD 807 series provides an extremely high level of strategic security which cannot be defeated by conventional techniques. It is also highly resistant to sophisticated electronic

Table 8-8. Table of Specifications for the 207/307 Series Including Interface Information (courtesy of TCC).

Codes in system	10^{16}
Codes selectable from front panel:	262,144
Code Selection	Six front panel thumb wheel switches plus internal jumper plugs.
Audio Channel Bandwidth	300 - 2300 Hz at ± 3 db
Front Panel Controls	Private/Clear Mode Switch ON/OFF Switch Fast/Slow Code repetition rate switch Manual/Push-to-talk synchronization selector switch Audio Volume Control Manual synchronization originate switch
Front Panel Accessories	Handset connector Fuse
Front Panel Indicator LED lamps	Off/Clear/Private Synchronization
Configurations	Vehicular, Manpack, Base, Telephone
Mode of Operation	207 is half-duplex (push-to-talk) 307 is full-duplex or half-duplex
Temperature range, operational	−30° to +60°C
Relative humidity	95%
Power Drain	DC = 3 watts AC = 12 watts

Size and Weight	Vehicular and Manpack 15″(l) × 11.5″ (w) × 4″ (h), 10 lbs. 38.1cm (l) × 29.2 cm(w) × 10.1(h),4.5Kg Base and Telephone 19″(l) × 11.5″ (w) × 4″ (h), 15 lbs. 48.2 cm(l) × 29.2 cm(w) × 10.1 cm(h)6.8 Kg
Model 37, Remote Control Head for 207/307	
Size and Weight	8″(l) × 6″(w) × 2″(h), 2 lbs. 20.3 cm(l) × 15.2 cm(w) × 5.0 cm(h) .9 Kg
Controls	ON/OFF Switch Clear/Private Switch Manual Synchronization Originate Switch Audio Volume Control (for specific interfaces only)
LED Indicators	Off/Clear/Private Synchronization
Accessories	Handset Connector Microphone Connector
Interfaces for 207/307	Microphone/Speaker Microphone/Volume Potentiometer Handset (in series with) 600 ohm OdBm, 2 wire and 4 wire (full duplex in 307 only) Telephone (in series with telephone's handset) Acoustically Coupled telephones IMTS

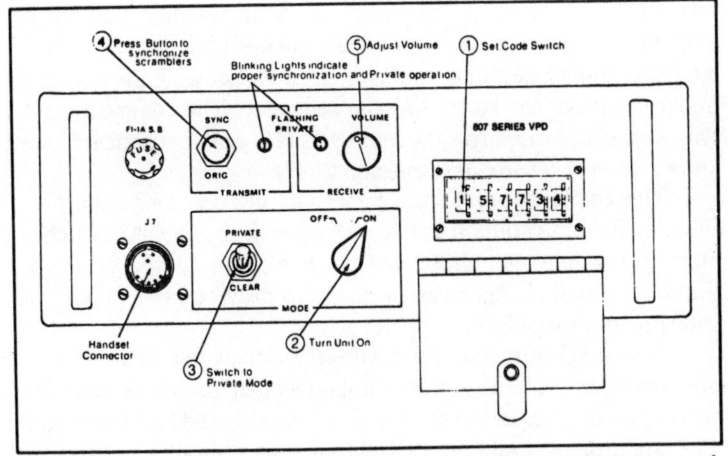

Fig. 8-7. Front face of the CSD 807 showing the various controls (courtesy of TCC).

deciphering methods, including computer analysis. The time and frequency processing is an analog audio cipher technique, controlled by digital codes and synchronized by an FSK code burst. Codes can be varied by the front panel code selector switches and the use of readily accessible internal switches and a coding plug unique to each customer, thereby providing an extremely wide range of code selections plus a high degree of built-in code protection. The simple combination of front panel and internal code selection provides control of three coding properties—synchronization, timing and frequency. A random number transmitted with each synchronization code burst changes the frequency and time codes with each synchronizing transmission. Maximum time delay is under 100 msec, and a VOX circuit is used to avoid objectionable scrambled side tone.

Frequency coding is accomplished by shifting the analog signal into various groups of frequency locations. These frequency slots are in turn shifted into 12 different frequencies having unequal increments ranging from 83 Hz to 157 Hz each at unequal time periods unrelated to the frequency coding. The shift from one frequency arrangement to another changes about 250 times a second, or 15,000 times a minute.

The independent time coding varies as well, and the maximum time delay at transmission is under 100 msec which will not impair two-way communication.

Code type and length is determined using a 31 bit shift register with a repetitive period of $2^{31}-1$ periods at a 61 Hz clock rate of 1.115 years. It is combined with another non-linear sequence in such a manner as to make the key immune from attack by conventional decoding methods such as knowing 2n bits of a linear sequence and computing the sequence. The combination of two sequences requires the entire 1.115 year sequence to be known by an eavesdropper to permit this type of decoding.

The combination of front panel and internal switching provides a total of 137 billion usable codes and the front panel switches alone permit any one of 262,144 codes at a time to be selected. Simple internal changes can be made to provide other blocks of code groups as required.

The CSD 807 contains a fail-safe circuit not found in less sophisticated systems that prevents uncoded transmission in the private mode in case the electronic circuits should fail, and warns the operator of a failure. In addition, the CSD 807 features an anti-tamper lock-out device which restricts unauthorized code changing during a transmission.

The 807 series is a full duplex system with the added capability of half duplex operation if required. This latter mode is selected by an internal switch. The system has a selectable dual-sync operation for long distance paths. This permits each receiving section to be synchronized by the other transmitting section to eliminate the effects of path delay and permit full duplex operation even when using a synchronous communication satellite.

A clear voice override permits the reception of unscrambled voice at the receiver even when the mode selector is set to the *private* position. When an operating communication channel is momentarily interrupted due to fading or other causes, synchronization will be maintained for forty minutes or longer by use of a built-in automatic synchronization feature. These latter features are especially useful for radiotelephone application. Recovered voice recognition is retained in channels with a frequency response of at least 300-2300 Hz. at ± 3 dB. This means that after the scrambling has been put into effect when transmitted, the voice received at the other end of the line will sound fairly natural after decoding. This is not the case with some security devices. Table 8-9 gives the specifications for the CSD 807 Series. A number of accessories are available for this series from TCC, and many of these are listed in Table 8-10.

FX 703 Series Facsimile Security System

Facsimile equipment has become a major carrier of electronic mail. It converts text, photographs, and graphics into electronic pulses which fall within the audio frequency spectrum. Since these are sounds which may be heard by the human ear, they may be transmitted over telephone lines through acoustic couplers. At the receiving end, these pulses are converted back into a facsimile of the original used for transmitting purposes. Thus, the name facsimile.

With the importance of facsimile transmissions today, facsimile security is vitally important and specially designed facsimile scramblers are now available. The FX 703 Series from TCC provides billions of code combinations, excellent recovered signal quality which is vital for a true facsimile reproduction, and is compatible with radio as well as telephone line channels.

The FX 703 is intended to be used with frequency modulation (FM) facsimile equipment. For facsimile machines using digital coding and modulation, TCC's Data Security Devices should be used to achieve encryption of the signal.

The FX 703 Facsimile Security System uses a unique digital processing technique controlled by non-linear codes and employing FSK synchronization to encrypt the FM Fax signal. The units are designed for facsimile transmissions over FM, SSB, and telephone communications systems having bandwidths as narrow as 2kHz (300 Hz to 2300 Hz).

Inherent flexibility allows the unit to easily adapt to a variety of interface parameters. State-of-the-art technology and stringent Mil-Spec testing insure adherence to high performance standards. Modular design allows easy access to internal circuitry for trouble-free maintenance.

The FX 703 is available in radio base station (RB), direct wired telephone (TW), and acoustically coupled/direct wired telephone (TA/TW) models. It is also available in a special configuration which is extremely cost effective as it permits the unit to be used for both voice and Fax transmissions. These models are designated with an SP suffix (i.e., FX703TWSP).

The FX 703 is a half-duplex system when used with facsimile transmissions. The voice portion of the SP models is full-duplex.

Interface parameters such as levels, impedances, and connections are easily changed by use of internal switches and interface programming plugs. A 115/230VAC NICAD battery power supply is provided with the unit. The FX 703 can also be powered from a 12VDC source or a 20-30VDC source on special order.

The scrambling or encryption technique used with the FX 703 is rather unique, in that it converts the analog signal to a digital

Specifications

Audio Modulation:

Dynamic time and frequency domain processing, controlled by digital codes, interjection of false noise and time delay, which is implemented with charge coupled devices, maximum delay is 100 MS.

Code Combinations:

262, 144 front panel selectable (six front panel thumb wheel switches)

137 billion total in system including internal switches and coding plug

Synchronization:

Type: Inband digital controlled FSK
Time: 1.6 Seconds
Duration: 10 minutes at temperature extremes or until next synchronization signal is received. At normal temperature approximately 40 minutes of sync duration can be expected.

Electrical Power Requirements

a. Basic Unit-11 to 15 VDC negative ground
b. 20 - 30 VDC to 12 VDC series regulator available as an accessory
c. AC supply 115-230 VAC 60 50 Hz to 12 VDC includes NICAD batteries and charger

Operation:

Full duplex, 2 wire or 4 wire
Half duplex, (push to talk), 2 wire or 4 wire

Operational Features:

Clear Voice Override
Automatic synchronization coasting in case of interrupted communication channel
Anti-tamper lock-out
Selective calling
Selectable single sync for local or dual sync for long distances
Protected against oscillator or logic failure
Fail-Safe indication

Current Drain:

0.2 A max. (up to 0.6 A with speaker output, Maximum voice received)

Push-to-Talk Signal:

Standard: Contact closure to ground on transmit, +30 VDC max for receive.

Frequency Control:

All timing frequencies are synthesized from a crystal controlled oscillator

Communication Channel Requirements:

The required channel bandwidth is 3 dB down at 300 and 2300 Hz for a 24 dB/octave rolloff rate.
The maximum tolerable channel non-linear distortion for good quality reconstituted speech is 10% for symmetric distortion and 5% for asymmetric distortion. These values apply to channels whose frequency response is equal to or better than specified above.

Frequency Offset:

±70 Hz maximum for SSB use

Front Panel Controls:
Clear/Private switch
On/Off switch
Audio Volume Control
Manual sync originate switch

Levels and Impedance:

Transmit:
600/150 ohm balanced or unbalanced center tap or split from – 50 dBm to +25 dBm
Other impedances at reduced power level, for dynamic, transistorized, and other microphone levels
For carbon microphones, unbalanced 100 mV to 2.2V p-p, up to 12 MA, positive bias
Telephone (direct-wired or acoustic coupler)
Receive:

Speaker, 4W, into 3-2 ohms unbalanced

Volume control potentiometer 2000 - 50,000Ω

High impedance, 0.1 to 8V p-p

50 mV to 5 V p-p 600/150 ohm balanced or unbalanced center tap or split – 50 dBm to 25 dBm

Telephone (direct wired or acoustic coupler)

Interface and I/O Characteristics:

Types of Interfaces:
- Mic/Spkr
- Mic/Vol Control
- Handset (in series)
- 600 ohm 4 wire
- 600 ohm 2 wire
- IMTS (vehicular or base)
- Telephone direct wired

Under Locked Cover (on front panel)

Thumbwheel Coding switch
Single/Dual Sync Selector switch
Man/PTT sync mode selector switch

Front Panel Accessories:

Handset Connector
Fuse

Front Panel Indicator LED lamps:
Transmit: Off/Clear/Sync/Private
Receive: Off/Clear/Sync/Private

Configurations:

Vehicular, Manpack, Base, Telephone, FAX

Temperature Range, Optional:

– 30°C to +70°C

Operational Humidity:

Up to 95%

Size and Weight:

Vehicular and Manpack: 11.5" W × 4" H × 14" L - 10 lbs. 29 cmW 10 cmH 35.5 cmL - 4.5kg.

Base and Telephone: 11.5" W × 4" H × 14" L - 10 lbs. 29 cmW 10 cmH 35.5 cmL - 4.5 kg.

Model 38 Remote Control Head

Controls:

On/Off Switch
Clear/Private Switch
Manual Synchronization Originate switch
Audio volume control (for specific interfaces only)
Single thumbwheel code selection switch (8 codes) not available with 600 ohm 2-wire and some 600 ohm 4-wire interfaces

LED Indicators:

Transmit: Off/Clear/Sync/Private
Receive: Off/Clear/Sync/Private

Accessories:
Handset connector
Microphone connector

Size and Weight:
6" W × 2" H × 8" L, 2lbs.
(15.2 cmW × 5.1 cmH × 20.3 cmL, .91kg.

Note: Model 38 Remote Control Head may be used for

Table 8-9. Table of Specifications for the CSD 807 Series (courtesy of TCC).

a. (RB) Terminal strip junction box and cable, AC power supply and cabinet or rack mounting for base station use.
b. (RV) Remote control unit with 20 ft cable and 38 remote control head, and mounting bracket for vehicular use.
c. (TW) Telephone interface cable, cabinet, and AC power supply.
d. (TA) Acoustic coupler for use with telephone handsets. Power supply with batteries standard for portable operation.
e. (RM) Handset, mounting bracket, cables.
f. (FA) Adaptor for use with facsimile machines.
g. 20/30V to 12 VDC Series regulator (Other voltages on special order)
h. 120/230 VAC power supply including NICAD battery backup with terminal strip connector and cabinet.
i. Spare PC board set (depot spares), full duplex.
j. Spare parts component kit (2 year field maintenance).
k. Model 75 back-to back test fixture.
l. Extended board for checking PC boards.
m. Set of extra interface programming plugs.
n. Model 38 Remote Control Head for complete remote control.
o. 20 ft. cable for use with Model 38 remote control head.
p. Stylized cabinet for telephone or base station installations.
q. Battery pack, dry or rechargeable.
r. Battery charger.
s. Rack mounting hardware.

Table 8-10. List of Accessories Available for the 807 Series (courtesy of TCC).

format. In this state, the digital signal is encrypted and then converted back to an analog signal for transmission. The FX 703's processing is controlled by non-linear digital codes and synchronized by an FSK code burst. Code settings are varied by thumbwheel switches located behind a locked hatch, internal switches, and by strapping options located on readily changeable jumper plugs. This three level coding provides a very wide range of code selections plus a high degree of built-in code protection. As a result of this three level coding, the code repetition cycle is extended to over 6 months.

When the FX 703SP is used in the voice mode, the analog signal is encrypted using what TCC calls pseudorandom, non-linear time and frequency domain processing.

The combination of front panel switches, internal switches, and internal strapping options provides over 54 billion usable codes. The front panel switches alone permit any one of 262,144 codes to be selected. Internal changes can be made to provide other blocks of code groups as required.

A clear signal override feature permits receipt of unscrambled signals at the receiver even when its mode selector is set to *private*. When a communication channel is momentarily interrupted due to fading or other causes, synchronization will be maintained for up to 40 minutes by use of a built-in synchronization coasting feature.

The FX 703 system provides a strategic level of security and is highly resistant to conventional or sophisticated electronic breaking techniques. Table 8-11 gives the specifications for the FX 703.

Table 8-11. Table of Specifications for the FX 703 Security System (courtesy of TCC).

Model FX703 Accessories

a. (RB) Terminal Strip junction box and cable, AC power supply and cabinet or rack mounting for base station use.
b. (TW) Telephone interface cable, cabinet, and AC power supply.
c. (TA) Acoustic coupler for use with telephone handsets.
d. 20/30V to 12V DC Series Regulator. (Other voltages on special order.)
e. 120/230V AC/DC Power Supply with Nicad Batteries, terminal strip connector and cabinet.
f. FX703 Spare PC Board Set (depot spares), full duplex.
g. FX703 Spare Parts Component Kit (2 years field maintenance), full duplex.
h. Model 75 Back to Back Test Fixture for FX703.
i. Extender Board for checking PC Boards for FX703.
j. Set of extra interface programming plugs.
k. Stylized cabinet for telephone or base station installations.
l. Rack mounting hardware.
m. Model 42 2 wire/4 wire hybrid.
n. Carrying case for portables.

SPECIFICATIONS

Signal Modulation:
Analog to digital conversion, digital processing, digital to analog conversion for analog transmission.

Code Combinations:
a. 262,144 front panel selectable (six front panel thumbwheel switches).
b. Over 54 billion

Synchronization:
Type: inband digital controlled FSK.
Time: 1.6 seconds.
Duration: Up to 40 minutes.

Frequency Control:
All timing frequencies are synthesized from a crystal controlled oscillator.

Communications Channel Requirements:
a. The required channel bandwidth is 3 dB down at 300 and 2300 Hz for 24 dB/octave rolloff rate.
b. The maximum tolerable channel non-linear distortion for good quality reconstituted signal is 10% for symmetric distortion and 5% for asymmetric distortion. These values apply to channels whose frequency response is equal to or better than specified above.
c. The FX703 can be used on any channel over which the fascimile unit will operate.

Frequency Offset:
±50 Hz maximum for SSB use.

Operation:
a. Full duplex, 2 or 4 wire on voice.
b. Half duplex (push-to-talk) 2 wire or 4 wire on fascimile.

Operational Features:
a. Clear signal override.
b. Automatic synchronization coasting in case of interrupted communication channel.
c. Anti-tamper lock out.

Front Panel Controls:
Private/Clear Switch.
ON/OFF Switch.
Signal Level Control.
Synchronization Originate Switch.
TX FAX/RX FAX
SP Models-TX FAX/VOICE/RX FAX.

Front Panel Indicator LED Lamps:
Transmit/Receive
Synchronization

Configurations:
Base, Telephone
Temperature Range, Operational:
−30°C to +60°C
Operational Humidity:
Up to 95%

Electrical Power Requirements:
a. Basic unit 11 to 15V DC negative ground
b. 20-30V DC to 12V DC series regulator available as an accessory
c. AC supply 115-230V AC 60/50 Hz to 12V DC
d. ACPDC Nicad Battery power supply available as an accessory
Current Drain:
0.2A
Power Drain:
3W DC, 12W AC
Size and Weight:
Base and Telephone:
11.75″W × 4″H × 15.5L-15lbs.
(29.8cmW × 10.1cmH × 39.2cmL-6.8 kg.
Interface and Output Characteristics:
Types of Interfaces:
a. Handset (in series with)
b. 600 ohm four wire
c. 600 ohm two wire
d. Telephone direct wired
e. Telephone acoustic coupled
Levels and Impedance:
a. Transmit
1. 600/150 ohm balanced or unbalanced center tap or split from −50 dBm to +25 dBm
2. Telephone (direct wired or acoustic coupler)
b. Receive
1. 600/150 ohm balanced or unbalanced center tap or split −50 dBm to 25 dBm
2. High impedance 0.1 to 8V p-p
3. Telephone (direct wired or acoustic coupler)
Specifications subject to change without notice

NOTE: When ordering, be sure to specify the interface parameters desired. TCC will set up the program plugs to meet the required characteristics. If additional interface plugs for different types of installation are desired, they should be ordered and their parameters specified. One may use the order form supplied by TCC or make a copy of the specifications and circle the specifications desired.

Specify RB for base radio, TW for direct wired, and TA for acoustically coupled to a telephone; also, indicate Half-Duplex or Full-Duplex operation.

Include: Electrical power, make and model of the facsimile machine the FX703 is to be used with, transmit interface, and receive interface. To facilitate a smooth and easy installation of the equipment, it is important to provide TCC with accurate interface information which will eliminate the need to make changes in the field during initial installation.

TCC has available an interface checklist which can be completed to facilitate compiling the required interface information.

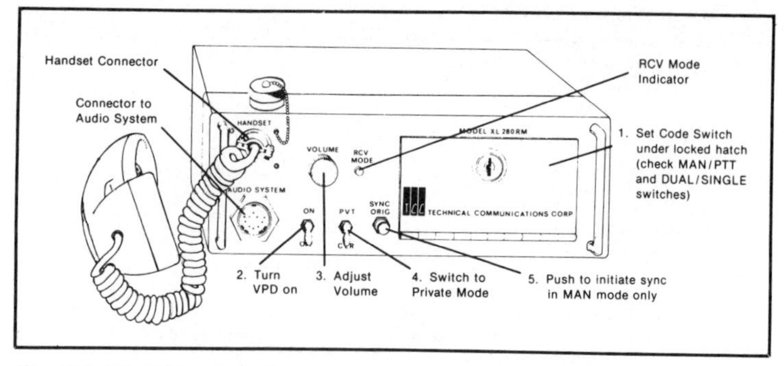

Fig. 8-8. XL 280 series of security devices uses a Pseudorandom time and frequency scrambling technique (courtesy of TCC).

XL-280 Series

Technical Communications Corporation's SX-280 series uses a unique pseudorandom time and frequency scrambling technique which provides a high degree of security that cannot be decoded by conventional techniques. Figure 8-8 shows the XL-280RM and indicates the various controls and connection points. Synchronization and scrambling are controlled by digital codes. The XL-280 series is designed for use over FM, AM, SSB, and telephone communication systems having bandwidths as narrow as 2.0 kHz (300 Hz to 2300 Hz). It is ideal for tactical use.

The XL-280 is a complete modular system, as is shown in Fig. 8-9. The units are small in size, have extremely low power drain, use modular hybrid construction, and are switchable to either full or half duplex for use with telephones, mobile radios, base station radios, manpack radios, and hand-held portable radios. Operation is simple; maintenance is easily accomplished by use of replaceable modules; and the units may be easily set up for use with different input/output characteristics.

The XL-280 series has the added operational feature of "clear voice override" permitting a user to receive a clear transmission when in the private mode of operation ("clear" is unscrambled; "private" is scrambled) when used in the push-to-talk mode.

The XL-280 units are small, permitting them to be located where convenient while remaining inconspicuous.

Various models are available permitting use for almost any type of application. For telephone use, the XL-280TW is direct wired, and the XL-280TA/TW is acoustically coupled and is also capable of hard wired connection. For radio applications, the XL-280RV is designed for vehicular mobile radios, the XL-280RB for use with base station radios, and the XL-280RP for portable radio use. The XL-280RM is for Manpack operation.

Coding is accomplished in both the frequency and time domains. A unique frequency translation and variable time technique modulates (scrambles) the voice under control of non-linear digital codes. There are 22 frequencies selected in a pseudorandom manner. The transition from one frequency to another is not continuous but uses a series selected from 85 frequency steps.

There are 8 sequences of 512 discrete frequencies each, stored in a memory. Sequences are selected at random, one at a time, under control of a pseudorandom code generator having almost 8.4 million states.

Frequency steps are changed every millisecond (mS), and the duration of a sequence, while random, is always over 0.5 seconds. The code repetition cycle is over 1165 hours. There are almost 18 billion codes in the system, of which over 1 million may be selected from front panel switches. Additionally, a 10-position internal switch increases user selectable codes to over 10 million.

Sync between two XL-280 series units is accomplished by transmission of a 1.02 second long 8-bit code preamble. The sync burst will be rejected by the receiving unit if the internal sync code (dictated by the thumbwheel switches) does not agree with the received sync code. The synchronization code and scrambling code are not the same.

XL-280 units have an optional dual sync feature which enables the transmit and receive sections of these units to be synchronized independently, thus accommodating long-path-delay channels. A

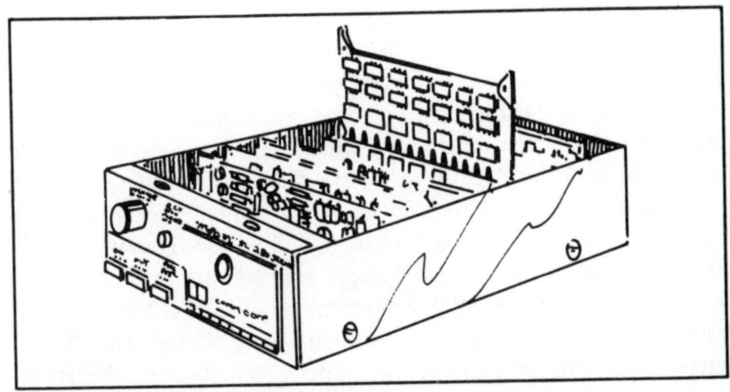

Fig. 8-9. The XL 280 system uses modular construction design for efficiency, durability, and ease of maintenance (courtesy of TCC).

switch which selects either single or dual sync is located under the front panel hatch. Units which do not incorporate the dual sync feature are always in the single sync mode, regardless of the single/dual switch position. Due to the modularity of the units, the dual sync feature may be easily added in the field in a matter of minutes.

DATA PROTECTION

With the high number of data terminal and data processing equipment in use today, an increasing amount of confidential, valuable, and proprietary information is being transmitted via phone lines and in digital form. Data terminals are vulnerable to data interception by many knowledgeable people with modest amounts of equipment. Therefore, persons with computer terminals are turning toward the security market to keep confidential the computer data sent over telephone lines. These terminals are involved with trade secrets, bank accounts, personal data, and market transactions.

TCC's DPD-72 and DPD-72A Data Privacy Devices ensure the security of encrypted transmitted digital information, as any intercepted data will be unintelligible. An enciphered message received by an data terminal not equipped with a Data Privacy Device (DPD) set for the same code will print out a stream of unrelated characters which cannot be deciphered. As the coding is extremely complex, no correlation can be drawn between clear data and the encrypted signal. An enciphered message received by a data terminal equipped with a DPD set for the same code will print out clear text.

A switch on the DPD-72/72A selects either clear or private operation. When the private mode is selected, the DPD enciphers in real time. The signal to be processed enters the DPD, is encrypted, regenerated, and then transmitted. The DPD requires little action by the terminal operator other than the need for a preamble of any five characters which must precede each new transmission. Table 8-12 gives the specifications for this series.

The DPD-72 can be set for any of 4.3 billion different codes, while the DPD-72A version has over 68 billion code combinations. All coding is digital, non-stationary, and non-linear. For every basic code set into the DPD, a different set of code changes result for each character, number, or word that is transmitted. The DPDs have a set of thumbwheel switches under a locked cover which are used to set the basic code. Codes can be easily changed daily or hourly.

Specifications

Audio Modulation
Wide deviation frequency displacement in both time and frequency domains controlled by digital codes.

Code Combinations:
1,048,576 front panel selectable (five front panel thumb wheel switches).
Almost 18 billion total in system including switches.

Synchronization:
Type: Inband digital controlled PSK
Time: 1.02 seconds
Duration: 10 minutes at temperature extremes or until next synchronization signal is received. At normal temperature approximately 1 hour of sync duration can be expected.

Frequency Control:
All timing frequencies are synthesized from a crystal controlled oscillator.

Communication Channel Requirements:
The required channel bandwidth is 3 dB down at 300 and 2300 Hz for a 24 dB octave rolloff rate. The maximum tolerable channel non-linear distortion for good quality recovered (decoded) speech is 10%, for symmetric distortion and 5% for asymmetric distortion. These values apply to channels whose frequency response is equal to or better than specified above.

Frequency Offset:
±90 Hz maximum for SSB use

Frequency Response:
300 Hz - 2300 Hz

Front Panel Controls

1. ON OFF (double action push button): In the OFF position, the VPD is switched off. All of the XL-280's input/output circuits are bypassed. The communication device operates as if the VPD is not there. In the ON position, the VPD is switched on.
2. CLR PVT (double action push button): In the CLR (clear) position, audio passes through the VPD circuits. However, no scrambling takes place. In the PVT (Private) position, transmission is scrambled (not necessarily synchronized.) Reception is "in the clear" until the receiving section is synchronized, if (and only if) the Sync Mode Switch is in the PTT position, the receive section will revert to clear operation each time the PTT push button of the radio is released and, therefore, resynchronization is necessary.
3. Sync Originate (momentary push button): This switch is operative when the MAN sync mode is selected. Activation will cause a sync burst to be transmitted.
4. Volume Control: This control functions in both CLEAR and PRIVATE modes.

Front Panel Indicator
A front panel LED indicator signifies the receive status of the unit as follows:

1. Blank - power off.
2. Continuously lit - receive section in CLEAR mode.
3. Flashing - receive section in PRIVATE mode.
4. Double Brightness - valid sync has just been received.

Control Behind Locked Hatch

1. Five thumb wheel switches - code setting.
2. Sync Mode switches:
   Single Synd/Dual Sync - only operative on units equipped with dual sync capability.
   PTT/MAN - determines the source of the sync command - either PTT switch actuation or SYNC ORIG push button actuation, and controls sync logic, as described above.

Rear Panel Components

1. Handset Connector - for use with "optional" TCC handset. (XL-280RM Handset Connector - on front panel.)
2. Audio System Connector - main interface IN/OUT connector.
3. TX IN/RX OUT (either from handset or from audio system) Selector Switch (optional).
4. TX OUT/RX IN level Selector Switch (optional).
5. DC Power Connector - for use with an external battery or the TCC AC/DC Power Supply.

Internal Controls

1. Half Duplex/Full Duplex selection switch.
2. "RX IN," "RX OUT," "TX IN," "TX OUT" level adjustments.
3. Two LED lights for "RX IN" and "TX IN" level adjustments.
4. Level indicator activate push button.
5. VOX IN/OUT activation switch.

Ambient Temperature
Operating – 30°C to +60°C
Storage – 40°C to +85°C

Operational Humidity
Up to 95%

Electrical Power Requirements:
RB, TW, TA/TW - 115/230 VAC, 48-63Hz using TCC AC/Battery supply
RV, RM - 7.2/14.4 VDC
RP - 7.2/14.4 VDC at 35 MA max.

Operation:
Full Duplex, 2-wire or 4-wire
Half Duplex, (Push-to-talk) 2-wire or 4-wire

Operational Features:
Clear Voice Override
Automatic synchronization coasting in case of interrupted communication channel
Anti-tamper lockout
Selectable single sync for local or dual sync for long distances
Protected against oscillator or logic failure

Current Drain:
50 mA (up to 0.6 A with speaker output, maximum voice received)

Push-to-talk Signal:
Standard: Contact closure to ground on transmit, +30 VDC max for receive. Up to ±100 VDC control voltage on special order

Size and Weight:
RB, TW, RV: 3-⅛"H × 8"W × 10¼D
(8cm H × 20cm W × 26cm D)
RP: 1-1/16"H × 3-⅛"W × 9"D
(2.69cm H × 7.39cm W × 22.86cm D)
TA/TW: 6½"H × 13"W × 21"D
(16.51cm H × 33.02cm W) × 53.34cm D)
Manpack: 3¼"H × 10¼"W × 10½"D
(8cm H × 26cm W × 27cm D)
RB, TW, RV, Manpack: 8 lbs. (3.6 Kg.)
RP: 1.5 lbs. (.85 Kg.)
TA/TW: 15 lbs. (8.5 Kg.)

Interfaces
All XL-280 units may be interfaced with any of the following devices or communication lines:

1. Mic/Volume Control Potentiometer
2. Mic/Speaker
3. Handset, Wired
4. Handset, Acoustic coupled
5. 600 ohm/0dB/2-Wire, Half Duplex
6. 600 ohm/0dB/4-Wire, Full Duplex

NOTE: All specifications subject to change without notice.

Table 8-12. DPD-72/72A Table Specifications (courtesy of TCC).

TCC's DPD devices are available in asynchronous or synchronous, half-duplex or full-duplex configurations. Connection is made usually between a data terminal and its transmission line or modulator-demodulator (modem), enciphering the data terminal output. The EIA (Rs232C) interface model plugs directly into any modem. The line model has standard teletype (current loop) input/outputs. Interfacing is effected by merely connecting the DPD between the output connector of the terminal and the modem or line. Transmission or reception of data can be accomplished as if the scrambler were not in the line.

Use of the DPD-72/72A for asynchronous operation tends to reduce transmission error because of the regenerating nature of the circuit. The baud rate of the regenerated signal is controlled by a quartz crystal oscillator which can eliminate telegraph distortion caused by the transmitting teletype and the line. Errors occurring due to noise may propagate only as far as five characters.

Any standard teletype rate is available: 50, 60, 66, 100 words per minute. Other rates are available on special order.

When the DPD-72/72A is used for synchronous operation, clock signals are always provided by the modem and the DPD is inserted between the terminal device and the modem using the EIA interface. Data rates for synchronous operation are determined by the modem. The DPD-72 can operate at speeds up to 1 MHz.

The DPDs can be easily adapted to work with any start-stop code with block length between 2 and 14 bits (including 5-level and 8-level teletype codes). Code level setting switches in the interior of the DPD are accessible by the user and eliminate the need for factory adjustment.

TCC manufactures the DPD-72 series in two models. The DPD-72A is a somewhat more secure version of the DPD-72, with more available codes and a greater memory capacity. Both models are available in half-duplex and full-duplex configurations. The DPD series is packaged in an aluminum case with black wrinkle finish. A ruggedized version is also manufactured which is designed for aircraft use.

PERSONAL AND BUSINESS DEVICES

The line of encoding scramblers already discussed is very sophisticated and is aimed mainly at the large corporate and military markets. This is not to say that they would have no application in the home or small business, but cost factors would play a very large role in these latter instances.

CCS Communication Control, Inc. manufactures several devices which are designed for the small corporate market. This company points out that as far back as 1955, business and industry have fallen prey to electronic eavesdropping. In 1955 alone, a telephone wiretap cost a large cosmetics firm over 30 million dollars in business which was lost to competitors. Throughout the years, books and movies have been written dramatizing industrial espionage activities and the subsequent wiretaps and electronic eavesdropping which play a major part in these covert undertakings.

While electronic eavesdropping was highly successful in the middle fifties, this science has been vastly improved with a wide range of inexpensive and easily accessible devices. CCS points out that what makes the practice of wiretapping even more attractive is a general lack of understanding on the subject and the threats that exist, along with the countermeasures available for protection. To quote from a CCS bulletin:

"Did you know, for example-

"A device as tiny as a kernel of corn can transmit your phone conversation over a range of several city blocks. To install one of the many radio transmitters, a wiretapper need never enter your premises. He simply places the device on a telephone pole, in a wirecloset, or anywhere along the line inside the building.

"Your telephone need never be picked up—need never even ring—to carry your private room conversation anywhere throughout the world. A wide variety of infinity transmitters are advertised in newspapers and magazines as inexpensive audio burglar alarms and come complete with manufacturer's instructions on installation.

"Certain modifications can convert your telephone handset into an open room microphone. The wiretapper simply short circuits or bypasses a hook switch mechanism on the instrument which allows an almost undetectable amount of current to flow through. This method does not require installation of a device nor does it affect normal operation of the phone."

Many simple devices can be purchased for as little as $10.00 at thousands of commercial locations which can be used either directly or through modification for telephone eavesdropping purposes. For the businessman who must conduct certain dealings by telephone, either at home or in the office, the ease with which phone lines can be successfully bugged may be a real concern.

The purpose of this discussion is not to make the reader

paranoid about illegal surveillance but to provide input on the possibility of phone taps and what can be done to prevent them. It can be assumed that the majority of phone taps attempted in this country are the illegal variety. Now, most of us do not carry on business affairs which are all that confidential or whose importance would prompt competitors to resort to illegal tactics. By the same token, most of our personal dealings are not on a level where surveillance might be used. However, if you are not an average individual, this discussion may have more import for you. Politicians on a local, state, and national basis often deal in areas of legislation which could have certain economic ramifications on property owners and businesses should confidential material get out. Under these circumstances, one might be more concerned than usual about phone conversations, where they're made, and who is present while the conversation is in progress. In these instances, it may be desirable or essential to explore some form of telephone security. It should be pointed out that the author is also an elected official on a local basis and has never considered telephone scramblers or other types of security devices. The only security measures taken involve making certain that confidential issues are not discussed in the presence of persons not directly connected with the group working on this particular project. True, it would be possible to obtain some confidential information by bugging the office telephone, but, quite frankly, most of the confidential issues are not of such great importance as to warrant such illegal activities. Other elected officials dealing with other types of matters may feel differently.

CCS manufactures a complete communications control center called the CC600, which is a union of sophisticated telephone security systems and provides virtually total telephone privacy. Housed in a briefcase, it includes a code phone scrambler which makes your conversation indecipherable to outside listeners. Only the party you call with a matching, encoding/decoding unit receives the clear, coherent speech pattern.

A telephone set replaces the conventional telephone unit so that even pay phones, hotel phones and mobile phones can be secured against eavesdropping. A built-in line voltage meter tests the phone lines for certain voltage changes which might indicate an illegal tap. A balance voltage meter determines that line voltage has been balanced to its proper level. Within this unit is also a telephone tap alert that provides 24-hour surveillance against wiretaps and telephone operated room bugs. When activated, the signal remains intact continuously and automatically screens your voice conversation from these taps.

A telephone monitor is included which can record actual evidence of a tap being installed. Used separately, it records your conversation for permanent proof of important verbal agreements. The CC600 goes with you wherever you travel due to its mounting configuration in a sturdy attache case.

Another model from CCS is really a voice disguise circuit called the Electronic Handkerchief. When you speak into this device, your voice is automatically disguised. A control allows you to deepen your voice or to distort your voice while all you do is speak naturally. The voice deepening control may be found useful by women who live alone as it can make soprano voices sound much more bassy, closely matching the voice of a man. The output from this device is clear speech and requires no decoding. The person you are talking with would probably not even be aware that anything unusual or electronic was taking place.

The Electronic Handkerchief is available in several different styles and serves as a standard telephone instrument until switched to the on position. It is available as a normal rotary dial telephone or a trim-line walnut cigar box. With one switch, the user's voice becomes anonymous while he or she has complete control of the depth and volume with a single rotary control.

SUMMARY

Security protection of phone conversations can be terribly important to that small segment of the population whose activities dictate complete privacy. One need not be involved in the James Bond world of espionage in order to require some form of business or personal communications security. Most of the time, an individual's needs will be discussed with the security systems manufacturer and the custom designed system installed. This is not to say that electronic circuits are especially built for an individual. Rather, the manufacturer may choose a specific combination of models to arrive at the desired security level for specific needs.

It should be pointed out that while all of the security systems discussed in this chapter are known to be excellent devices with varying levels of proficiency, there is no such thing as an absolutely perfect security system. All of them can be deciphered given the proper amount of time and talent. It is safe to say that all of the devices mentioned in this chapter will keep the user's conversation secure from the average snoop. Some of the more complex systems go much further than this and provide a high degree of security even when professional eavesdropping is attempted. Fortunately, each manufacturer provides complete details on the levels of security offered by their individual devices. Naturally, as the decipherability of a device increases in complexity, so does the price. Some very inexpensive security devices may sell for a few hundred dollars, while elaborate systems cost tens of thousands.

Chapter 9

Alternative Communications

When most of us think of telephone communications, we think of telephone lines made from copper wires. We also think of all of the telephone instruments being directly linked by these wires and electrical impulses traveling from phone to phone. While this is true to an extent, in actuality, the telephone network is much more complex than this and involves the use of microwave stations which transmit communications information through the air instead of over copper wire conductors. Satellites are also brought into play, and even fiberoptics may be used to transfer light signals which correspond to the audio input at a telephone. So far, we have mentioned three ways in which telephone conversations may be transmitted—by hard wire, by radio waves, and by light rays. Indeed, in one simple, long distance phone communication, all of these methods of transmission may be used.

The telephone channel which is connected to your home today receives a continuous stream of frequencies, both from your phone and from all the others. This is called an *analog* channel. The standard phone line is responsive to audio signals of from about 300 to 3400 Hz. This is the range which makes up the normal human speech pattern. While humans can hear frequencies of from 20 to over 15,000 Hz, those frequencies which lie above and below the telephone line range frequencies are not essential to voice communication.

The limitation of audio bandwidth in the telephone channel is easily detected, in that your stereo system or FM radio sounds more natural in its broadcast of the human voice than does the telephone. This is because you're hearing a truer reproduction of all of the frequencies of the voice than you would hear on the phone line. The phone line frequency range does not encompass all of the human voice frequencies, only the major portion of them. Since your stereo has a much broader audio bandwidth than the phone line, it is a high fidelity device. Fidelity is a word whose root means faithful; therefore, a high fidelity system is one which has a high degree of faithfulness of reproduction. Due to its limited audio bandwidth, the standard telephone line is not a high fidelity channel.

Once telephone signals are transmitted into the phone channel, they all travel together. This is called *multiplexing*, which means that a single channel may carry thousands of phone conversations. To accomplish this, the audio signals are usually multiplied so that an original frequency of 1000 Hz might be transmitted through the channel at 61,000 Hz, while the same tone transmitted by another telephone might be at 71,000 Hz. Since each signal or, to put it in practical terms, each user's voice is multiplied to a different frequency range, it is possible to extract the audio information by detecting the multiplied frequencies by receivers tuned to the correct frequency. This is very similar to commercial radio broadcasting, where the announcer's voice is multiplied in frequency and transmitted through the air to your receiver, which is tuned to the same frequency and extracts the audio information, playing it through a speaker. On the phone channel, the multiplex conversations are transmitted in an analog manner in a continuous wave of frequencies. Figure 9-1 shows what this wave might look like on an oscilloscope. Analog transmissions have been used since the start of telephone service, and most telephone company equipment is designed to handle this type of signal. For this reason, analog systems will be with us for quite some time to come.

Another system of transmission is rapidly progressing in modern telephone communications. This is called *digital* transmission and offers major advantages over analog systems. This new method uses a technique called pulse code modulation, where the human voice is converted into a stream of pulses which look like computer data. Whereas analog transmission is a continuous flow of frequencies, digital transmission is a stream of on-and-off pulses. This is the same type of transmission method used by computer terminals. The pulses do not vary in the same method that analog signals do. The pulses are either full on or full off. The rate at which they are turned on and off is dictated by the audio information contained. The major advantage of this system is that line noise is almost wholly eliminated. Using the former method, the amplification of the audio information also involves amplification of line noise which is made up of spuriously induced unwanted signals. With digital transmission, the voice is simulated, and each transfer station generates its own pulses based upon the input from another station. A call from the other side of the world using digital transmission would consist of demodulated pulses which were generated by a station near you and responds to the coded digital information which originated at the caller's station. Figure 9-2 shows a representation of a digital signal, which may be compared to the analog version shown in Fig. 9-1.

If the entire telephone system today converted to digital transmission, then computer data would no longer need to be converted to an analog form for transmission by phone line. It must be pointed out, however, that the human voice is an analog system and must be converted to a digital system so that it can be transmitted in the form of pulses. These pulses must then be converted back to analog to be understood.

It's no wonder that an analog method of transmission was first developed for phone company transmissions, since this system required no conversion for human voice relaying. With the advent of the computer era, however, many advancements have been made through the use of computer language instead of human language. For phone communications purposes, analog systems are common to human language, while digital systems are common to computer language. Computers can disseminate more information in the same period of time than can humans. This is one of the reasons why a conversion from analog to digital, or from human to computer, is advantageous.

One of the most important advantages when comparing digital to analog transmissions is that using the former system, all signals become a stream of similar pulses. Consequently, information bits

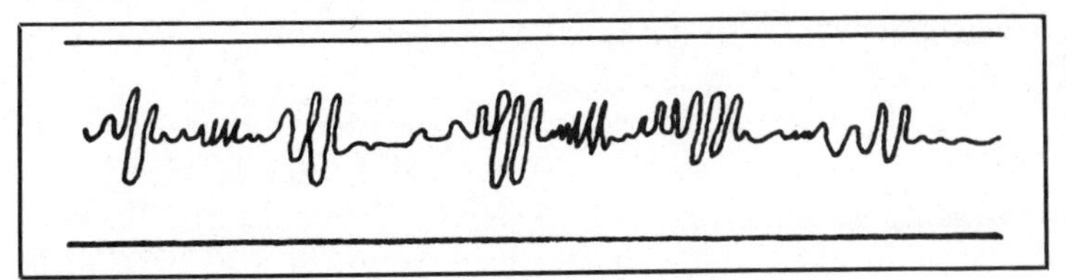

Fig. 9-1. Waveform of an analog communications transimission.

will not interfere with each other and will not make differing demands on engineering of the channels. In other words, the manufacturing process becomes more standardized. Analog transmissions involve constantly varying continuous streams of frequencies. This places a greater stress on system reactions and tolerances and requires that the telephone channel exhibit a greater fidelity factor than is required for the same signal to be transmitted digitally.

COMMUNICATIONS WITH LIGHT

It has been known for quite some time that a beam of light may be used to carry audio information just like a radio wave. Using the latter system, the audio information is superimposed upon a transmitted radio carrier. This means that the carrier will be changed in some manner at the same rate as the changer frequency from the audio input. AM modulation varies the amplitude of the transmitter carrier, while FM, or frequency modulation, varies the frequency of the carrier. Intensity modulation may be used with a beam of light whereby the output brightness of the light varies with the audio input. The light beam is modulated. For instance, a signal of 300 Hz may cause the light beam to be ten times brighter than it is when a signal input of 3000 Hz is used. There is a ten to one ratio in frequency between 3000 Hz and 300 Hz. Likewise, there is a ten to one ratio in intensity of the light beam between these two frequencies. As long as the ratio is maintained, the audio information used to modulate the light beam may be retrieved as the same audio information by a light-to-sound receiver.

Telephone companies are researching and actually using light waves to transmit audio information. Most people think of radio waves and light waves as being completely different, but they are very similar in that they are both forms of electromagnetic radiation. They differ only in the rate at which they oscillate. Electromagnetic waves include all forms of radiation which span the spectrum from visible light to radio waves and below.

This conversation on light wave communciations has, so far, involved only visible light or electromagnetic radiation which can be detected by the human eye.

The lighting source or carrier is generated by the sun. Upon striking objects on the earth, it is reflected at different intensities, depending upon the materials it strikes. Some material absorbs light, while others reflect it. The material which absorbs reflects back to a receiver less light than does highly reflected material. In

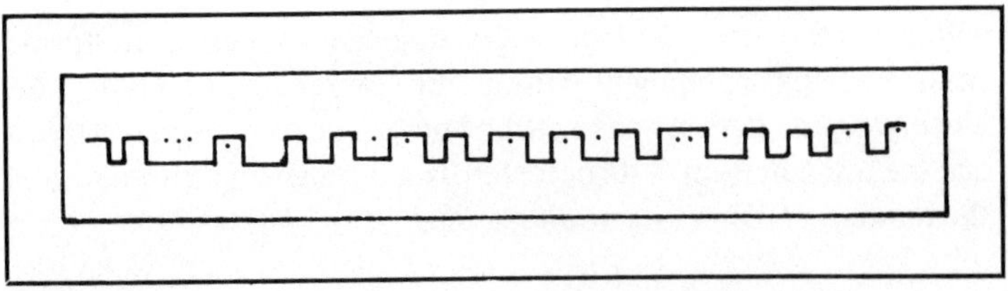

Fig. 9-2. Waveform of a digital communications transmission.

other words, the original light source has been modulated or superimposed with information. Much of this information is contained in the intensity ratios of the light waves which are picked up by the receiver.

In this case, the receiver is the human mind. The antenna is the human eye. Just as a radio receiver antenna has the ability to respond to the passing radio waves and to pass this information along to the receiver which processes it and converts it into an energy system which may be readily deciphered, the human eye also responds. Its response is to light rays. When it reacts, this information is passed directly to the human brain, where the processing occurs. Thus it can be seen that the act of visual identification is part of a complex transmitting and receiving system.

Taking this comparison one step further, we know that to receive a radio signal which can be heard by the human ear, it is necessary to have a transmitter as well as a receiver. If either component is not working, then the communication cannot be maintained. Likewise, should the transmitter in our natural light transmitting system be rendered ineffective, no communication can be had. For example, if you were deep in an underground cavern, the sun's rays could not reach the receiver (your eyes). The receiver is functioning perfectly; it simply has no transmitted information to detect. Therefore, you see nothing. Persons who are blind can be likened to those with a defective receiver. The transmitted information is always present, but there is no means to convert this information into conventional points of reference.

This simplistic explanation of light wave communication is presented to give the reader a very basic understanding of how these communications take place. Science has developed electronic components which take the place of the human eye to differentiate between the various light reflections. A television camera is a good example. The camera tubes or vidicons have specially treated surfaces which react to light intensity and to colors. As the tube's surface reacts, it sends out electronic signals to processing equipment within the camera's circuitry. This closely copies the response of the human eye to the same criteria and the transmission of its detected data through electric signals to the brain.

Laser Communications

Everyone has heard of a laser, but most persons do not really know what it is. Light is made up of excited atoms. In an incandescent light bulb, electric current flowing through a conductor or element causes heating effects, thus exciting the atoms and causing them to move. In a fraction of a second, each individual atom unloads this extra energy as a pulse of light. Each energized atom acts independently, as if the other billions of atoms did not exist. Since each atom behaves in this fashion, the overall light radiation is not orderly. Since there may be many different kinds of atoms in the wire element, each type may radiate at a different frequency.

A laser produces ordered light, also called *coherent* light. Figure 9-3 shows a basic laser design for practical discussion. The device is a hollow glass cylinder with a mirror at each end. The cylinder is sealed from the normal atmosphere and contains a substance which is composed of atoms that will give off waves of a single frequency when excited. The atoms are kept in a continual state of excitation and, as in an electric lamp, emit their waves at random and in every direction. Due to the construction of the laser, any random radiation which moves along the axis of the enclosed cylinder is bounced back and forth between the two end mirrors.

Whenever this radiation hits an excited atom, it is stimulated to emit its stored energy as a new pulse of waves which follow the same path of the energy that triggered its reaction. This establishes an ever-building radiation flow along the cylinder axis. This is a single-frequency beam of coherent waves which is ever-building in intensity.

In order to utilize this energy, one of the mirrors is made slightly transparent which allows the intense beam to escape into the open. This beam may be further reflected using a series of mirrors to bring it to the proper point of detection or utilization.

It is possible to purchase experimental lasers for as little as a few hundred dollars from some hobby electronics outlets. These are usually extremely low-powered devices, but they can be used

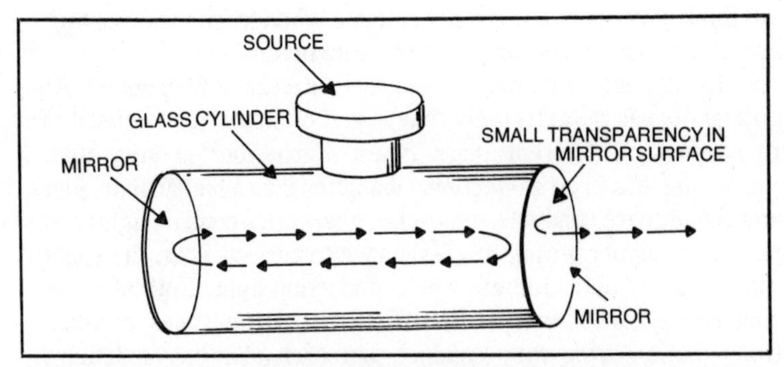

Fig. 9-3. A basic laser system consisting of a hollow glass cylinder with a mirror at each end.

for voice modulation provided the proper modulation and detection circuitry can be built. Lasers especially suited for communciations purposes are in a constant state of development. Many major companies are experimenting with different ways of generating the light beam. One method uses crystals made from a yttrium aluminum garnet treated with neodymium. This is known as the YAG laser, and it generates a powerful, continuous beam at a light frequency which is excellent for communications purposes. Experimentation is also going on with *laser diodes*. These are semiconductor diodes usually made from gallium-arsenide. They emit coherent light when a voltage is applied to their terminals.

At this point, it is good to realize that the types of lasers used for communciations purposes are not of the same intensity or, at least, focused intensity, of lasers used for welding and defense purposes. These latter devices can cut through quarter-inch steel plate. Communications lasers do not concentrate their intensity as do the other devices.

Along with suitable communications generators, there must be a means for modulating the beam with audio information. To do this, it is usually necessary for the laser beam to be divided into pulses. Each pulse must follow the other in close procession, but each must also remain distinct from the other. One system that has been accomplished by combining the YAG laser with another special crystal allows a series of short pulses to be transmitted, each of which begins and ends within 30 billionths of a second.

Light Wave Communications Cable

Standard telephone systems require cables to interconnect many of the devices on the circuit. The same is true of light wave communications, but copper wire and other types of conductors, as we think of conductors, are not suitable when frequencies as high as those which light is composed of are involved.

Light wave communications cables are called fiberoptics. An optical fiber is an extremely thin strand of material composed of two layers of different glass types. Each fiber is very thin, measuring less in cross-sectional diameter than a human hair. Just as a copper wire conducts electricity, fiberoptics conduct light; and just as all copper wiring has resistance to current flow, the same can be said of optical fibers when conducting light. Both of these conductors exhibit very low resistance to the types of electromagnetic waves they conduct, but each also has a definite resistance factor. Using a copper wire, the resistance factor will be determined by the cross-sectional density of the wire, the ambient temperature, and the impurities found within the wiring structure. The creation of efficient optical fibers depends upon controlling the purity, uniformity, and composition of microscopic glass layers to very high degrees of precision. Highest precision allows for the lowest resistivity factor in regard to the conductance of light. Often, these thin fibers are combined in a single cabling package with each fiber having the ability to conduct light which is separate from all of the other fibers. A single fiberoptic cable an eighth of an inch in cross-sectional diameter may contain many thousands of separate fibers, allowing for thousands of separate light channels.

Since May of 1977, the Bell System has been studying a light wave communications system in the city of Chicago. This system covers an area of about 1½ miles and the cable is buried beneath the streets, carrying voice, data, and video signals as pulses of light transmitted through the optic fiber.

Overall studies indicate that the light wave communications system is more highly efficient than a standard telephone network. The light wave system has surpassed the 0.02% outage rate, with a projected annual outage rate of 0.0001% which breaks down to a total of 30 seconds downtime in a year's period of operation. On the average, less than 1 second per day has contained a transmission error, making the system rate 99.999% error-free.

It should be pointed out that while this is a discrete light wave communciation system, it is also interfaced with the standard telephone network and the high efficiency factor is maintained. The interface devices must convert the light wave pulses into analog signals for the standard phone line and vice versa when going from the major network to the light wave system.

The science of fiberoptics has produced many revolutionary communication tools which have extended into other areas of human services. For example, a fiberscope which is a flexible bundle of optical fibers having a lens at each end is used for medical work. This allows doctors to view the interior of the human body by inserting the fiberscope through a natural opening or through a small incision. One channel may be used to carry a light beam applied externally to one end of the cable, while another fiber channel will be used to view what is reflected by the projected light source.

Fiberoptics also opened up the world of direct video communication between subscribers without having to convert from audio or radio wavelengths to those of the video spectrum. This has long been a problem with development of the video telephones which, in the 1950's, were projected to be a commonplace feature in the 1970's and 80's.

VIDEO COMMUNICATIONS

Video communications may indeed be a thing of the late 1980's. Some experts believe that it will catch on as an economically viable consumer item during the latter part of this decade due to advances in modern electronic engineering. Figure 9-4 shows a pictorial diagram of AT&T's experimental picturephone which consists of three components: a conventional tone dial telephone instrument; a picturephone set with a screen, video pickup and loudspeaker; and a control unit containing a microphone and permitting adjustment of the video image and audio volume. There is also a separate control unit which interfaces the line and powers the electronic circuits. System specifications are:

—Bandwidth:	1 MHz
—Screen size:	5½″ × 5″
—Frames per second:	30
—Number of lines per frame:	250
—Normal viewing distance:	36″
—Normal area of view:	17½″ × 16″ to 28½″ × 26″

All control functions are basically the same as for conventional phone systems, but wires are added to carry the picture in parallel with the existing local loops from the central office which carry the audio information to the instrument. New switching facilities are required which operate in parallel with the telephone switching system, but both are under the same control. Telephone trunks must be expanded upon to add the capability of carrying the video signals as well as the normal audio information.

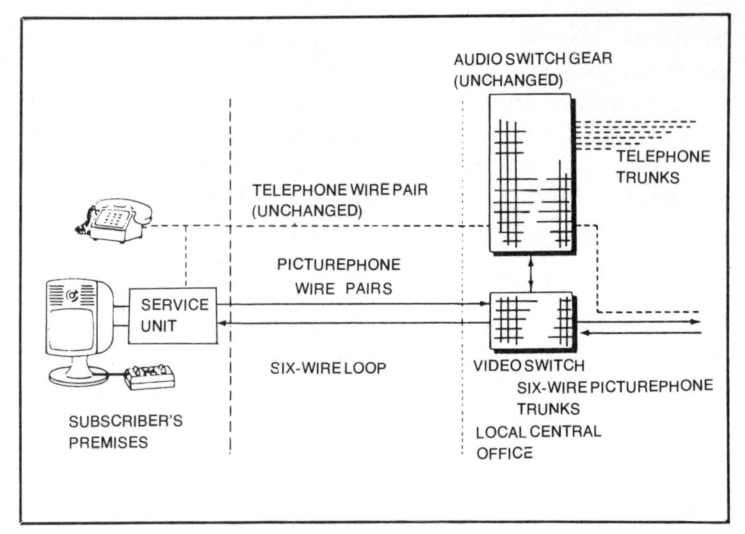

Fig. 9-4. Pictorial diagram of AT&T's experimental picturephone.

Video transmissions are specially adapted to be transmitted in digital format. All picturephone transmissions between central offices will be handled in this manner. Once these signals are digitally encoded, they will remain in that form until they reach the central office of their destination. These signals will then be converted from digital to analog form.

This picturephone system is already in use in several large cities across the United States. These devices are normally found at large centers and have not caught on with the average subscriber. This is mainly due to the fact that this service is applicable only to certain areas of the country. Of course, a subscriber can only communicate with another subscriber in the same area or in another large city where picturephone communications are also operative. Naturally, cost plays a big factor in taking this elective service. At present, a rather phenomenal price tag goes along with this rather phenomenal communications device. As is the case with all new electronics instrumentation, the majority of the price of initial units is tied up in research. Once this research has been paid for, the per unit cost can be cut by several hundred percent. It is possible that by the end of this decade, it may be practical for almost every home to lease a picturephone from the central office or, perhaps, purchase one from a manufacturer. Figure 9-5 provides an overall picture of the conversion from digital to analog and of the various interconnections.

FACSIMILE COMMUNICATIONS

Facsimile communications involve the use of visual scanning machines which send graphical information over telephone lines. Facsimile can efficiently send reproductions of letters, pictures, maps, and other graphic subjects by scanning the material with a light beam and photoelectric cell. The variation in the transmitted material causes fluctuations of electric current within the photoelectric cell. These signals are converted into signal waves which lie within the audio frequency spectrum. These are connected to the phone line by means of a line coupler and sent in the same form as a normal conversation.

On the receiving end, the facsimile recorder converts the audio signals into current pulses which drive a specialized stylus. This action moves the stylus across a piece of paper, reproducing what was originally transmitted.

Due to the nature of facsimile transmission, the heaviest requirement of any type of communication is placed on the overall telephone system. Special consideration must be given to level change, low-level hum which is most often transmitted by the frequency of the AC line, minute impulse noise interference and small variations in transmission delay. Other conditions which must be minimized or completely overcome include echoes, multi-path transmissions, envelope delay distortion, lack of synchronization, and undesirable audio compression techniques. In the transmission of a million elements comprising a typical

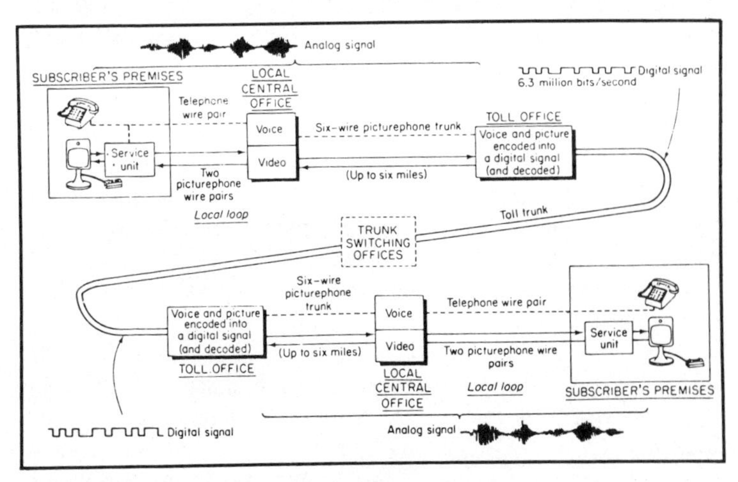

Fig. 9-5. Overall picture of the conversion process from digital to analog transmission.

facsimile picture, one pulse received considerably above or below its correct gain level, position, or sharpness becomes an outstanding character in the received picture. Since each pulse of information is so vitally important, it can be seen why much more line distortion can be tolerated when the phone lines carry a normal conversation. Distortion on a phone line which would be almost unnoticeable during a voice communication might completely prohibit the reception of readable facsimile copy.

Modern facsimile machines make use of the photoelectric cell which is a solid-state device that converts the energy stored in light into electric currents. In a facsimile system, a focused, intense beam of light scans the picture and reflects into a photoelectric cell. This light beam is very narrow and is concentrated at any one time on only a very tiny portion of the overall picture to be transmitted. The intensity of the beam which ultimately strikes the surface of the photoelectric cell is directly dependent upon the material being transmitted. Dark areas of a picture reflect back less light than do the brighter areas. The cell reacts to this by generating more current when the light beam is scanning a bright area than when the portion of the picture is dark.

The output current pulses are processed and end up providing the driving power for an audio oscillator which changes frequency in response to the current pulse input. The output of the oscillator is coupled to the telephone line and sent to the desired number. At the receiving end, the phone is answered and is then coupled to the facsimile converter, which changes the audio tone back into current pulses in order to drive the stylus, as was previously discussed.

There are many types of facsimile devices. Figure 9-6 shows a facsimile machine which is used by a newspaper field office to transmit picture information and script back to the central office and printers. This device features an acoustic coupler which accepts the handset of the telephone instrument. The copy to be sent is placed in position and scanned by the light beam. At the other end, the converter takes over and changes the audio pulses back into hard copy.

The use of facsimile communications is heavily relied upon and increases each year. The industrial applications have been taken advantage of for many years, but smaller businesses and individuals are now beginning to realize specific advantages in using this system. The cost factor has decreased considerably and small units may be purchased new at reasonable prices. Some very

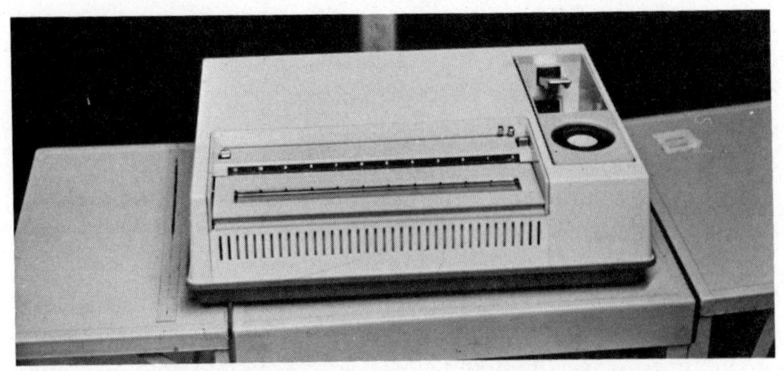

Fig. 9-6. Small facsimile device made by Xerox (courtesy of Xerox Corp.).

good bargains may be obtained on the industrial surplus market. These anticipated cost reductions, along with the continuing refinement of facsimile technology, will probably lead to the heavy use of electronic mail by the end of this decade.

Facsimile equipment is a major carrier of electronic mail. This is due to its speed, accuracy, and ease of operation. Electronic mail is just what it sounds like, mail which is sent in an electronic manner. Instead of a picture, a letter is placed on the scanning plate of a special facsimile machine. A drum spins around its axis, and when in the scanning mode, a light beam is played across the material. The same machine at the receiving end is set for the recording mode and a stylus is driven across the paper on the spinning drum. This effectively recreates what was originally scanned. Since facsimile is completely compatible with the telephone line, it can be sent to any part of the world which is equipped with telephone service. Of course, you must have one unit at your end and another at the number you are calling. The scanning line frequencies of the scanner and recorder must be essentially equal.

Not all facsimile machines operate in the same manner nor transmit their audio pulses in identical configurations. While steps are being taken to establish certain set functioning parameters, at present it is often necessary to transmit to machines made by the same manufacturer as the one doing the scanning.

The facsimile unit in Fig. 9-6 and shown in detail in Fig. 9-7 is called the Xerox 400 Telecopier. It is a fascimile transceiver that can be used anywhere a telephone and an electric outlet are accessible. It is ready for immediate use anytime an operator loads a document or recording paper and inserts the telephone handset into the acoustic coupler.

In a matter of minutes, the operator can send or have sent a document, chart, photograph, or almost anything else that can be put on a piece of paper. The 400 Telecopier scans the original, converts the light and dark areas into audible signals, and transmits these signals over the telephone line. When receiving, the 400 Telecopier converts audible signals into electric impulses to drive the stylus. A clear facsimile of the original copy is the result.

This particular model, when using a normal telephone, can transmit and receive copy to a maximum size of 8½″ × 11″ in a time period of four minutes. A 4/6 speed switch provides a six-minute capability to make the 400 Telecopier compatible with all other telecopier equipment manufactured by Xerox.

Some of the device features include an end-of-message signal, self-test mode, carrier position control, and multi-speed settings. The end-of-message signal alerts the operator when the transmission has ended. The signal will be heard when the operator wishes to send, if the carriage has not been moved away from its extreme left travel. In the receive mode, the signal will be heard until the carriage is moved all the way to the right. With the self-test mode, the telecopier may be tested at any time to insure that the send and receive components are operational. This is a valuable feature on any facsimile device, since it ascertains that the machine is functioning as designed when the telephone line or other telecopier station may be initiating the problem.

A carriage position control permits the sending operator to edit the transmission when it becomes necessary to shorten transmission time. For instance, to send only a middle portion of a document, the carriage position control is set to the right of the copy to be sent. When the indicator has moved past the copy, the drum door is opened, stopping the transmission and setting off the end-of-transmission signal at the receive operator's position.

The Model 400 Telecopier, while designed primarily for general office use, is rugged enough to be truly portable. A carrying case is available for this application and resembles a large briefcase. It securely holds the 400 Telecopier, the power cord, and a supply of 150 sheets of recording paper.

The electrical requirements are 107 to 127 volts AC at a current of about one-half ampere. The input line is terminated in a two-pole, three-wire receptacle with one grounded terminal. Maximum size of documents to be transmitted is 8½″ × 11″, while thickness may range between .002″ to .007″. The document stiffness is not to exceed that of a computer card and documents

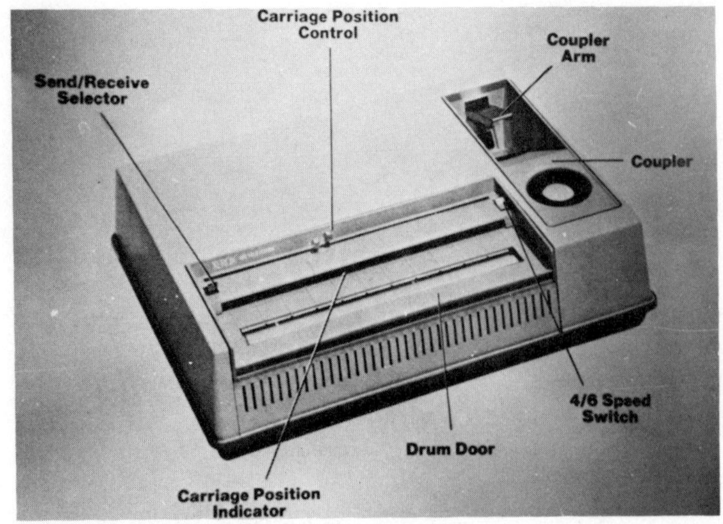

Fig. 9-7. Closeup view of the Xerox 400 Telecopier.

with minor damage such as tears, folds, and wrinkles can be sent without fear of jamming the machine. Reasonable care in making the transmission will insure that no further damage to the documents occurs. All attaching devices such as paper clips, staples, rubber bands, etc., must be removed before loading into the machine.

Figure 9-8 shows the 400 Telecopier connected to the telephone line through the acoustic coupler. The sending operator calls the appropriate phone number, establishes contact with the receive operator who positions the phone instrument to the voice coupler of another telecopier set for receive operations. At the transmitting end, the scanning circuit is activated and in four or six minutes (depending on which operation speed has been chosen), all information has been transferred.

Operator maintenance is very simple and involves cleaning of the main cover, control panel and hinge doors with a clean, moist cloth. A mild detergent is recommended for buildups which cannot be removed with a damp cloth alone. The drum is also cleaned with a moist cloth. All this involves normally is brushing off loose dirt and paper particles. An auxiliary dust cover is used to protect the mechanical portions of the device when not in use.

SATELLITE COMMUNICATIONS

Digital technology as it applies to telephone communications has enabled satellies in stationary earth orbits to be used for

Fig. 9-8. Xerox model 400 telecopier connected to telephone instrument by means of an acoustic coupler.

worldwide hookups. The first commercially available digital satellite service for voice, facsimile, and data transmission was introduced several years ago by American Satellite, a domestic satellite communications company. This system demonstrated simultaneous transmission and reception of voice and high speed data transmitted digitally over a live satellite channel.

There are two types of basic telephone communications satellite. One is *passive* and simply reflects microwave signals, focusing them to a specific area. The surface of the satellite is focused upon by a ground transmitting station and the striking angle adjusted for the desired deflection to a distant point back on the earth. This type of satellite has been with us for over twenty years with the first successful coast to coast telephone calls in the United States being made in the latter part of 1960.

Passive satellites are extremely cheap and uncomplicated when compared to other types of satellites, which contain highly sophisticated and numerous electronic circuits. The passive satellite is really a large metallic sphere and little else. The real expense with this type of orbiter comes from the technology needed to place it into orbit.

Active satellites contain microwave receivers and transmitters, along with amplification and processing circuits. An active satellite is a repeater, in that it takes a weak signal and retransmits it as a strong one. Figure 9-9 shows a comparison of an active and a passive satellite. It can be seen that a radio signal is transmitted from the earth, strikes the surface of the passive satellite, is reflected back to earth where it is received at a distant site. The communications signal is stronger at the earthside transmitting antenna. It loses strength from this point on. By the time it has traveled the vast distance to the orbiting satellite, it has become many times weaker. This is due to absorption and certain space losses. Another loss occurs when it strikes the satellite body and is reflected. This highly weakened signal must now travel another long distance back to earth. This distance can be longer, shorter, or the same as the uplink path depending on the reflection angle. By the time the signal reaches the earth again, a very complex receiving system is necessary to boost the level to a value which is usable.

Looking at the active satellite portion of this figure, it can be seen that the first step is the same as with the passive satellite. The signal which is transmitted earthside travels to the orbiter. Here is where the difference in the two types of communications satellite is plainly recognized. A receiving antenna on the satellite detects the transmitted signal, converts it into digital information,

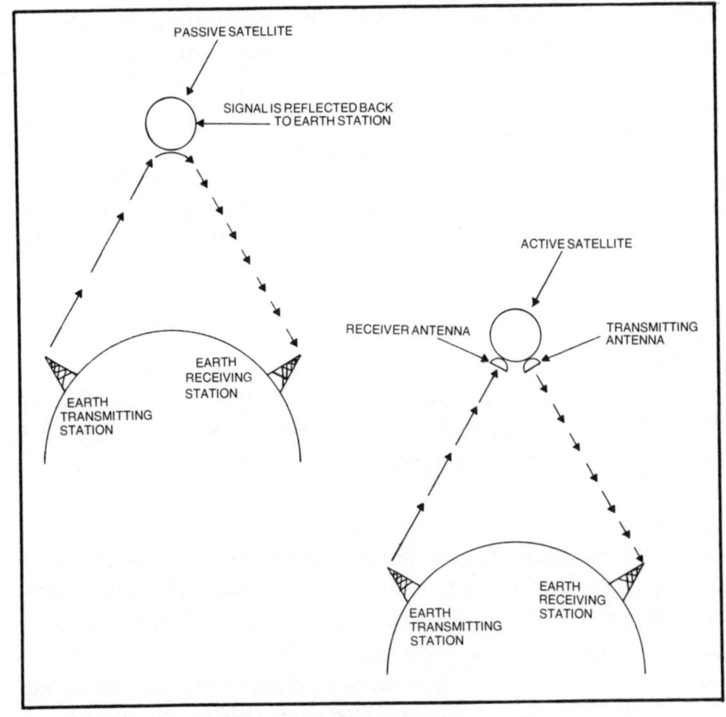

Fig. 9-9. Comparison of active and passive satellite operations.

amplifies this data, and then feeds it to the input of a self-contained transmitter which re-transmits the original information back to the earth station. Upon arriving at the earthside receiving antenna, the signal is much stronger than with the previous system and more reliable communications are obtained.

In addition to being classified as active or passive, communications satellites are also categorized by their altitudes. Certain altitude satellites are locked in a stationary position above the earth. These are said to be *synchronous* with the earth. This means that the satellite speed is matched to the rotational speed of the earth on its axis. The earth turns, but the satellite keeps up with this motion and can always be located at the same compass point on the horizon. A synchronous satellite is very easy to find when using radio signals, because the antennas on earth can be locked into stationary positions. They do not have to track the satellite as it orbits from horizon to horizon.

Medium-range satellites are not synchronous and must be tracked. However, their maximum altitudes above the earth are usually less than 10,000 miles, whereas fixed satellites are more than twice this distance. While medium-range satellites must be tracked, they can only be used when not behind the horizon as is determined by the placement of the transmitter and antenna used on the earth. Transmitting signal strengths do not have to be powerful due to the lower orbital altitudes.

Communications satellites are placed in orbit from time to time and receive very little news coverage because they are just not that unusual anymore. Many of the satellites are privately owned by such companies as AT&T, Western Union, and others. Some are used for relaying computer data, while others may be used for standard telephone communications. There are even amateur radio satellites which have been built by experienced teams of amateur radio operators. These are placed into orbit as "riders" with commercial or government satellites. Their sole purpose is to serve as repeaters for amateur radio communications.

Without satellite communications, the conveniences afforded by the telephone companies in the United States and abroad would not be of the caliber which we presently enjoy. In truth, these conveniences are being augmented on an almost weekly basis and new features are often disclosed monthly. A satellite used for communications is a costly piece of equipment, but the expanded uses it offers is easily worth the added pennies to our phone bills.

INTERFACING COMPUTERS WITH THE PHONE LINE

It has already been learned that computers speak a digital language, while we humans communicate through analog methods. However, the advancements in digital interfacing with the phone line make computer tie-ins an important facet of telephone line use.

This seems to be the age of the home computer. You can buy one at almost any hobby store, along with the hardware and software needed to take full advantage of your purchase and to expand its capabilities almost without limit. Computers can play chess with you, keep your checkbook balanced, help forecast the weather, and even provide interesting electronic games which are played through your television set.

In recent years Radio Shack has become a leader in low-cost home computer systems. They even offer an acoustic coupler which will allow you to attach one of their computers to a phone line for exchange of information with another computer across town or on the other side of the world.

The TRS-80 Telephone Interface is shown in Fig. 9-10. It features a cradle to which the phone handset attaches and may be switched to receive data or to transmit it. This device makes it

Fig. 9-10. Radio Shack TRS-80 acoustic coupler (courtesy of Allied Electronics).

possible to transfer data from a branch office to headquarters or to exchange programs with a friend. All the operator need do is dial the remote location and the interface can put two TRS-80 Computers in touch with each other. It also permits coupling between the TRS-80 and many other compatible computer types. It features a baud rate of up to 300, a receive sensitivity of – 45 dBm., and requires a Radio Shack RS-232 Serial Interface Card. It is powered from an AC adaptor which provides 24 volts AC at 150 milliamperes. This interface sells for about $200.00.

The Serial Interface Card is shown in Fig. 9-11 and is a general purpose interfacing device which allows the TRS-80 computer to be attached to many different types of systems. It sells for about $100.00.

Many large volumes of text could be written on computer-to-computer linkups using the telephone system. For the purpose of this book, it will suffice to say that these linkups are presently possible within certain limitations, and telephone systems will be designed in the future to provide a full range of linkup possibilities. The main limitation at the present time is the analog method of telephone communications systems. With the industry slowly switching to digital methods, computers will eventually be connected to systems which directly speak their own language.

In the future, when everyone owns a home computer, it may be possible to order your groceries, make a bank deposit or withdrawal, or even buy a new car simply by typing this request on the computer keyboard. The address and computer linkup number at the location where you are ordering from, or doing business with, would also be entered and might be obtained from a directory very similar to the one where you look up phone numbers today. The computer linkup number might be processed through a central computer terminal and then channeled directly to a computer at the location you have chosen. The two computers would speak with each other in a digital language and the business conducted. If you ordered a new car, after entering the data you might, almost instantly, see a printout at your computer terminal asking what color you desire. When you type this information into the system, another question might be fired right back at you in less than a second. Computers can talk with each other much faster than humans can. As a matter of fact, computers think in millionths of a second while humans often think several million times slower. With the speed which is available when using computers, the only problem encountered is finding a method for them to talk with each

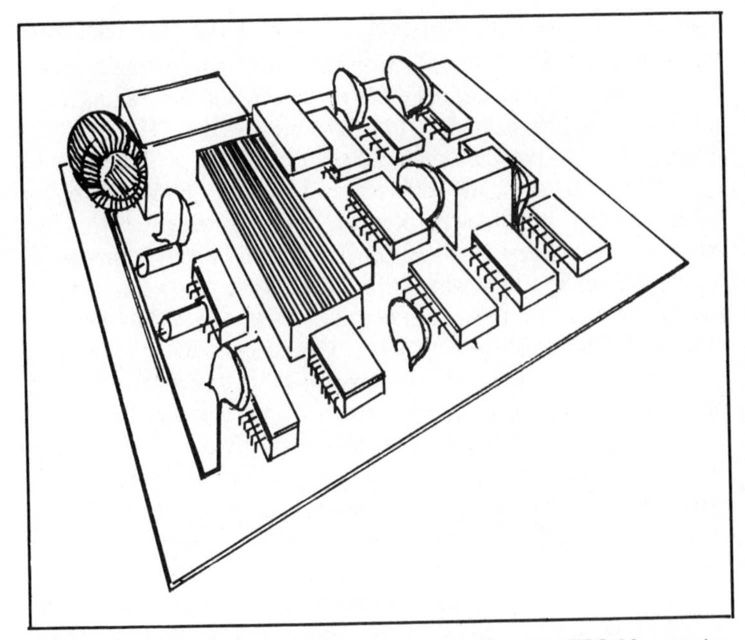

Fig. 9-11. Serial interface card which allows connection of the TRS-80 computer to a variety of devices (courtesy of Allied Electronics).

other. The telephone system would seem to be the most obvious solution to this problem.

SLOW SCAN TELEVISION

Another specialized form of communications by phone line involves the use of true television. As we normally know it, television transmissions require vast bandwidths millions of times larger than those available on the standard phone lines. However, slow scan television works on a much slower basis than the fast scan version which we view in our homes. This latter type makes 30 pictures every second, while slow scan television makes only one picture in eight seconds or more. This means that slow scan television is similar to facsimile transmissions discussed earlier, although slow scan methods involve electronic "painting" on a cathode ray tube and, therefore, is true television.

SSTV (slow scan television) uses cameras and video receivers just like the more conventional types but is limited to broadcasting still materials only. Some of the high resolution transmitters may take a half minute or more to scan a single picture. SSTV scans the material to be sent in an electronically equivalent manner to facsimile. The scanned information is converted to audio pulses which are coupled to the phone line and received at a distant location. Upon reception, the audio signals are converted into driving pulses for the video receiver. The information then paints a picture on the cathode ray tube.

Robot Research, Inc. first entered into the SSTV market by manufacturing medium resolution devices designed for the amateur radio operator. They are now heavily involved in phone line television systems which make it possible, and economically feasible to transmit television pictures anywhere in the world by telephone. Phone line television (PLTV) means that visual data can now be transmitted almost instantaneously to any area which can be reached by telephone. Many time, distance, and cost limitations are eliminated using this method. Subject matter may be large or small from check signatures or gauges to buildings or power line structures. With very few exceptions, anything that can be televised with closed circuit television can be televised with PLTV.

PLTV circumvents the limitations of broadcast and closed circuit television by converting the video image to audio tones that can be transmitted over any audio communications system, including the telephone network.

PLTV employs standard CCTV cameras and monitors in connection with Robot Series 600 Phone Line Television transmitters, receivers, and/or transceivers. In order to transmit an image, the Robot PLTV system "grabs" and frame-freezes a video image from the CCTV camera and converts this stored image to audio tones for transmission. At the receiving end of the transmission, the audio tones are then fed into the Robot PLTV Receiver, which converts the signal back to video for display on any CCTV monitor.

Figure 9-12 shows the Robot Model 530 Transceiver with direct interface. The 500 series of PLTV equipment is designed for fast, economical picture transmission in applications where fine picture detail is not required. The PLTV format consists of a 128 by 128 (square) array of picture elements transmitted in 8.5 seconds. In viewing the pictures on a CCTV monitor, the left-hand and right-hand edges of the video picture are blanked to convert the TV's 4:3 aspect ratio to 1:1. In the Robot 500 Series equipment each picture element is represented by one of 16 grey shades in a solid-state digital memory.

When sending or preparing to send PLTV pictures, Robot transceiver models frame grab and store single standard TV fields

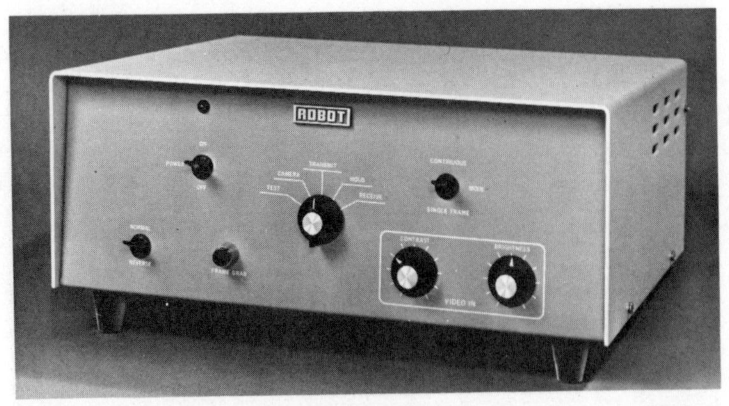

Fig. 9-12. Robot model 530 transceiver with direct interface (courtesy of Robot Research, Inc.).

(1/60 second) from a CCTV camera or other video source. This may be automatic, in which case frame-freeze occurs at the beginning of each PLTV picture. Or it may be manual, in which case a video field is grabbed whenever the operator presses a switch. Transceiver models include controls for setting the stored grey level range of the TV pictures grabbed for transmission. The grey scale mode allows the operator to frame grab a precision internally generated grey scale for comparison purposes. In the camera mode, the real time digitized camera video is displayed on the CCTV monitor for viewfinding purposes. In the receive mode, the incoming picture information replaces the previously stored picture in the memory, picture element by picture element, without intervening gaps or loss of picture information. At any time, reception may be discontinued by means of a front panel hold switch and the currently stored picture is retained and displayed as long as power remains on.

Model 600 Series

The Robot 600 Series Phone Line TV (PLTV) equipment is designed for a wide variety of applications where picture quality is important. Resolution (and therefore transmission time) can be selected to suit the needs of your application. The 600 Series equipment features a full screen 4:3 aspect ratio display (except in signature verification models). Six bits of memory are used for each picture element, which provides up to 64 levels of grey for the displayed picture. With this number of shades, the picture appears to be much smoother (grey scale steps are not obvious), and details

GENERAL

INPUT

Video Input—Standard 525-line TV video signal (625 in 50 Hz countries). 1.0 volt p-p white positive into 75 ohms; BNC connector. Positive or random interlace.

PLTV Input—Accepts high resolution PLTV signals with input sensitivity of -40 dBm. Input impedance 600 ohms. Modulation standards are*: white—2300 Hz, black—1500 Hz, sync—1200 Hz.

OUTPUT

Video Output—Standard TV Video signal 1.4 volt p-p white positive into 75 ohms, BNC connector, compatible with standard 525 line TV receiver (625 lines in 50 Hz countries).

PLTV Output—High resolution PLTV modulation 1200 Hz to 2300 Hz audio-FM, -9 dBm nominally into 600 ohm load.*

CONTROLS, CONNECTORS AND INDICATORS

Line Switching Models—Two front panel buttons and indicators are used to determine telephone line status. In the VOICE mode (VOICE indicator illuminated) the attached telephone set is active for dialing, ringing or talking. In the PICTURE mode, reached by pressing the PICTURE pushbutton while the attached telephone set is off-hook, the PLTV is connected to the telephone line. In the PICTURE mode the telephone set is inactive.

Rear Panel Connectors and Controls—FROM CAMERA VIDEO is a BNC connector for input of a standard CCTV video signal. TO VIDEO MONITOR is a BNC connector which supplies standard video selected by the front-panel switch for viewing on a CCTV monitor. TELEPHONE SET is standard 6 lead modular jack for connection of any FCC type accepted single line telephone set. TELEPHONE LINE is a pendant cord terminated in a modular plug for connection to a telephone company installed modular jack. REMOTE is a 25-pin "D" type connector for connection of an optional (not supplied) remote control box which enables the PICTURE and VOICE control functions to be remotely duplicated. REMOTE also contains leads to permit picture black/white reversal and remote control of FRAME GRAB. TAPE jack provides for recording transmitted or received PLTV picture signals on an audio or cassette tape recorder.

DIRECT INTERFACE MODELS

Rear Panel Connectors and Controls—FROM CAMERA VIDEO is a BNC connector for input of a standard CCTV video signal. TO VIDEO MONITOR is a BNC connector which supplies standard video selected by the front-panel switch for viewing on a CCTV monitor. TELEPHONE receptacle is a standard 2-wire ¼" phone jack fully isolated and d.c.-blocking for transmitting or receiving PLTV audio-FM signals to or from the telephone line. REMOTE FRAME GRAB jack parallels FRAME GRAB switch on front panel, for plugging in a cord and remote switch (not supplied). TAPE jack provides for recording transmitted or received PLTV picture signal on audio or cassette tape recorder.

*This specification may be preset to a different value from that described above, in order to meet certain foreign regulations.

MODEL 500 SERIES

Picture: 128 line by 128 element resolution; 1:1 aspect ratio; 16 shades of grey.

Transmission Time: 8.5 seconds over dial telephone network.

OTHER CHARACTERISTICS

Power Input—Line voltage range is 105 to 125 volts AC or 210 to 250 volts AC (specify), and 50 or 60 Hz (specify). Power consumption is 20 watts.

Mechanical—Width: 12½ in. (31.75 cm); depth: 11 in. (27.94 cm); height: 7 in. (17.78 cm); weight: 12 pounds (5.44 kg); shipping weight: 15 pounds (6.8 kg).

Construction—All solid state circuits on one glass epoxy printed circuit board. Two-tone brown all-aluminum cabinet.

MODEL 600 SERIES

MODEL 601 and 606

Picture: 64 line by 256 element resolution; 4:1 aspect ratio (¼ screen display); 64 shades of grey.

Transmission Time: 8.5 seconds over dial telephone network.

MODEL 603 and 608

Picture: 128 line by 128 element resolution; 4:3 aspect ratio; 64 shades of grey.

Transmission Time: 8.5 seconds over dial telephone network.

MODEL 612 and 617

Picture: 128 line by 256 element resolution; 4:3 aspect ratio; 64 shades of grey.

Transmission Time: 17 seconds over dial telephone network.

OTHER CHARACTERISTICS

Power Input—Line voltage range is 105 to 125 volts AC or 210 to 250 volts AC (specify), and 50 or 60 Hz (specify). Power consumption is 25 watts.

Mechanical—Width: 14½ in. (36.8 cm); depth: 13 in. (33 cm); height: 7 in. (17.8 cm); weight: 15 pounds (6.8 kg); shipping weight: 18 pounds (8.2 kg).

Construction—All solid state circuits on glass epoxy circuit boards. Two-tone blue all-aluminum cabinet.

Fig. 9-13. Table of specifications for the Robot 500/600 transceiver series (courtesy of Robot Research, Inc.).

are not lost due to contouring. Minor variations in video input (caused by switching cameras or changing light levels) will have much less effect on the picture and will not require readjustment of controls. The 600 Series comes in transceiver models only and offers a range of picture formats to fit different applications. Figure 9-13 provides general specifications for the 500/600 Series.

Robot Model 635

The Robot Model 635 Transceiver grabs and frame-freezes video fields from any CCTV camera, converts the stored picture to audio tones, and transmits the picture tones any distance over the dial telephone network. It also receives these audio PLTV picture tones transmitted over any distance by telephone and converts the picture information to video for display on any TV monitor.

The standard PLTV signal consists of audio-FM. This signal is insensitive to wide variations in line attenuation and has a "capture" effect in the presence of noise that insures good reception over most telephone connections.

The standard high resolution PLTV format consists of a 256 by 256 array of picture elements transmitted and received in 35 seconds. A multiple speed option is available which allows selection of lower resolution, faster frame rate modes by means of a rear panel switch. These alternate modes are a 128 line by 256 element 17 second picture element, or a 128 line by 128 element 8.5 second picture. In the model 635 each picture element is represented by one of 64 grey shades in all solid-state digital memory.

The PLTV signal produced by the Model 635 Transceiver is −9 dBm nominally into a 600 ohm load impedance. The signal can be stored on any audio reel or cassette tape recorder.

The Model 635's connection to the telephone line serves for both receiving and transmitting pictures. The device is registered with the FCC for direct connection to the dial telephone network in the U.S. The connector for the telephone, which is supplied with the Model 635, is the telephone industry standard Modular plug. A modular jack is provided on the rear panel for connection of a telephone set. During PLTV transmission or reception, the telephone set is disconnected from the line by the Model 635 so that unwanted noise will not interfere with the PLTV signal. If the operator should happen to hang up the telephone set or turn off the power, it will "drop" the telephone line. This is to prevent an unintentional "tie-up" of the line. The method of connection to the telephone line varies in different countries depending on government regulations. In some countries, specific modifications have been made to the 635 to meet these requirements; and in others, the use of an approved coupling may be required. Details of specific requirements are available on request. Simultaneous voice and PLTV over the same audio channel is not possible.

SPECIFICATIONS

INPUT

Video Input—Standard 525-line TV video signal (625 in 50 Hz countries). 1.0 volt p-p white positive into 75 ohms; BNC connector. Positive or random interlace.

PLTV Input—Accepts high resolution PLTV signals with input sensitivity of -40 dBm. Input impedance 600 ohms. Modulation standards are*: white—2300 Hz, black—1500 Hz, sync—1200 Hz.

OUTPUT

Video Output—Standard TV Video signal 1.4 volt p-p white positive into 75 ohms, BNC connector, compatible with standard 525 line TV receiver (625 lines in 50 Hertz countries).

PLTV Output—High resolution PLTV modulation 1200 Hz to 2300 Hz audio-FM, -9dBm nominally into 600 ohm load.*

CONTROLS, CONNECTORS AND INDICATORS

Front Panel Controls—A front panel selector switch chooses among five possible operating modes. In **RECEIVE,** the unit is set to receive PLTV picture signals from the telephone line and produce video for TV display. In **HOLD,** memory up-dating is discontinued and the currently stored picture is retained for video display as long as power remains on. In **TRANSMIT,** the unit grabs TV fields from the CCTV camera input and produces PLTV picture signals on the rear-panel **TELEPHONE LINE** cord; video supplied to the rear-panel **TO VIDEO MONITOR** jack is from the stored picture in memory. In **CAMERA** position, used primarily for TV camera adjustment, no transmission or reception occurs; the video supplied to the rear-panel **TO VIDEO MONITOR** jack is the camera's real-time video after being quantized to 256 by 256 picture elements and 64 grey shades (exactly as it is presented to the memory for storage and transmission). In **TEST**, an internally generated grey scale replaces the TV camera's input to memory; the grey scale appears on the video display and can be converted to PLTV and presented for transmission on the rear-panel **TELEPHONE LINE** cord in the **TRANSMIT** position. The front panel **FRAME GRAB** switch is activated in **TEST, CAMERA** or **TRANSMIT** modes. When in the **AUTOMATIC** position, it frame-grabs (1/60 second) from the camera video or grey scale test pattern at the beginning of each high resolution PLTV picture (every 35 seconds). In the **MANUAL** mode, it frame-grabs whenever the operator actuates the **FRAME GRAB** push button. The two front panel **VIDEO IN** controls are effective in **CAMERA** and **TRANSMIT** modes. They adjust the input video **CONTRAST** and **BRIGHTNESS** to match the memory input range; the effect of these controls can be viewed in the **CAMERA** position of the selector switch. In the **VOICE** mode (**VOICE** indicator illuminated) the attached telephone set is active for dialing, ringing or talking. In the **PICTURE** mode, reached by pressing the **PICTURE** push-button while the attached telephone set is off-hook, the PLTV is connected to the telephone line. In the **PICTURE** mode the telephone set is inactive.

Rear Panel Connectors and Controls—**FROM CAMERA VIDEO** is a **BNC** connector for input of standard CCTV video signal to Model 635. **TO VIDEO MONITOR BNC** connector supplies standard video selected by front-panel switch for viewing on CCTV monitor. **TELEPHONE SET** is standard 6 lead modular jack for connection of any FCC type accepted single line telephone set. **TELEPHONE LINE** is a pendant cord terminated in a modular plug for connection to a telephone company installed modular jack. **REMOTE** is a 25 pin "D" type connector for connection of an optional (not supplied) remote control box which enables the **PICTURE** and **VOICE** control functions to be remotely duplicated. **REMOTE** also contains leads to permit picture black/white reversal and remote control of **FRAME GRAB. TAPE** jack provides for recording transmitted or received PLTV picture signal on audio or cassette tape recorder.

OTHER CHARACTERISTICS

Power Input—Line voltage range is 105 to 125 volts AC or 210 to 250 volts AC (specify), and 50 or 60 Hz (specify). Power consumption is 25 watts.

Mechanical—Width: 14½ in.; Depth: 13 in.; Height: 7 in.;

Weight: 15 pounds (shipping weight: 18 pounds).

Construction—All solid state circuits on glass epoxy circuit boards. Two-tone blue all-aluminum cabinet.

Fig. 9-14. Specification chart for the Robot model 635 high resolution PLTV (courtesy of Robot Research, Inc.).

The Model 635 includes controls for setting the stored grey level range of the TV pictures grabbed for transmission, for inserting a grey scale pattern for test purposes, and for receiving blacks and whites in the displayed picture. Figure 9-14 gives the specifications for the Model 635.

As was stated previously, SSTV might be considered as facsimile handled in a completely electronic manner without involving any mechanical or moving parts. The SSTV received image looks similar to facsimile copy. Figure 9-15 shows an example of a high resolution SSTV image. By looking closely, you may be able to pick out the individual lines making up the overall image.

SSTV offers many possibilities for sending video information through a standard phone system. The period of time required to scan and paint the graphic data is reduced from that required by most facsimile systems. The main advantage, however, lies in the fact that almost any object which may be televised by a close circuit television camera can be sent through the telephone system using SSTV. Graphic material no longer needs to be the only transmittable data, as is the case with facsimile. A friend or relative sitting across the room from you may have his or her image transmitted to a party hundreds of miles away on the other end of the telephone line. It is not necessary to place data on a drum or easel to be scanned; the camera is simply pointed and focused upon the object to be transmitted. Remember, due to the time required to paint a single frame, moving objects cannot be properly transmitted. They will appear at the receiving end as a series of visual distortions.

The SSTV transceivers discussed here are portable, easily attached to the phone line, and are relatively inexpensive considering all of the technical advantages and conveniences they offer the user.

SUMMARY

Too many of us tend to think of the telephone system in a very similar manner to the Dixie Cup telephones we might have experimented with as children. This consisted of a small hole punched in the bottom of a Dixie Cup through which the end of a long string was slipped and often tied to a shirt button to hold it in place. This process was repeated at the other end of the string and when this connecting line was held taut, you could communicate by having one person speak into one cup while the second person listened with his ear to the other.

Fig. 9-15. High resolution SSTV image resembles fascimile copy in many ways (courtesy of Robot Research, Inc.).

The telephone system which we enjoy almost every day of our lives can be likened to this child's toy just as an anti-aircraft gun is likened to a slingshot. In other words, there is really no comparison.

Today's telephone system offers a multitude of services and conveniences, all of which are utilized by everyone from the private citizen to the giant conglomerate. Indeed, there are many services which are not even dreamed of by the general public at present which may be reality in less than a year.

Versatility is built-in to the telephone system assuring that it will be a precious commodity for many years to come. Certainly, there are aspects of this communication system which are outmoded. One of these might be analog transmissions. These are being replaced, where possible, by digital interfaces. Sooner or later, the digital method of transmission may be outmoded and another method will take its place. Constant research, development and updating have made the telephone system a most valuable necessity for the housewife who wants to call a friend or neighbor, the handyman who needs to locate a hardware store which carries a product he needs, the radio station that needs to broadcast a football game, and to the giant computer terminal which must talk to a computer on the other side of the world.

Chapter 10

Wireless Telephone Systems

In recent years, the cordless hand-held telephone has become a practical alternative to locating many extension phones throughout a large home or area. With the wireless telephone, the user may take or place telephone calls through his own telephone system without the necessity for any attachment wires. This means that you can go wherever you want to go and still not risk the possibility of missing an important phone call provided you are within the range of your system. Likewise, if you should desire to make a telephone call while relaxing at the pool or even while mowing the grass in the backyard, you can do so with the wireless telephone.

The wireless telephone should not be confused with the radiotelephone although they both work on the same basic principles. Whereas the radio telephone is a system unto itself, the wireless telephone is simply a means of tapping into your present hard-wired telephone system. Radio transmissions are used to send the caller's voice to the remote unit and to send the voice of the person at the remote location to the central telephone system. Figure 10-1 shows the Cobra Model CP-200S Cordless Electronic Telephone system which can be used up to several hundred feet away from your telephone system.

Figure 10-2 shows just how the cordless system is integrated with the standard system in your home. A cordless telephone normally consists of two units. The base unit is attached directly to the telephone line through a mating connector. The telephone

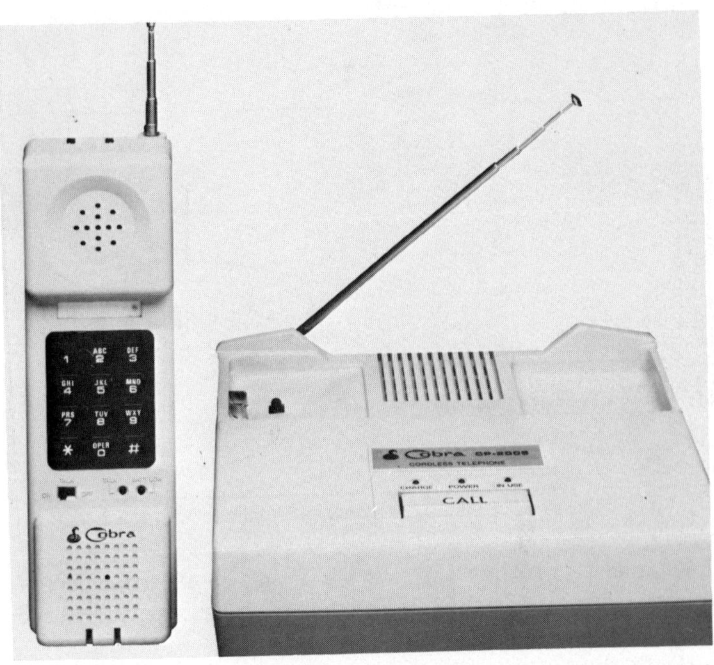

Fig. 10-1. Cobra model CP-200S cordless electronic telephone system which has a range of at least 300 feet (courtesy of Dynascan Corp.).

which is normally used in the home is usually attached to the base unit as well to allow for local phone calls to be made without the necessity of using remote systems. The portable unit may be carried with the owner as he goes about his daily household routines. It is light enough and small enough to fit in a coat pocket.

Figure 10-3 shows the connection of the base unit to the modular phone jack and to the hard-wired telephone. This can normally be a accomplished in just a few seconds as the telephone connection cable is fitted with a modular plug designed to insert into the modular phone jack which is most commonly installed by the telephone companies today. An identical phone jack is mounted in the back of the base unit case and receives the connector cable from the wired telephone. So, the installation of the base unit simply involves the removal of the connector cable from the standard telephone which is inserted into the back of the base unit. Then the connector cable from the base unit is attached to the wall jack where the wired telephone was formerly connected. Inserting the AC power cord from the base unit into a 110-volt wall outlet causes the unit to be activated. Most base units have one and

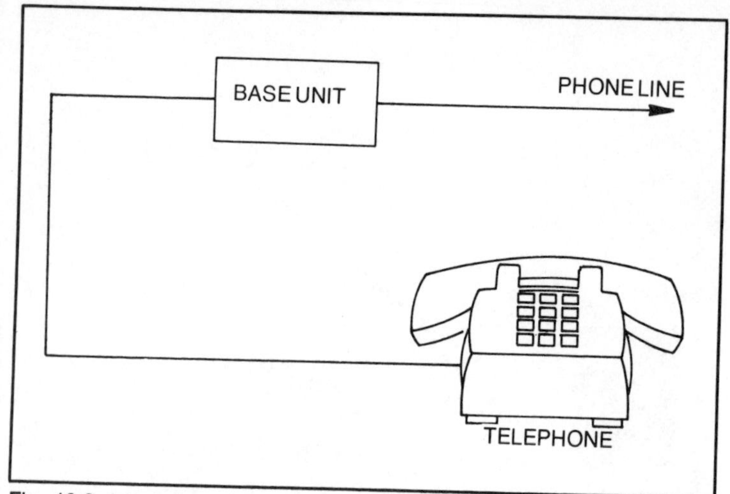

Fig. 10-2. Integration of the Cobra cordless telephone with the telephone instrument and line.

sometimes two vertical antennas which may be telescoped out of the base. Any antennas should be extended to maximum height and not located near any large metallic objects.

For the Cobra CP-200S and most other units like it, the installation of the base unit is the most complicated procedure in getting this system operational. This, of course, is not really a complicated procedure at all, but the operation of the portable unit is simpler yet. All that one needs to do to operate the portable unit once the base unit has been installed is to extend the portable antenna, throw the on-off switch to the "on" position, activate the "talk" switch, and you're in business. Dynascan Corporation, who makes the Model 200S, indicates a range of about 300 feet distance between the base unit and the portable one; however, tests have shown that unit to be effective up to a half mile or more, depending upon the amount of metal within the home where the base unit is placed and the terrain between the base unit and portable unit.

Figure 10-4 shows the circuitry of the cordless telephone system in block diagram form. The base unit transmits the audio information from the telephone line to which it is directly attached at a low frequency of about 1.7 MHz. While the base unit is equipped with an antenna, this is used for receiving only and picks up the transmission from the portable unit which are at about 49.8 MHz in the radio spectrum. The base unit feeds its transmitted signal directly into the AC line from which it receives operating current. The output of the 1.7 MHz amplifier is split through a

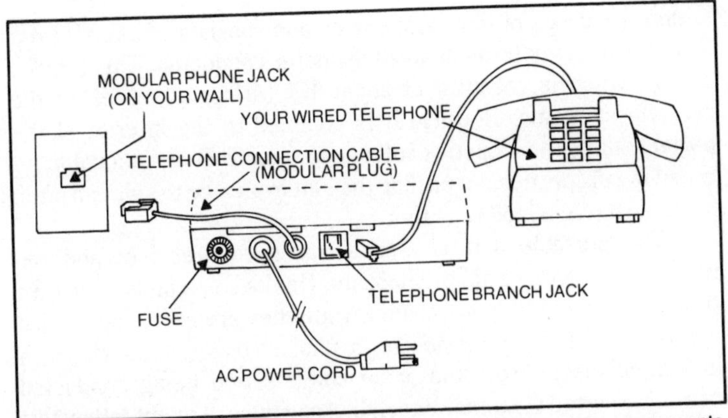

Fig. 10-3. Modular plug connection between the various system components (courtesy of Dynascan Corp.).

transformer and each side of the AC line is connected to each output through a .001 μF capacitor. Each capacitor effectively blocks the AC line current from entering the base unit circuitry and allows the RF output to travel through the home electrical system and to be radiated into space. The house wiring system, then, becomes a very complex antenna system. A trap filter is built into the base unit where the AC line enters the circuitry to prevent the 1.7 MHz energy from re-entering the base unit circuitry. This trap makes the line a one-way circuit regarding the transmitted RF. It can only travel into the house wiring circuit and not into the base unit circuitry.

THE ANTENNA

Because house wiring is used to form an antenna system when the portable unit is used within the home, reception is excellent because it should always be in fairly close proximity to a portion of the house wiring antenna. It was noted that signal strength was maintained to a high degree outside of the home when the portable unit was operated beneath the electrical cables suspended from poles on the street. Some of this RF power from the base unit was travelling on these wires as well.

The base unit, as was noted before, transmits its output near 50 MHz. The telescoping antenna located within its case is used for transmitting purposes only. It is far too short to be effective at the low frequency transmitted by the base station. For reception of the base signals, the portable unit has a "loopstick" antenna located within its case. This is very similar to the antennas which are found within the bases of most AM radios and consists of a ferrite rod would with many turns of small diameter conductor. This forms a circuit which is resonant at about 1.7 MHz and serves as the receiving antenna. Because it is confined to the interior of the portable unit, the loopstick antenna is somewhat directional in its receiving properties, so turning the unit from side to side can often increase the received sensitivity.

The portable unit is a true transmitter-receiver and can transmit and receive simultaneously. Unlike some units which are transceivers and cannot transmit while they are receiving or vice versa, the CP-200S carries on a phone conversation which is indistinguishable from the same conversation being conducted over hard-wired systems. Receiving a phone call might follow this procedure:

—Caller dials your number.

—Hard-wired phone begins to ring.

—Base unit sends ringing tone to portable unit which beings to oscillate.

—User with portable unit extends antenna to full height, pushes "talk on" switch, and carries on a conversation in a normal manner.

—When conversation is ended, user pushes talk switch to "off" position and collapses antenna.

—System is now ready to receive another phone call.

An internal battery supplies power to the portable unit and is recharged by placing the entire portable unit in its cradle formed in the base unit. Mating connectors are automatically engaged when this is done, and an overnight charge will provide a full day's operation. When the talk switch is in the "off" position, the portable unit draws very little operating current and is activated only when a call comes in. At this point, the talk switch is placed in the "on" position and maximum current drain is realized during this phase of the operation. The nickel-cadmium batteries typically will last for at least 3 years under conditions of maximum daily usage. This much usage is fairly rare, and batteries in most typical operations will last far longer.

The following procedures are used to place an outgoing call:

—Extend the antenna of the remote unit to full height;

—Place the talk switch in the "on" position.

—After hearing a dial tone, punch in the desired number on the touch tone pad.

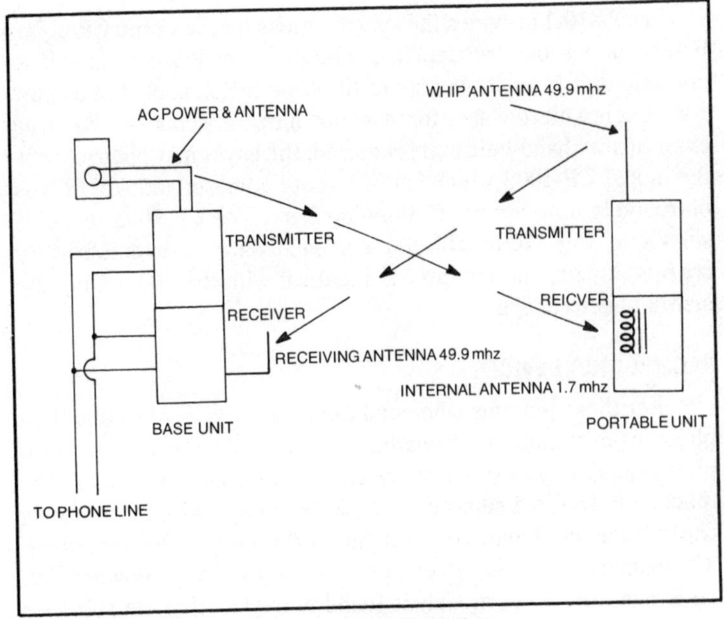

Fig. 10-4. Block diagram of cordless telephone system.

—If you make an error in dialing, hang up the phone by returning the talk switch to the "off" position and then back on again to redial.

—After your number has been punched in, you will hear ringing as the phone on the other end is dialed up.

One convenient feature of the Cobra unit tested was the built-in memory which can be used to redial numbers which were busy on the first try. If the user punches in a number which proves to be busy, the same number may be redialed minutes later by pushing the (#) key to the right of zero. The memory is stored at the base unit and released by pressing the numbers key. It is interesting to note that even though a touchtone pad is used to place the phone call from the portable unit, the wired phone line is directly dialed. A rotary switch which corresponds to telephone dial mechanism is located within the base unit and is triggered by the tones from the portable unit pad. When the number 4 is punched at the portable unit, the number 4 is dialed by the base. This is the only mechanical movement which takes place in the cordless telephone system and it can be heard when listening in close proximity to the base unit as a series of audible clicks.

Table 10-1 provides the specifications for the Cobra CP-200S, listing the various frequencies, tolerances, and dimensions. The portable unit is small enough to fit in the inside pocket of a sport coat, but is a bit too large for insertion into a shirt pocket. When an even smaller hand-held unit is desired, the buyer may opt to choose the model CP-100S which is much more compact and will fit less obtrusively into various clothing pockets. The CP-100S model is shown in Fig. 10-5 with its corresponding base unit/battery charger. Again, the circuitry is identical with the 200S. Only the size has been changed.

RECEIVE ONLY DEVICES

For those persons who need the capability of answering their phone from remote locations but do not need to be able to place phone calls, a less expensive system is available from Cobra. The model CP-15S is a remote cordless telephone with receive only capabilities and looks very similar to the other units described. The main difference between this latter unit which is shown in Fig. 10-6, and the earlier models is found in the fact that the 15S does not contain a touch tone pad on the remote unit. Therefore, it is used only to answer telephone calls and the owner who needs this capability does not have to go to the extra expense of purchasing a portion of the circuitry for which he has no need. The frequencies for transmitting and receiving are identical with the former units.

The circuitry involved in the Cobra phones is not highly unique and is probably built by a foreign manufacturer and distributed to the United States where it is packaged by the various electronic companies for sale under their brand names. Another fine cordless telephone is the Duofone ET-300 which is sold by Radio Shack. Pictured in Fig. 10-7, it can be seen that the Radio Shack unit closely resembles the Cobraphone Model 100S. An examination of the schematics also shows them to be almost exactly identical. Many other wireless telephones are available on today's market. Some of them are very expensive, while others, depending upon the features offered, may be relatively low in price.

One of the lowest priced wireless telephones on today's market is the Muraphone which was purchased for research purposes from J S & A Marketing Group in Northbrook, Illinois for only $89.95. The Muraphone represents a very simple approach to a rather complex telephonic function. Unlike the other wireless phones discussed, the Muraphone is a transceiver. The portable

Table 10-1. Table of Specifications for the Cobra CP-200S (courtesy of Dynascan Corp.).

X. SPECIFICATIONS

GENERAL

The COBRA ModelCP-200S Cordless Electronic Telephone complies with FCC regulations, Parts 15 and 68, and is registered and approved for direct connection to your telephone extension outlet without installation or monthly service charges. NOTE: Specifications are subject to change without notice.

Frequency Control	Crystal-controlled, except 1.7 MHz transmission.
Modulation	FM.
Frequency Stability	±2.5 KHz
Operating Temperature	−20° C to +50° C.

BASE UNIT - Model CP-200B

Transmit Frequency	1.6 to 1.8 MHz (1 channel within this range).
Receive Frequency	49.8 to 49.9 MHz (1 channel within this range).
Power Requirements	120 V, 60 Hz, 10 Watts.
Size (Including Antenna)	8-5/8″ W × 2½″ H × 7¼″ D. (220 mm × 64 mm × 184 mm.)
Weight	Approximately 3¼ lbs. (1475 grams).

REMOTE UNIT - Model CP-200R

Transmit Frequency	49.8 to 49.9 MHz (1 channel within this range).
Receive Frequency	1.6 to 1.8 MHz (1 channel within this range).
Power Requirements	Rechargeable nickel-cadmium batteries.
Size	2½″ W × 8″ H × 1¾″ D. (64 mm × 203 mm × 44 mm.)
Weight	Approximately 15 oz. (425 grams).

XI. CORDLESS TELEPHONE CHANNELS AND GUARD TONES

To prevent your COBRA cordless electronic telephone from being interfered with by other cordless telephone systems, a variety of radio channels is used. As a further security precaution, to prevent unauthorized use of your phone line a number of guard tones are used.

For example, if your COBRA cordless telephone is marked CH 1 and 5.3 KHz, this means it utilizes 49.830 MHz for the remote transmitter and base receiver, with a 5.3 KHz guard tone and 1.665 MHz for the remote receiver and base transmitter.

RADIO CHANNELS

Channel No.	Remote Transmit & Base Receive Frequency	Remote Receive & Base Transmit Frequency
1	49.830 MHz	1.665 MHz
7	49.845 MHz	1.695 MHz
13	49.860 MHz	1.725 MHz
19	49.875 MHz	1.755 MHz

GUARD TONES

The guard tones are: 5.3 KHz, 6.0 KHz and 6.7 KHz.

unit resembles a walkie-talkie, and indeed, functions in a very similar manner. Voice quality as heard by the caller and by the person with the remote unit is not of standard telephone quality and sounds more like a conversation by radio. However, it should be noted that within the 400-foot range of this device, speech is highly intelligible and adequate communications can be maintained. This is a receive only unit which is one of the reasons for the low price. Recently, the Mura people have added a touch-tone pad to the basic circuitry to provide a unit through which you can place calls as well as receive them.

Transmitting and receiving by both units is on the same frequency which lies near 50 MHz. Again, this is a transceive system which will not transmit and receive simultaneously. One might ask how the base unit switches from transmit to receive. The remote unit is switched between the two modes by a thumb activated control, but the base unit naturally cannot be directly controlled by the human hand. The answer is the remote unit determines the mode of operation of the base unit. The base unit's normal state is the transmitting mode. When the phone rings, the base unit switches to this mode and keys an audible sound at the remote unit. The base unit, however, does not transmit continuously. Rather, it transmits a pulsed signal which switches from transmit to receive during different fractions of a second. Indeed, this pulsing can actually be heard by the person with the remote unit. He will hear the voice on the other end of the line plainly, but the voice will have a fluttering effect which is the result of the rapid switching between transmit and receive.

Now, when the person with the remote unit desires to speak, he depresses the push-to-talk button, which causes a signal to be transmitted to the base unit. If the base unit is in the transmitted portion of its cycle, it will not receive the transmitted signal from the portable unit until it switches to the receive mode. Since the base unit switches to receive several times a second, the user does not notice this lag. When the transmitted signal from the portable unit is finally received, this triggers a relay at the base unit which switches the base circuitry into a receive only mode. Now, the voice of the remote operator is received from the portable unit and fed directly into the telephone line. When the remote operator desires to listen again, the push-to-talk switch is released and the base unit returns to the transmit mode due to the lack of the transmitted signal from the portable unit.

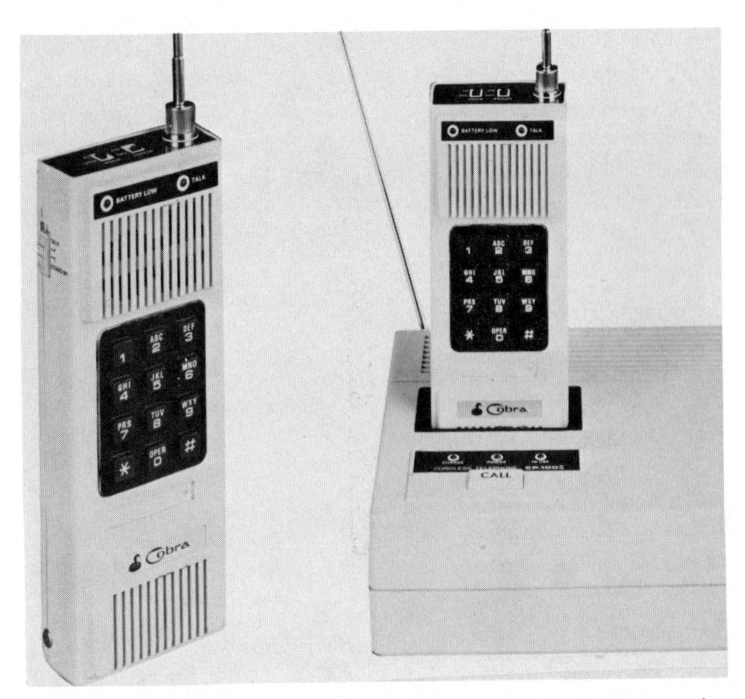

Fig. 10-5. Cobra model CP-100S consists of the same basic circuitry as the earlier model but is built into a more compact case (courtesy of Dynascan Corp.).

It should be understood that the Muraphone is not as complex nor as efficient a device as were some of the other cordless designs discussed in this chapter. The audio quality is just not there, but good communication abilities are possessed by this system. The main advantage of the Muraphone is its simplicity of design and operation and, of course, its low cost. The former units sell somewhere between the $200 to $300 range while the Muraphone costs less than $100. For most applications, 90% of the usage of a cordless telephone will be for the receiving of calls. The other 10% of the time it is necessary for calls to be placed through the average cordless system. This would mean that, for the average user, the Muraphone might be perfectly adequate 90% of the time. In some situations, its range may be even longer than some of the more expensive models. This is mainly due to the fact that only a single high frequency is used to make the connection between the base unit and the remote. The antennas on each unit are used for both transmit and receive, and these antennas are more efficient at their operating frequencies than the house wiring system and the

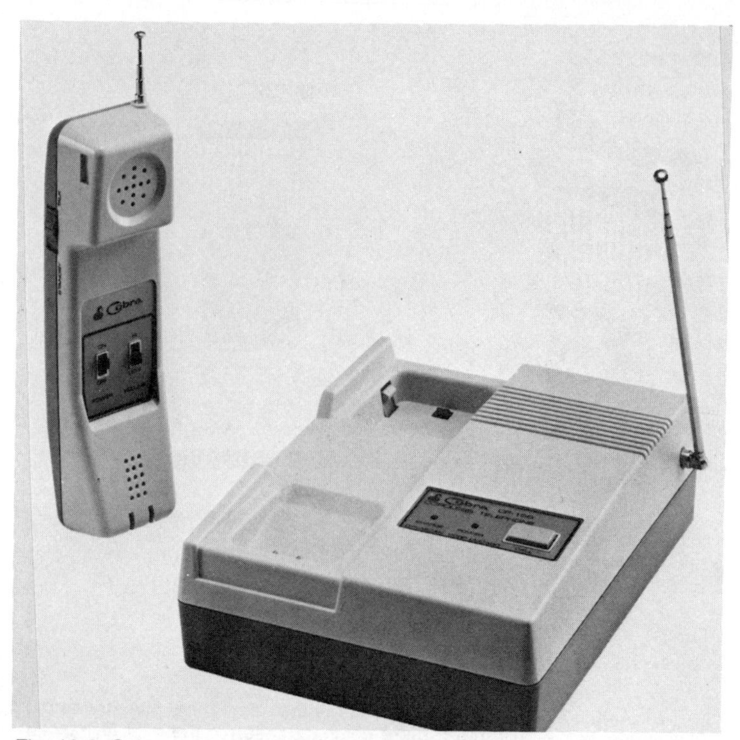

Fig. 10-6. Cobra model CP-15S receive-only cordless telephone (courtesy of Dynascan Corp.).

internal loopstick are for the cordless phones which utilize the latter. An efficient antenna at 1.7 MHz must be very long (usually over 200 feet), whereas an identically efficient antenna at a frequency near 50 MHz would be less than 5 feet long. The short whip antennas located on the cordless telephone for high frequency transmission and reception are much closer to the required length than are the loopstick and house wiring systems.

THE INTERCOM

The Muraphone is shown in Fig. 10-8 and also features an intercom button which can allow for local communications between the base and mobile unit. Pressing and holding down the intercom button on the base unit broadcasts the ringing signal to the remote unit. It also disconnects the cordless system from the wired phone system. Should a regular phone instrument be attached to the jack in the Muraphone, you can use the wireless system as a pager or call interceptor. This is accomplished by picking up the phone attached to the base unit and pressing the intercom button. When the remote station answers, you can carry on a two-way conversation. During the whole conversation, the intercom button must be depressed. If it is released, it will connect the phone system to the Muraphone and a dial tone will be heard by both parties. The hold-in method is an important safety feature, as it prevents the user from leaving the wireless system in the intercom position which disconnects the phone instrument from the line. At the end of the intercom conversation, the "line release" button must be pressed. The line may be released at either end of the wireless system, either by the base or mobile unit. If this is forgotten, however, the timer will drop out the Muraphone in two minutes. This internal timing circuit assures that the wireless system will be armed and ready to take a phone call at almost any time regardless of whether or not the correct arming procedures have been carried out by the user.

If a phone call is answered by the regular phone instrument attached to the base unit jack, pressing the intercom button puts this call on hold. You can then talk to the remote without the caller hearing what you are saying. Again, the intercom button must be depressed during the entire length of this intercom conversation. If desired, you can release the intercom button and hold a 3-way conversation. Either party can hang-up his phone. The party on the remote end does this by simply depressing the line release button while the party with the wired instrument simply hangs up in the usual manner.

Fig. 10-7. Radio Shack ET-300 cordless telephone system.

The portable unit of the Muraphone system is operated from a 9-volt battery which is of the transistor radio type. Some units are supplied with a rechargeable battery which is recharged by connecting a mating plug between the portable and the base unit. An overnight charge will supply enough charging for many hours of use. The number of hours will depend upon how many calls are received and on whether or not the portable unit is kept in the "standby" position. This standby position allows the portable unit to be keyed by the ringing phone and to emit the ringing pulse. When this is heard, the user simply turns the volume up to an acceptable level and carries on a conversation. When in the standby position, the portable unit is drawing some battery current though not alot. Maximum current is drawn only during normal transmit and receive periods in the midst of a conversation.

When using the system around the house or yard where you can hear the regular phone ring, it is not necessary to keep the portable unit on standby. When the phone rings, the portable unit is then turned on, the antenna extended, and the call answered. By keeping the portable unit completely off until a call comes in, the rechargeable battery will last much longer because it is used less. Recharging time, too, will be reduced and the life of the battery extended. The only reason to keep the Muraphone on standby is to be able to hear the phone ring through the remote unit. This would be most advantageous in situations where the user is removed from the normal phone by such a distance as to render it impossible for him to hear the ringing bell.

The rechargeable battery in the Muraphone system represents a substantial portion of the cost of the entire system. The stock battery delivers 12 to 15 hours of operation per charge and should be adequate for the majority of users; however, some people may want to have a more powerful battery for extended operational periods. Mura offers a special optional battery which is rated at 3 times the capacity of the stock battery. To get this option, it is necessary to send both the remote and base units back to the factory where the high storage battery is installed and the charging current rate provided by the base unit increased to handle the demands of the new battery. The cost for this battery and service is less than $20.

Figure 10-9 gives the complete specifications on the Muraphone system. It can be seen that the frequency of operation is within the same high frequency range as the previous system discussed; however, only one frequency is used to both transmit

Fig. 10-8. Muraphone answer-only cordless telephone system is inexpensive and efficient over short ranges (courtesy of J S & A Group).

and receive. Also, it can be seen that the transmission ratio for the base unit includes 450 milliseconds of transmit time to 40 milliseconds of received time. The Muraphone is manufactured by the Mura Corporation of Westbury, New York and represents a low-cost alternative to cordless telephone systems.

USEFUL RANGE

As has already been pointed out, the useful range of the cordless telephone system between the portable unit and the base and vice versa will depend upon many different factors. These include frequency of transmission and reception, receiver sensitivity, transmitter power output and modulation characteristics, antenna design, antenna height above surrounding objects, weather conditions, and general terrain. All of these conditions can have a pronounced effect on range. Probably the most important of all the range-determining factors mentioned will be antenna design and height above surrounding objects. Receiver sensitivity will be next in line, closely followed by transmitter power output and frequency of transmission and reception.

Most of the newer cordless telephone designs utilize the 49.8 to 49.9 MHz frequencies for transmission from the portable unit to the base. Likewise, many of the newer units use the frequency spectrum just above the AM broadcast band for transmission from the base unit to the portable one. The reasons for these choices are obvious once a closer examination is made of the requirements of the circuitry. First of all, to have truly natural and efficient communication it is necessary to be able to transmit and receive simultaneously as is the case with the standard wired telephone in your home. When transceive operation is used, it is possible to miss certain portions of the conversation because when the person with the transceiver is transmitting, he has no way of knowing if the party on the other end is listening or talking. On the standard telephone system, this is not the case because the talker can also hear the listener should he decide to interrupt or ask a question. Transceive operation usually necessitates ending all transmission with the word "over" to let the party on the other end know that he is switching to the receive mode.

When simultaneous transmit and receive operations are desired, it is necessary to build a circuit which contains a separate transmitter and separate receiver. The whole time the unit is activated, the transmitter is broadcasting its signal and the receiver is constantly detecting an incoming signal. If the frequency of the transmitter is very close to that of the receiver, then the close proximity of the two antennas will cause much of the transmitted signal to be fed into the front end of the receiver, causing overload and even circuit damage. For this reason, it is essential that the transmit and receive frequencies be separated by as great a distance as possible, as it is impossible to separate their respective antennas when used for wireless telephone purposes. The greater the separation between frequencies, the less the transmitter will interfere with the receiver. This is the reason for the utilization of the frequency spectrum near 50 MHz for transmission on a portable unit and the spectrum above the broadcast band at about 1.75 MHz for broadcasting from the base unit. There is a separation here of about 47 MHz which is more than adequate.

Federal Communications Commission rules and regulations play a most definite role in frequency choice. Cordless telephones are restricted to the two frequencies just mentioned and a few others. Another frequency which was often used for these purposes was in the 27 MHz region of the radio spectrum. This lies within the citizens band channels and is no longer desirable because of the massive interference encountered in this area.

It would be just as easy to have designed the cordless telephone system to have the base transmit at the high frequency and the portable unit transmit at 1.75 MHz; however, the antenna system becomes an immediate problem when this is contemplated. As was pointed out earlier, an antenna system designed to be

SPECIFICATIONS:

1) Remote Handset Model No. MP-100
   Carrier Frequency: 49.860 MHz.
   Power Output: 10,000 u volts at 3 meters (nominally 80 milliwatts).
   Receiver Sensitivity: 3 u volts typical.
   Control Frequencies: talk control: 7000 ± 200 Hz
   line release: 5000 ± 200 Hz
   Power Supply: 7 cell stack NICAD batter. 8.4 volts, one watt hour.
   Speaker/Microphone: alnico 2-¼ inch speaker.
2) Base Unit Model No. MP-101
   Carrier Frequency: 49.860 MHz
   Power Output: 10,000 u volts at 3 meters (nominally 80 milliwatts)
   Transmission Ratio: 90% duty cycle (450 milliseconds transmit, 40 milliseconds)
   Receiver Sensitivity: 3 u volts typical
   Power Supply: regulated to 9.5 volts ±5% from 110 to 125V AC.

Phone Connections: a) 5 foot cord with modular plug to plug into the phone system lines
b) USOC: RJ11C to accept a telephone instrument

Charger Output: 5 ma into the 8.4 volt battery stack of the model MP-100 remote unit.

Disconnect Timer: 2 minutes after no receipt of any signal from the remote unit the phone line is automatically released.

Other Controls: a) line release button
b) intercom switch

Fig. 10-9. Specifications of the Muraphone cordless telephone system (courtesy of Mura Corp.).

efficient at 50 MHz is tiny in physical size compared to one which is efficient at 1.75 MHz. The house wiring system in the home normally comprises a very great, although complex, length. This system is better suited as a radiator for the 1.75 MHz frequency than would be a simple whip antenna which is all that could be mounted on a portable unit. This latter antenna would be so inefficient as to be next to useless. Therefore, a wise choice was made in designing the portable unit to transmit at the higher frequencies where a short whip antenna would be much closer to optimum dimensions. It would be even more ideal if the loopstick antenna located within the case of the portable unit could be attached through a length of wire at least 200 feet long. Range would be greatly extended, but of course this is not a practical improvement to make.

At the base unit, the external whip antenna is used purely as a receiving antenna and should be kept as free and clear from large metallic objects as possible. The same applies to the antenna of the portable unit. Even the case should be kept away from large metallic structures if possible, as the loopstick antenna can be affected by the capacitance to ground exhibited by many of these objects.

In an effort to increase effective range, many persons have tried extending the factory installed antennas on their wireless phone systems. This is not to be recommended and in many cases is illegal. According to FCC regulations, the transmitting antenna (the one located on the portable unit) may not exceed five feet. A similar requirement is placed on antennas connected to the base unit which operates at 1.75 MHz; however, the direct attachment to the electrical system apparently precludes this regulation. It should be pointed out, however, that should an external antenna somehow be attached to the base unit transmitter, FCC regulations do determine its length.

Another factor which enters the picture when changing the length of transmitting antennas is the circuit loading factor. The output circuit of the various transmitters is adjusted to operate into an antenna load which is usually firmly fixed. By varying the length of the transmitting antenna, the impedance changes at the feed point, and the circuit may operate outside of its ratings. This could quickly cause damage. Changing antennas necessitates readjustment of the output circuit, and this will probably void the FCC certification placarded on each unit.

Receiving Antennas

Receiving antennas do not come under FCC regulations as long as they are used for receiving purposes only. Here it may be possible to extend the antenna to a more appropriate length and obtain better results. On modern units this would apply only to the base station which usually contains a whip antenna about 18 to 24 inches in length. This is a bit short for a quarter wavelength at the 50 MHz frequency. The ideal height would be about 4′9″. This antenna may be extended to this height without any bad effects and may cause a slight increase in range to be had between the portable unit and the base regarding transmissions from the portable unit. This, however, is not where the problem in shortened range usually lies. In experiments run by the author, it was noted that the distance the portable unit was able to cover between it and the base was greater than the transmission distance between the base and portable unit. In other words, the 50 MHz transmission will travel a greater distance to be received by the base than the 1.75 MHz transmission from the base to the portable unit. This is due to several factors, the main one being the poor efficiency of the receiving antenna located within the portable unit. The lower frequency, too, has more of a tendency to be absorbed by wiring systems in other homes, AM radio receivers, and a myriad of other metallic objects and electronic devices. From a practical

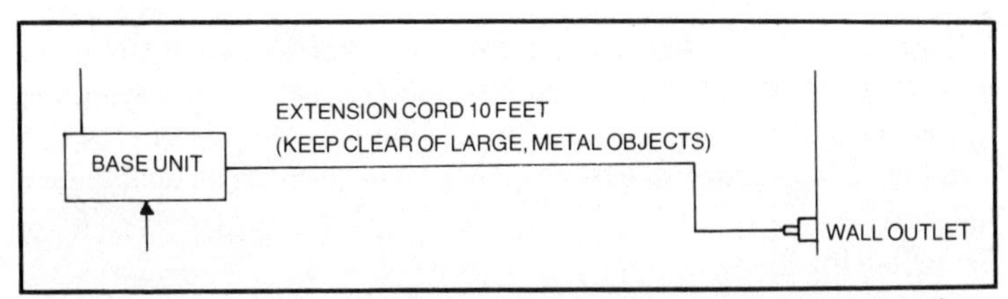

Fig. 10-10. Installation of extension cord to provide greater range in locations where electrical wiring is run through metal conduit.

standpoint, there is no easy way to extend the effective range of these units. Any sizable increase in range extension would probably be in violation of the FCC rules and regulations. Anyway, the distances available are adequate for the purpose these units were designed to serve. If you want more distance, you will have to resort to a mobile telephone system, which is discussed elsewhere in this book.

This best approach to system range is to make the most out of what you've got. If the remote unit can be attached to the phone line on an upstairs floor of a building and the unit and antenna located near an open window, longer range will probably result. Poor locations for placement of the base unit would include basement, steel closets, or on top of any metal object. The basement location would not significantly affect transmitting range from the base unit to the portable, but transmissions from the portable unit to the base might be severely suppressed. The height advantage obtained with placement in the upper stories of the home will not significantly increase broadcast distance between the base unit and the portable but should increase the transmission distance from the portable to the base unit.

Since most of the modern systems utilize the home wiring system as a transmitting antenna, a close look at your electrical system may be in order. Transmission distance will be determined to a large extent on the size of your electrical system, whether or not you have a one or two story home, and the amount of steel and other metals used in home construction. For example, a home with aluminum siding will probably severely limit the range of the base unit. Another severe range-shortening condition is found when all of the home wiring is run through aluminum conduit. Here it will be nearly impossible to get any range at all out of the base unit because the antenna (the home wiring system) is completely encased in a grounded metal shell.

Fortunately, this later situation can be partially overcome by installing a standard extension cord ten or more feet in length between the power plug of the base unit and the wall outlet as is shown in Fig. 10-10. This extension cord can even be strung in the upper corners of the room near the ceiling or even run outside the window. This section of extension cord will serve as an unshielded antenna.

Extending the Range

Extending the range of the portable unit does not offer as many possibilities as does the base unit. The loopstick antenna cannot be improved upon in a practical manner. Extending a whip from this antenna for a distance of three of four feet would have little or no effect on the range. While it is possible to extend the whip antenna used for transmitting, this is not highly practical either and involves internal circuitry readjustment and adherence to the FCC regulation stipulating a length of no more than 5 feet.

It would seem, then, that there is little to be done to extend the range from the portable unit to the base. This is not exactly true and one method of operation, simple though it may be, may provide you with maximum range from your system. This can be had if you always use the portable unit with the antenna fully extended and held in a completely vertical position. Since the antenna is attached directly to the case, tilting your head to the side can cause the antenna to be situated almost horizontally with the ground. The vertical antenna at the base will receive signals best which are transmitted from an antenna situated in the same vertical configuration as it is. So, when the transmitting antenna goes substantially off the vertical, transmitting range can be decreased. Figure 10-11 shows the proper way to hold the portable unit. Notice that the antenna is almost perfectly vertical in relationship to the ground below it. This method of transmitting will insure maximum range during most conditions. Figure 10-12 shows an improper method of transmitting. The antenna has been brought into a near-horizontal configuration with the ground below it and there will be a substantial signal loss at the receiving end of this system.

Since the loopstick antenna used for receiving at the portable unit is installed within the case, as was pointed out earlier, some directional characteristics will often be noted. This means that the antenna will receive better when pointed in a specific direction. The direction of strongest reception can be found by listening to the received dial tone or voice in the portable unit and turning your body in a circle until the strongest reception is noted.

By using the positioning method described, effective communications can be had far outside of the advertised maximum ranges for these units. True, at the range fringes, the transmitted and received signal may contain a high degree of static, but effective communication can still be maintained. This will not sound exactly like a normal telephone call because of the decreased voice quality, but each party should be able to get his or her message through.

Those users who are fortunate enough to live in flat areas of the country or near the water may find the ranges of their unit to be doubled, tripled, or multipled by more times than the advertised maximum range. This is where terrain plays a part. When radio signals intersect hills, trees, metal objects, etc., a portion of the signal is absorbed and the signal strength drops at the received end. In hillier, mountainous areas radio signals must often pass through the earth where these objects penetrate the radio path. Flat areas do not offer nearly as many obstructions to radio signals, with transmission over water usually producing the maximum ranges. There is little or nothing to be done regarding surrounding terrain and use of wireless telephone systems, so persons in less than ideal operational areas must resort to maximizing what they have and putting up with much shorter transmitting and receiving ranges.

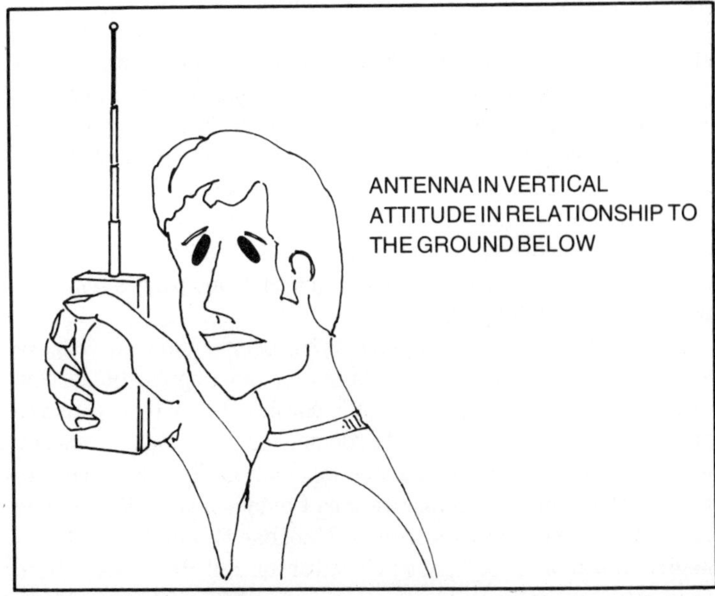

Fig. 10-11. Correct method of holding portable transmit/receive unit. Note antenna is held in vertical position.

Fig. 10-12. Improper method of using remote unit with antenna in slanted position.

SUMMARY

It can be seen that the cordless telephones discussed offer many advantages over the wired instruments more commonly associated with telephone systems. They are ideal for the busy executive who may be confined to different areas of a large office building and who may need to place and receive telephone calls at times when he is not near a standard telephone. With the wireless telephone, anyone may take or place calls through the standard phone system at any place or time within the effective range area.

The cordless telephone may be thought of as an extension to the standard system, one which does not require cumbersome cables and which allows the user the convenience of maintaining a telephone station which is completely portable and can be relocated to any point at any time. The truly nice thing about the finer cordless systems lies in the fact that once the base units are installed into the phone line, normal use and operation are almost identical to that of a standard instrument. Neither the caller nor the person answering the call need be limited in any way as long as a portable unit is within range.

Installation of the phone line, as has been pointed out, is very simple and takes only a few minutes for most applications. One important note: before installing the cordless telephone system to any phone line, it is necessary to contact your local phone company and advise them of your intended installation. This is a requirement of the Federal Communications Commission. Tell them you intend to connect an FCC cordless telephone to their lines. Do not neglect this most important and legally required procedure. There is no extra charge by the phone company for this installation as long as no service call is needed.

Chapter 11
Mobile Telephone Systems

While the cordless telephones of the previous chapter provide a great deal of convenience for their owners, their uses involve "close-in" operation to present phone systems. The cordless telephones are actually adjuncts to the standard hard-wired systems and merely extend rather than replace the home phones. On the other hand, a similar telephone system can be considered as discrete from the standard home system. The mobile telephone is a unit unto itself with a separate number and billing. Not really a short range device, the mobile telephone allows direct calls to be made from distances of up to several hundred miles. Actual range depends upon the channel frequency or frequencies, antenna installations, receiver sensitivity and transmitter power output.

RADIOTELEPHONE

The true name of the devices under discussion is *radiotelephone*. This accurately describes the combination which results in telephone communications via radio transmissions. Without going into great detail on basic radio theory, the radiotelephone works in a similar manner to any other radio transmitter. The voice of the operator is superimposed on the radio carrier. This means that changes in the human voice result in identical changes to the very high frequency output of the transmitter. Higher frequencies will travel a much greater distance than those of substantially lower frequencies. This can be proven very simply by speaking to someone across a sizable room. Generally, the voices of women, which are higher pitched, travel farther or are more easily understood than are those of men which are pitched much lower. At a concert, the tones of the flute are heard above the sounds of a tuba. Both of these comparisons assume that both types of tones are transmitted at the same volume or output level.

Knowing this, it can be easily seen just how radio transmissions carry sound much greater distances than the unaided human voice. Whereas, the normal voice frequencies range from between 300 and 3,000 hertz, the frequency output of the radiotelephone is usually around 160,000,000 hertz or may even be as high as 450,000,000 hertz in some systems. These radically higher outputs carry much farther through the earth's atmosphere.

Of course, the frequencies mentioned are completely out of the range of human hearing and require receivers to convert the modulation contained on the carrier back to the human audible range. Frequency modulation is used for radiotelephone transmissions. *Modulation* is an electronic term which means to change. In this case, the frequency of the transmitter output is changed to correspond with the changes in the tones which form the human voice. Taking the human voice frequency spread as an example, we can simply show how modulation works. Taking a single tone of 2,500 hertz (or cycles per second) and feeding it to the microphone input of a radiotelephone which operates at 150 Megahertz (150,000,000 hertz), the following results will be obtained:

☐ Before the tone is fed to the microphone, the output of the transmitter will be 150 MHz.

☐ When the tone is entered, the modulation circuitry affects the frequency of the transmitter output, lowering it to 149,997,500 hertz. This figure is 2,500 hertz lower than the non-input signal. If the tone changes to 3,000 hertz, then the transmitter output frequency is 149,997,000 hertz or 3,000 hertz lower than non-input normals.

☐ The transmitter output frequency or carrier will exactly follow the input frequencies fed to the microphone, so the complex make-up of the human voice can be duplicated by varying the frequency of the transmitter carrier.

Admittedly, this is a simplistic explanation of how the radiotelephone transmitter operates, but it does serve as a proper basis for understanding frequency modulation. Again, to modulate means to change, and the frequency of the carrier is changed in a manner which corresponds to the changing voice frequencies supplied at the microphone input.

Mobile telephone systems have been in use in the United States for nearly 35 years. The modern systems of today far outweigh the earlier devices and offer advantages in communication ease as well as in size, weight and portability. The early radiotelephones were, of course, tube-type. Today, almost all radiotelephones in use feature completely solid-state circuit design with integrated circuits used wherever possible instead of discrete devices. This compact packaging means that these units, which often are subject to vibration stress, large operation temperature fluctuations, and other road conditions, operate more reliably and are much less subject to physical damage and electronic failure.

Early radiotelephone operations necessitated the calling of a mobile operator. Basically, the user simply transmitted to a central operator station. Once in contact with the mobile operator, the various connections were made by hand, patching the mobile transmissions into the standard phone line. The voice from the person being called was patched from the phone line into a transmitter at the operator's station which broadcast this information back to the mobile receiver.

Today this principle of radiotelephone communications is very similar, but the mobile operator is sometimes done away with. In some systems, a dial tone is heard by the mobile station when a channel is clear, and a tone encoder (push-button pad) is punched with the proper number combination. All of this is done by the mobile caller. Electronic circuitry at the central office automatically handles the patching work. It can be seen that mobile radiotelephone operation is about as simple from a user's point of view as dialing a number from your home.

In these modern times, there are two types of radiotelephone systems. The *Mobile Telephone Service*, abbreviated MTS, is composed of mobile units which operate on one or more channels. The radiotelephone user must manually select the intended channel of communication by means of a rotary channel-select switch on the panel of his unit. This switch is turned until a clear channel is found. Remember, radiotelephone operation is a shared service. It's like one big party line. Owners of single-channel units will often have to wait in order to obtain a free channel as other users will be operating on the same frequency from time to time.

Multi-channel units offer their owners the convenience of searching through several channels until a clear frequency is found. These channels are selected by internal crystals or frequency synthesizers which also control the transmitting frequency. The transmit and receive frequencies are usually separated by a fixed frequency range to provide two-way communications. Radiotelephone communications necessitate push-to-talk operations where the user must press a microphone switch to transmit and release it to receive. This can be a little confusing when talking to persons not familiar with this type of communication procedure. It usually means that each person will have to say "over" when they are ready for the other party to speak. It is also impossible to interrupt the person talking using this communication system, because he is not receiving while he is transmitting.

The second, basic radiotelephone system is abbreviated IMTS and stands for *Improved Mobile Telephone System*. This system offers multi-channel mobile units which automatically scan the many available channels and select one which is clear. Touch-tone or rotary dialing is also standard with this latter system. The user never needs to manually select a clear channel. This is done automatically by the internal scanning circuitry which monitors each channel. When a channel is in use, the circuitry skips over it and seeks another. When the clear channel is located, the receive circuitry locks onto it, and the user may dial the desired number. This scanning feature usually begins when the handset is lifted from the master unit. This action releases a switch which sets the scanning circuit into operation.

INSTALLATION

Figure 11-1 depicts a typical radiotelephone installation in an automobile. While the actual location of the communications circuitry and the control unit will vary from vehicle to vehicle, this is the most popular installation practice and has been carried over from other mobile radio installations used for many, many years.

The transmitter/receiver is often housed in a discrete unit. Trunk-mounting of this piece of equipment is the rule rather than the exception. Variations will be found when rear-engine automobiles or trucks are outfitted for radiotelephone service. Some of the newer units may be mounted completely below the dash or beneath the driver's seat. One necessary requirement for mounting is adequate ventilation. The output transistors used when the unit is transmitting get rid of component package heat through a

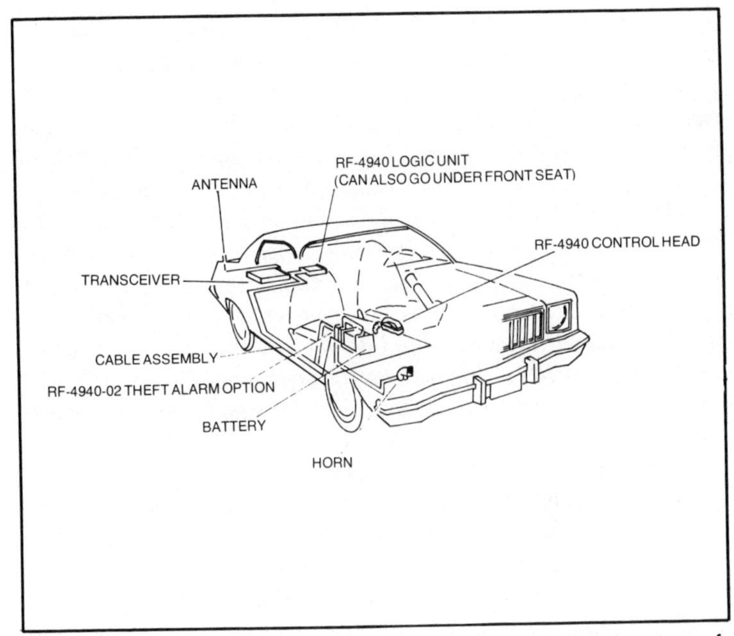

Fig. 11-1. Typical radiotelephone installation in an automobile (courtesy of Harris Corp. RF Communciations Division).

large, external heat sink. This is a heavy finned metal casing which is mounted to the outside of the transmitter/receiver case. The heat from the transistors is conducted to the sink which transfers this build-up to the air. Should air circulation be inadequate, enough heat cannot be exchanged, and the unit will shut-down or worse, the transistors may become overheated and be destroyed. Most manufacturers suggest horizontal mounting of the transmitter/receiver package, but vertical mounting is often used where space is at a premium. Trunk-mounted units should be placed in an area removed from the heating effects which can be present directly over the tailpipe or muffler. Modern cars with catalytic converters must be closely examined. These converters produce a tremendous amount of heat, and mobile units should not be mounted in the trunk area directly over them. Normal installation calls for the unit to be mounted on the right (or passenger side) of the trunk.

In order to control the transmitter/receiver circuitry, a control head is placed on the front dash within reach of the driver. This is connected by means of a multi-conductor cable which must be routed through the back seat area and up under the dash. Again, the passenger side of the automobile is normally chosen for this cable path. Access through the trunk is found by removing the back seat and pulling the cable through a small channel. The cable is then run beneath the floor mats or carpet through to the front seat and exits the front mat on the passenger side beneath the dash. Here, it can be connected to the control head which is mounted nearby.

Power for the radiotelephone is obtained from the automotive electric system (the battery and alternator). Cables of adequate size to carry the current are run from the battery terminals to the rear compartment where they are connected to the transmitter/receiver unit. The other cable (between the unit and the control head) supplies power for the control head mounted on the dash. This current demand is much lower and very small conductors are used.

In order to transmit and receive efficiently, an external antenna which is designed for the operating frequencies is mounted on the roof or trunk lid of the automobile. A connecting cable attaches between the antenna base input and the output of the transmitter or input to the receiver. During transmit operations from the mobile unit, the antenna is automatically switched to the transmitter output. When the push-to-talk switch is released, the transmitter/receiver unit switches to receive, and the antenna is internally connected to the receiver input.

CALLING FROM A MOBILE UNIT

As was previously stated, the operation of a modern radiotelephone system is very similar to calling from your home. When you lift the handset from its cradle to make a call from your automobile, the transmitter sends a short, seeking tone which is received at the central office. If a line is available, the central office transmits back to you a tone which indicates the equipment is ready to receive your dialed or punched-in number sequence. This is accomplished through a rather complicated process of short tone sequences which identify your radiotelephone for billing purposes. These sequences occur in tiny fractions of a second and are sent and received automatically by programming in the radio telephone mobile and central office equipment.

When the dial tone is received at the mobile end, the desired phone number is dialed. This procedure transmits other short tones which indicate the number you desire to reach. This operation is similar in many ways to a pure tone dialing system using a push-button telephone in your home or office.

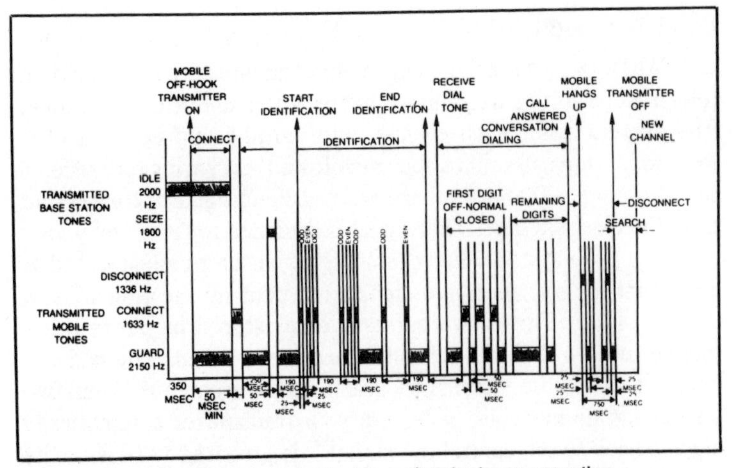

Fig. 11-2. Sample tone sequence chart for radiotelephone operation.

Once the central mobile office has detected and read the dialing sequence, the desired number is called over a standard phone line. When the called party answers, a normal conversation ensues. Again, with push-to-talk operation, it is not possible to hear the party you are in communication with while you are talking. Likewise, they cannot hear you until you interrupt them by pressing your push-to-talk switch. You can interrupt them, but they can't interrupt you. In other words, the mobile caller controls the phone conversation. When the push-to-talk button is pressed in the automobile, a signal is automatically sent to the central receiver which patches your transmission into the called party's instrument.

When the call you initiated is finished, the handset is placed back in its cradle. This transmits a tone code which indicates disconnect to the central office. The called party's line is freed and the communication has ended. A new call is initiated by lifting the handset again after disconnection has occurred (a matter of less than a second), and the entire process is repeated again for the new call.

This has been a very simplistic explanation of, basically, how a mobile radiotelephone call is initiated. The switching, tone and sequencing processes are highly complex but are handled automatically between the mobile and central transmit/receive stations by the electronic circuits of both systems. Figure 11-2 shows a chart of the actual, complex tone sequence. A step-by-step calling procedure is discussed later in this chapter.

CALLS TO A MOBILE UNIT

When a party calls your mobile number from a standard telephone, a similar sequence is set up at the mobile central office. The system scans each channel until an idle or free channel is located. A signal is sent to your mobile unit, preparing it to receive the incoming call. Shortly thereafter, a calling signal or tone is sent from the central office. This signal is designed to "ring" only your unit by a pre-set code. For example, if your unit is designated as unit number five, then five signals are sent by the transmitting central office. These five bursts are registered by every other mobile unit on the same channels and are counted. The count is compared with the internal programming of each unit. When five pulses and no more are received, your transmitter automatically signals back to the central office that it is prepared to receive the call. Upon reception of this signal, the central station sends a series of tones which causes the call indicator on your mobile control head to ring or buzz. You would then pickup the handset and begin the conversation which is handled in the same manner as when initiating the call.

When the call is completed, you hang up the handset, and a disconnect signal is sent to the central office. The central transmitter/receiver is returned to its standby position, awaiting another incoming call.

THE HARRIS MOBILE TELEPHONE SYSTEMS

Harris Communications and Information Handling has long been a trusted name in electronics. They are very well known in the commercial broadcasting industry as makers of AM, FM, and short-wave communications systems. It is not surprising that they offer a very sophisticated line of radiotelephone apparatus.

Figure 11-3 shows the Alpha 2000 Series Mobile Telephone and Control Unit. On the bottom is the transmitter/receiver while the control head and handset are on top. Harris offers a choice of three mobile systems which feature high performance solid-state components for excellent reliability and low operational cost. Each system can be equipped with a wide variety of options and accessories to function as private radio networks, or public radiotelephone networks complete with customer billing. They are designed for either simplex (same channel transmit and receive) or duplex (split channel transmit and receive) operations in the VHF or UHF frequency bands. The VHF band lies in the 150 MHz. portion of the frequency spectrum, while UHF operations occur around 450 MHz. Each Harris system has the capacity to incorporate additional channels to serve many mobile telephone subscribers.

Harris makes not only the mobile units but the central switching equipment as well. They point out that mobile systems are limited in their capacities to deliver a large number of calls at the same time. Each mobile channel, however, can effectively service between 30 and 70 radio units.

When care is taken regarding base and mobile station installations (especially antenna height and placement), Harris advertises an effective communications range of between 15 and 30 miles. Again, terrain will play an important factor in actual communications range. In the Midwest where land features are flat, longer ranges can be expected over what might be had in the mountainous East.

The Alpha 2000 System pictured is frequency-controlled through two electronic synthesizers. This provides a great deal of system flexibility in channel assignment without compromising performance. The synthesizing circuits take the place of separate

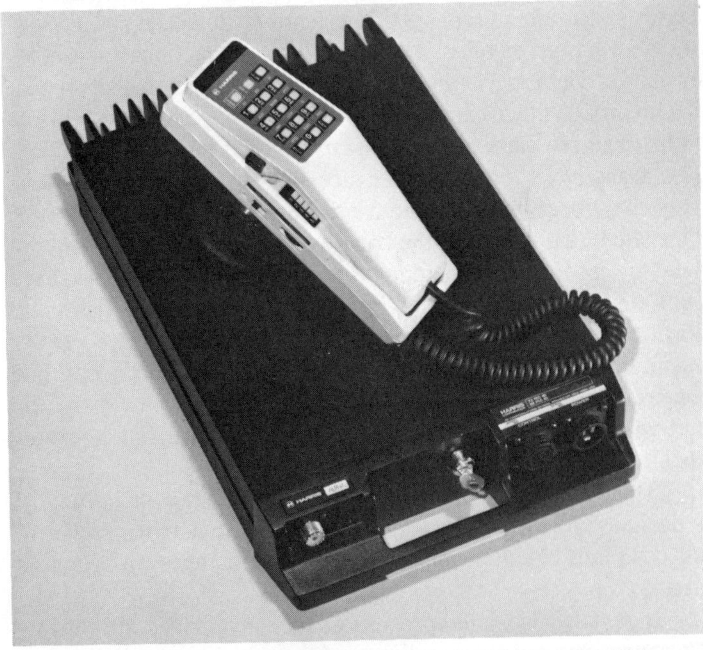

Fig. 11-3. Alpha 2000 series mobile telephone and control unit (courtesy of Harris Corp. RF Communciations Division).

crystals for each channel of operation. With two synthesizers, transmit and receive frequencies are controlled independently, allowing operation in any channel spacing plan. All common channel spacings are easily accommodated even at 12.5 kHz. Future channel spacing plans can also be incorporated for expansion purposes.

Coverage frequency is 450 to 512 MHz in the UHF spectrum and 150 to 174 MHz in the VHF band. Power output is 50 watts on VHF and 30 watts on UHF, and the unit is rated for continuous duty. This power output is rather high for the frequencies of operation and provides greater talk-back range for users who live in rural areas. Power output is adjustable or the units can be ordered with fixed output levels in areas which require output power to be held to a certain limit.

Figure 11-4 shows a top view of the internal circuitry of the transmitter/receiver. Modular construction is used throughout, which facilitates maintenance and repair operations. Circuit sub-assemblies can be removed quickly for ease of repair. Also, replacement sub-assemblies can be quickly snapped into place for a minimum of equipment down-time. Later, the malfunctioning circuit board can be repaired. There are no soldered connections between sub-assemblies. All inter-circuit connections are handled by plug-in contacts and cables. The design is so simple that the unit can be broken down completely with a phillips screwdriver. As an added feature, there is a self-test mode built into the equipment. The transmitter and receiver are set up for simplex (same frequency) operation, and the control head generates a test tone. This allows for checks of the basic systems operation without the need for external test equipment. Some of the modules even have light-emitting diode indicators which glow when a particular portion of the circuit becomes inoperational. This allows a faulty module to be immediately identified by visual inspection and repaired or replaced. Figure 11-5 shows the bottom view of the unit with the cover removed. The module construction is plainly visible along with the snap-on connectors.

The control head of the Alpha 2000 is a rather complex unit. It is designed to work in conjunction with the transmitter/receiver and combines circuit sophistication and convenience with compact size.

It contains dual microprocessors for convenience and simplicity approaching that of a home telephone instrument. An alphanumeric display gives a visual display of the called number as

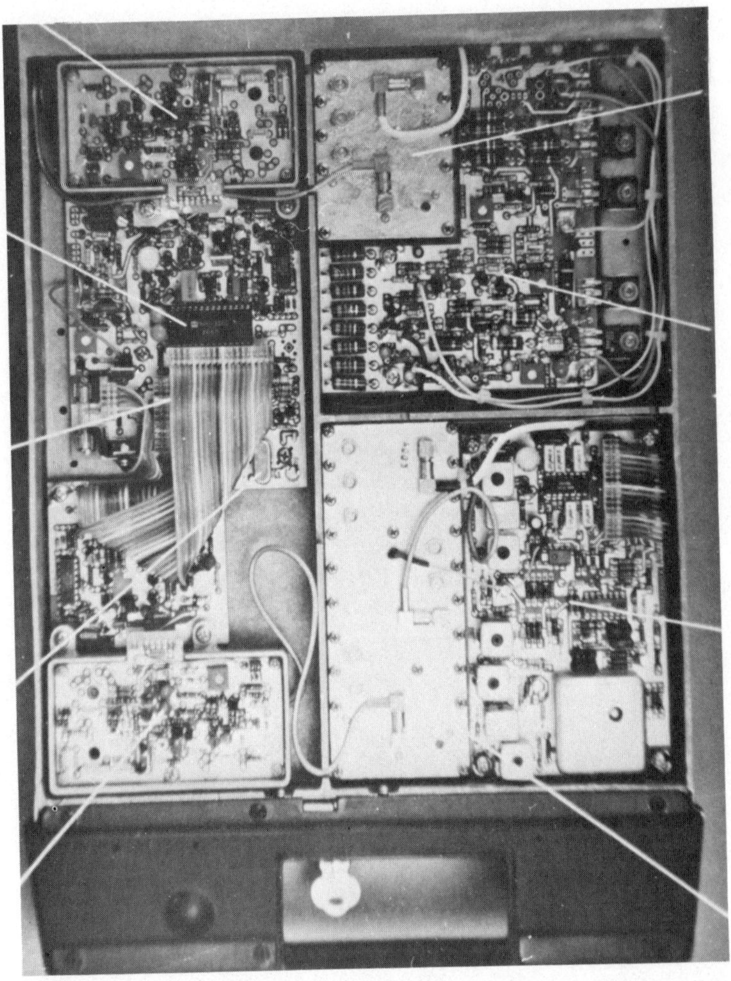

Fig. 11-4 Top view of circuit boards used in radiotelephone transmitter/receiver (courtesy of Harris Corp. RF Communciations Division).

well as unit status and mode. A speed-dialing circuit allows for up to ten numbers to be stored in memory and recalled with the push of a button. One problem with radiotelephone operation in crowded areas is obtaining a clear channel when you need one. This unit offers a circuit which has been named "call-queing." If all channels are active when you desire to place a call, this circuit is activated. It will automatically scan the busy channels and lock onto the first one which becomes clear. When the clear channel is indicated, you may then pick up the handset and initiate a call. This saves the user from

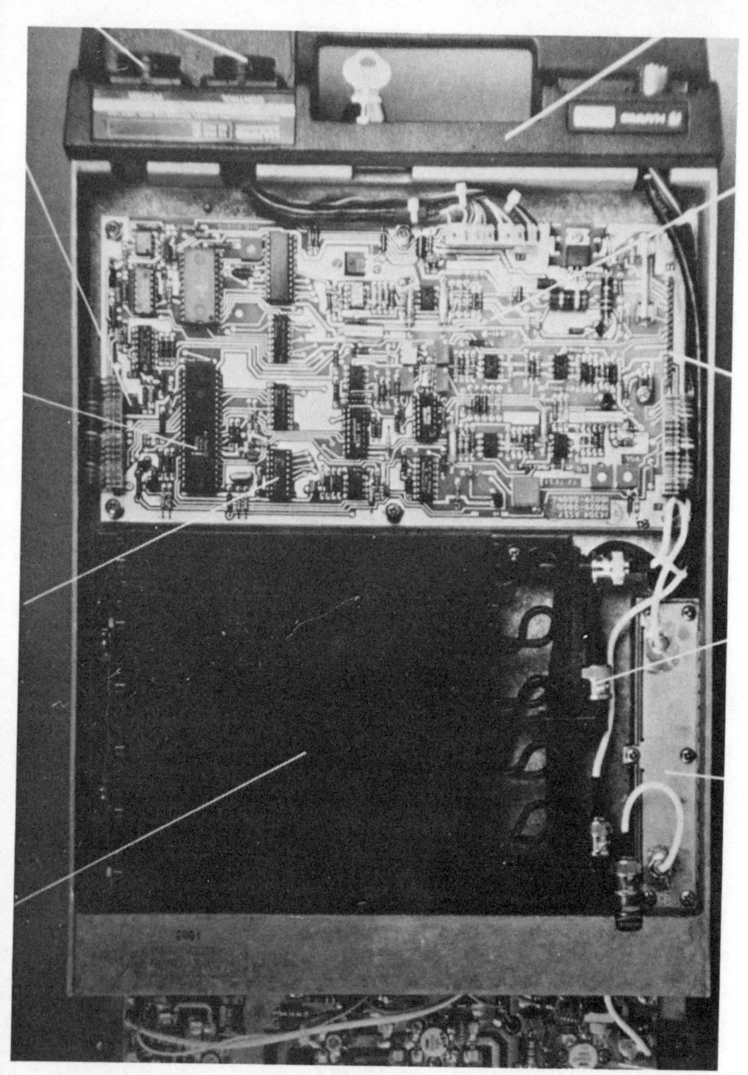

Fig. 11-5. Bottom view of radiotelephone unit with cover removed (courtesy of Harris Corp. RF Communications Division).

the inconvenience of constantly having to check to see when the channel has cleared. It also allows him to keep his mind on driving the automobile which adds to the road safety factor. Additionally, there is an electronic lock with a three-digit input code which prevents unauthorized calls from being made.

The packaging of the control head is rather unique, accentuating compactness. Even the smallest automobile should easily accommodate this unit which mounts in an incredibly small space. Horizontal or vertical mounting configurations may be used or the control head may simply be attached to the transmission hump in the front passenger section. The packaging is ultra-modern and should fit in well with the space-age console decor of modern automobiles.

Many other radiotelephone units are offered by Harris. Figure 11-6 shows their Dial-In Mobile Control Unit which takes on the appearance of a slightly modified multi-line desk phone. This conventional styling arrangement is carried over to the IMTS Mobile Control Unit (Fig. 11-7), which is far less complex regarding user functions. The IMTS unit is more complex electronically, however, with automatic scanning features actuated by lifting the handset. Figure 11-8 shows these units with a VHF duplex transceiver and with the UHF transceiver in Fig. 11-9.

It should be pointed out again that Harris makes complete systems for business purposes. Here, all of the central switching equipment may be contained at a company office which feeds a multi-line standard phone system. The latter is operated by the phone company while the company handles the former with their own equipment. Figure 11-10 shows a styled fixed station being used at company headquarters to talk with one of the mobile units. Called the Credenza, the wooden cabinet looks like a piece of furrniture but contains all base station components. The access

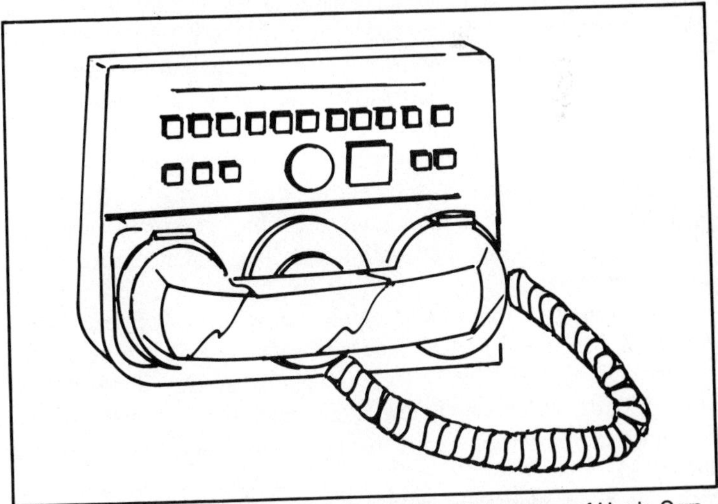

Fig. 11-6. Dial-in mobile control radiotelephone unit (courtesy of Harris Corp. RF Communications Division).

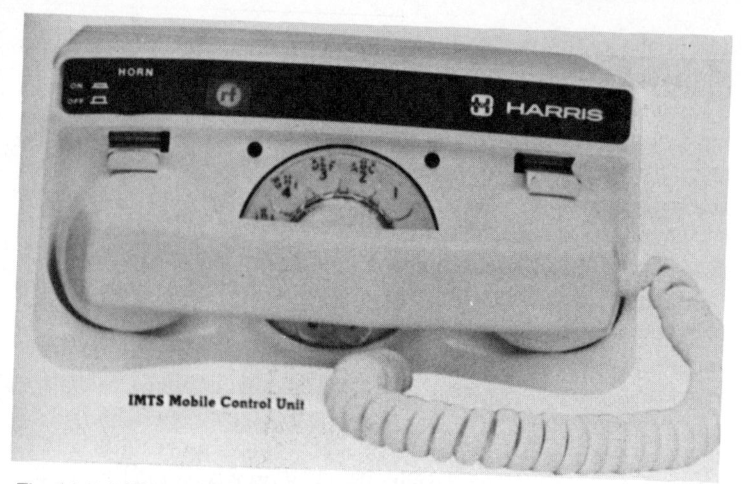

Fig. 11-7. IMTS mobile control unit for operatorless communications (courtesy of Harris Corp. RF Communciations Division).

instrument to this equipment is the RF-4912 Slim Line control unit which resembles a standard slim-line telephone instrument. This unit is the same type which may be used in mobile installations as a control head.

THE IMPROVED TELEPHONE SERVICE (IMTS)

The Improved Mobile Telephone Service is a radiotelephone system which is designed to connect mobile units with standard dial telephone exchanges. It usually provides fully automatic operation and avoids the necessity of operator patching which is prevalent much of the time in the previous system discussed. It provides fully automatic operation and means that the mobile radiotelephone user has almost the same convenience as the fixed subscriber. Each mobile unit using IMTS is assigned a conventional telephone number in the central office and is given the same treatment as the land telephone regarding automatic number identification, message accounting (for long distance numbers), and dialing privileges.

According to Harris Communications, the IMTS technique was first developed in 1961 and since about 1964 has been the standard automatic mobile radiotelephone system used in the United States. It has also found use in many other countries and is currently the most universally used system of its type throughout the world.

The main advantage of IMTS over other systems is that it increases the effectiveness of the mobile radiotelephone service by providing completely automatic dialing facilities. This assures better utilization of channel capacity, eliminates the necessity of operator-assisted calling and gives to the mobile subscriber the same facilities and convenience found in the home or office telephone hookup. Figure 11-11 shows the features of IMTS and a brief explanation of each.

Figure 11-12 depicts a basic IMTS network. Each element of this complex system is shown along with its interrelationship with other component systems. The terminal unit performs the necessary control, signaling, and switching functions to interface the local telephone exchange with the radio base station equipment. In addition to its connection to the dial office, the control terminal may also have a trunk connection to a switchboard for operator assistance on certain types of calls.

The radio base station includes the transmitter, receiver, antenna units, and duplexer along with any other necessary control equipment for connection with the terminal unit over wire line or radio carrier facilities.

Depending on the average terrain, antenna height and required operating range, the required output of a transmitter unit may vary from 20 to 250 watts. Where several channels are to be utilized, it may become necessary to add a more complex duplexer/circulator/combiner network to reduce antenna requirements and at the same time protect the system from intermodulation interference which can occur when the transmitting antenna is very close to or the same as the receiving antenna and when the transmit and receive frequency are relatively close to

Fig. 11-8. IMTS unit with a VHF Duplex transceiver (courtesy of Harris Corp. RF Communications Division).

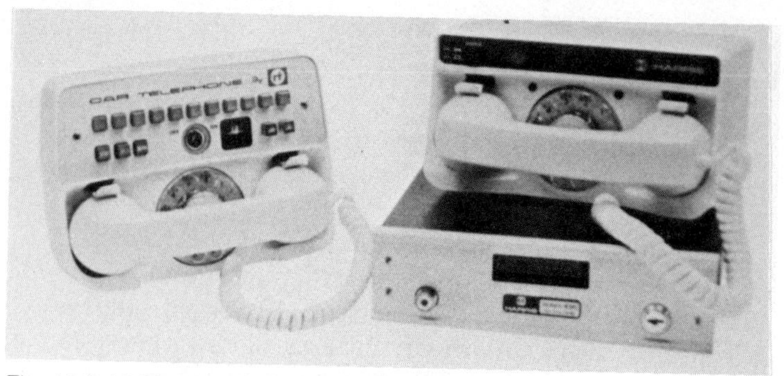

Fig. 11-9. UHF transceiver and IMTS mobile control unit (courtesy of Harris Corp. RF Communications Division).

each other. In some cases, it may be necessary to locate receiver units at a different site than the transmitting units.

The mobile radiotelephone includes a small integrated control/logic unit which can be located within convenient reach of the vehicle driver. There is also a compact receiver/transmitter package which can be located at any out-of-the-way place. The mobile installation also includes the antenna and duplexer(s). Transmitter output power and antenna type are chosen to assure effective communications throughout the anticipated or desired coverage area. As was previously mentioned, the local terrain will determine these last two factors.

The most basic feature of the IMTS system is the automatic selection and marking of a radio channel for each phone call. Whenever channels are available, the terminal unit selects one and activates the associated base station transmitter by modulating an idle marking tone. All idle mobile units scan their available channels until an idle tone is detected. The mobile unit then locks

Fig. 11-10. Fixed, base station used to communicate with mobile radiotelephone units (courtesy of Harris Corp. RF Communications Division).

DIAL-IN/DIAL OUT SERVICE. Calls to and from mobile units are dialed in the same manner as normal telephone calls.
AUTOMATIC CHANNEL SELECTION eliminates the need for a mobile user to search for a radio channel on which to make a call.
FULL CHANNEL PRIVACY. Only a mobile unit participating in a conversation has access to a working channel.
FULL DUPLEX OPERATION permits simultaneous two-way conservations without the inconvenience of "push-to-talk" operation.
BUSY CHANNEL INDICATORS. When all channels are occupied, the mobile unit's "busy lamp" will illuminate and an engaged tone will be heard in the mobile handset while off-hook.
FULL LAND CUSTOMER SIGNALING. Land customers are provided with all normal ring-back and engaged signals as well as an "out-of-service" voice recording if a mobile unit does not automatically acknowledge receipt-of-a-call indication.

Fig. 11-11. Operational chart of IMTS features (courtesy of Harris Corp. RF Communications Division).

onto the marked channel. The next call in either direction, incoming or outgoing, is then established on this channel with the participating mobile unit locked on. In completing the call, the terminal unit moves the idle marker to another free channel, again locking idle mobile units to a common channel in readiness for the next call. This eliminates the necessity of monitoring the available channels through manual switching in order to find a clear frequency, or to maintain watch on a separate channel used for calling purposes only.

Should all of the available channels be in use, a busy lamp will be illuminated by the circuits when the telephone handset is removed from its latch or cradle. Additionally, the Harris RF Communications IMTS unit provides the conventional "busy" signal at the handset earpiece when all channels are actively engaged. When a person calls a mobile unit from a fixed installation, he or she may hear the all-trunks-busy tone or an optional voice recording when all of the channels are engaged.

To maintain system order, each mobile unit is assigned a unique seven-digit identification and selective calling number. Mobile units will then respond only upon receipt of their assigned numbers. During mobile-originated calls, the unit transmits this same number back to the terminal unit as identification, so the call will be routed to the proper line circuit for completion. When the full capacity of the seven digits is not required, the number of digits can be reduced accordingly. This results in a savings of the time required for selective calling and identification.

The Harris IMTS system uses several tone-and-code signalling formats to provide the necessary calling, identification, command and control functions between the mobile units and the IMTS terminal. Figure 11-13 gives these tone codes and their initiating functions.

IMTS STEP-BY-STEP SYSTEM OPERATIONS

Using the Harris-RF Communications IMTS system as an example, it is possible to describe the step-by-step processes which take place. Each mobile radio scans its complement of radio channels until it detects the channel designated by the IMTS terminal as the idle channel.

When a wired subscriber dials an exchange number corresponding to the mobile unit, the exchange supplies a ringing signal on the corresponding mobile subscriber line into the IMTS terminal. The following sequence of events occurs:

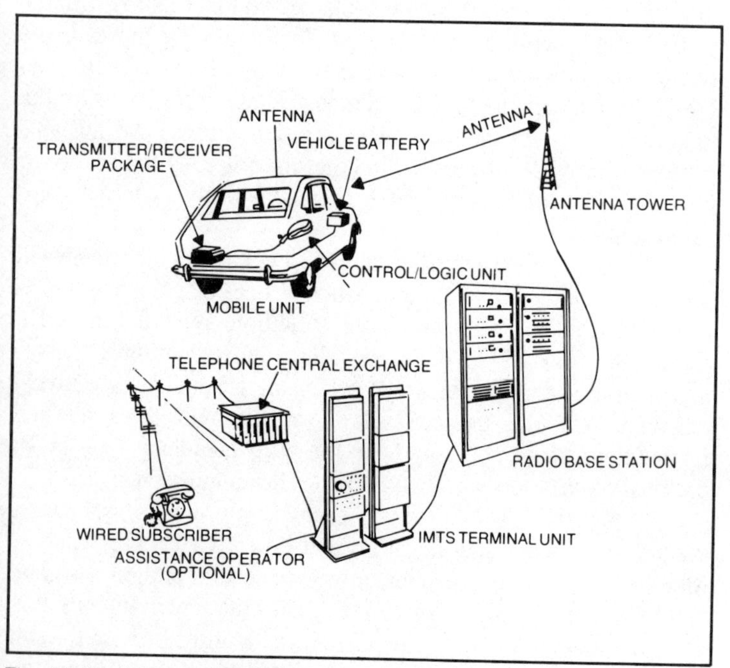

Fig. 11-12. A typical IMTS systems network (courtesy of Harris Corp. RF Communciations Division).

Base To Mobile	
Idle Channel Marker	2000 Hz
Channel Seizure	1800 Hz
Mobile Calling	1800/2000 Hz FSK 10 pps
Mobile Ringing	1800/2000 Hz FSK 20 pps
Mobile To Base	
Guard Tone	2150 Hz
Connect	1633 Hz
Automatic Number Identification (ANI)	on/off pulses of 1633 Hz with 2150 Hz even parity at 20 pps
Dialing Impulses	1633/2150 Hz FSK 10 pps
Disconnect	1336/2150 Hz FSK 20 pps

(FSK—Frequency Shift Keying; pps—pulses per second)

Fig. 11-13. Table of tone codes and initiating functions for the Harris IMTS system (courtesy of Harris Corp. RF Communications Division).

☐ The idle tone (2000Hz) (which has been marking the channel designated for the next call and holding idle mobile units to that channel) is dropped.

☐ Seizure of the channel is marked by the seize tone (1800 Hz). A mobile unit which attempts a call on this channel will receive a busy light and will be blocked from making a call on this channel.

The seize tone (1800 Hz) is transmitted for approximately 110 ms. This period is used to give the IMTS terminal time to receive and recognize a guard tone (2150 Hz) if a mobile attempts to initiate a call just before the seize tone was applied. All mobiles are locked to the channel by the seize tone since at this point the call could be for any one of them.

☐ After 110 ms of seize tone, the IMTS terminal selectively calls the requested mobile unit using the mobile calling sequence 1800/2000 FSK at 10 pps with 225 ms interdigit time. This selective call is transmitted once and is received by all the idle mobile units in service.

As the selective call is transmitted, each unit except the one being called will unlock from the channel as it detects a digit that does not match its precoded identification. Unlocked mobile units automatically search for the newly designated idle channel.

☐ Upon receipt of a correctly coded signal, the called mobile unit automatically acknowledges receipt of the call with a guard tone (2150 Hz) for 750 ms.

If the called mobile was not in range of the base station or was not in service, it will not acknowledge the call. If the IMTS terminal receives no such acknowledgment within 3 seconds, it abandons the call and the channel is free to accept a different call. The IMTS terminal can be supplied with facilities to activate a voice recording to announce that the called party cannot be reached.

☐ After the IMTS terminal receives the acknowledgment signal from the mobile, it knows that the called mobile is in service so that it then transmits the ringing signal to the mobile. The ringing signal consists of 1800/2000 Hz FSK at 20 pps for 2 seconds with 4 seconds between rings. The ringing continues until the mobile answers or until the ringing signal is transmitted for 45 seconds (7 rings). Receipt of the ringing signal in the mobile activates both an alert tone and a latching "call lamp" that remains illuminated until reset by removing the handset from its cradle. If the "H" button on the mobile's control/logic unit is activated, the vehicle horn or another annunciator will sound with each ringing signal.

☐ If the mobile operator answers the call by removing his handset from its cradle in the control/logic unit (also referred to as being off-hook), the mobile unit transmits the connect tone called the "answer burst" (1633 Hz) for 400 ms. This tells the IMTS terminal that the mobile has answered and instructs it to connect to the incoming call. When the mobile unit's handset comes off-hook the handset is muted for 400 ms so that the tone burst will not be heard. In addition, the unit is inhibited from searching for an idle tone. The mobile unit continues to transmit until its handset is replaced.

☐ At the end of the conversation when the handset is replaced the mobile unit transmits a disconnect signal, 1336/2150 Hz FSK for 750 ms. The IMTS terminal detects this signal and breaks the connection to the central telephone office. After the mobile unit completes transmission of the disconnect signal, it resumes scanning for the idle-channel marker.

If the mobile user fails to replace the handset in its holder after the call has been completed, the terminal unit will automatically disconnect the call after a predetermined time period.

LAND-TO-MOBILE CALL (MOBILE IN "ROAM" MODE)

When a wired subscriber wishes to contact a mobile unit that is outside its "home" area, but is in an area covered by a compatible IMTS system, the sequence of events is similar with the following two exceptions:

☐ Since the mobile unit's exchange number will not be valid in the area's IMTS terminal, the land subscriber must first contact the mobile operator of the area in which the mobile unit is traveling.

The operator must manually seize the designated idle channel (the idle tone is replaced by the seize tone) and selectively call the mobile unit.

☐ At the completion of the call the mobile subscriber will replace his handset, and his unit will emit the normal disconnect tone.

The operator will be alerted that the call has been completed and will disconnect the wired connection. At that time a record of the length of the conversation can be made for billing purposes.

Each unoccupied mobile unit in the system is locked to the radio channel marked by the IMTS terminal as the designated idle channel.

When the mobile subscriber takes his handset off-hook, the following sequence of events occurs:

☐ The mobile unit transmits a guard tone (2150 Hz) for 350 ms.

If, after going off-hook the mobile fails to continue to receive an idle tone from the IMTS terminal for 350 ms (which would indicate that the channel was seized by another mobile), the call attempt is blocked.

☐ After 350 ms of guard tone (2150 Hz) under the condition above, 50 ms of connect tone (1633 Hz) is transmitted by the mobile unit. The mobile unit is inhibited from attempting to seek another marked idle channel.

☐ Upon receipt of the connect tone from the mobile, the IMTS terminal drops the idle channel marker (2000 Hz).

After about 10 ms without the idle channel marker, the mobiles are blocked from making a call. After about 175 ms without the idle tone all mobiles except the one attempting the call will start to search for the newly designated idle channel.

If the IMTS terminal does not respond to the calling unit by dropping its idle channel marker (indicating that it did not receive the connect tone from the mobile), the call attempt is blocked.

☐ Approximately 250 ms after the idle tone is dropped the IMTS terminal transmits the seize tone (1800 Hz). The seize tone signals the mobile unit to prepare to transmit its Automatic Number Identification (ANI).

☐ Removal of the seize tone (1800 Hz) is the command for the mobile to transmit its ANI. It takes approximately 200 ms for the mobile unit to recognize this command. The ANI consists of an FSK format between no tone and the connect tone (1633 Hz). A parity tone or guard frequency (2150 Hz) replaces the "no tone" after each cumulatively totaled even pulse. ANI is transmitted at 20 pps with an interdigit time of approximately 200 ms.

The ANI consists of a seven-digit number (or fewer in some applications) used to identify the mobile unit to the IMTS terminal.

☐ The IMTS terminal connects the corresponding mobile subscriber's line to the central exchange upon receipt of a valid mobile identification. This results in the return of the normal exchange dial tone to the mobile unit. Once the mobile user hears the tone he is alerted that he can proceed with dialing.

☐ When the mobile user dials, the digits are outpulsed to the IMTS terminal using FSK format between the connect tone (1633 Hz) and the guard tone (2150 Hz) at 10 pps. This information is translated to standard dial pulsing at the IMTS terminal and is applied to the central exchange which completes the connection.

The mobile unit continues to transmit until its handset is replaced.

☐ When the mobile user completes the call and returns his handset to its cradle (goes back on-hook), the control/logic unit generates an FSK tone (1336 Hz) and the guard tone (2150 Hz) at 20 pps for about 750 ms. The IMTS terminal recognizes this signal and disconnects the corresponding mobile subscriber line to the central exchange.

☐ At the end of the 750-ms disconnect signal, the mobile unlocks from the channel and begins to search for the currently designated idle channel.

If the mobile user fails to replace the handset in its holder after the call has been completed, the terminal unit will automatically disconnect the call after a predetermined time period.

Figure 11-14 shows the complete sequence of events for an IMTS Land-To-Mobile call. The reader may easily trace the events which take place in the completion of such a call. Figure 11-14 shows the sequence of events for an IMTS Mobile-To-Land call. Systems operation for this latter mode is similar to the former and is described below.

MOBILE-TO-LAND CALL (MOBILE IN "ROAM" MODE)

When a mobile unit is outside its normal "home" area and wishes to initiate a call, the sequence of events is similar to the procedure for "home" area calls up to the completion of the ANI.

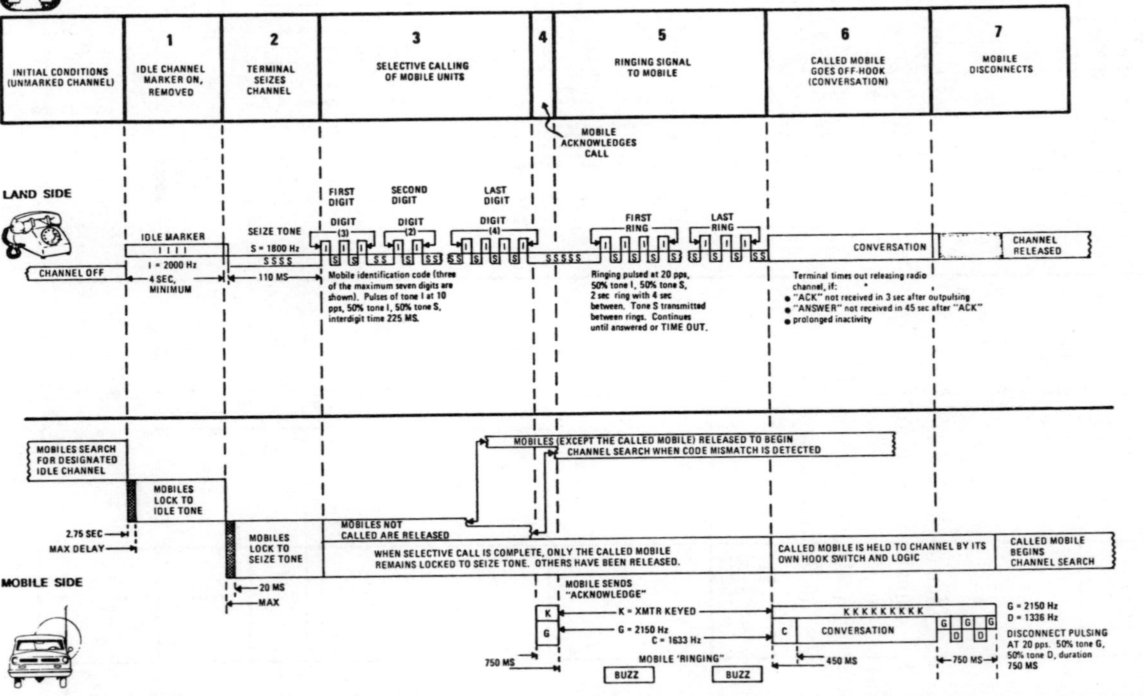

Fig.11-14. Chart of the complete sequence of events for an IMTS land-to-mobile call (courtesy of Harris Corp. RF Communciations Division).

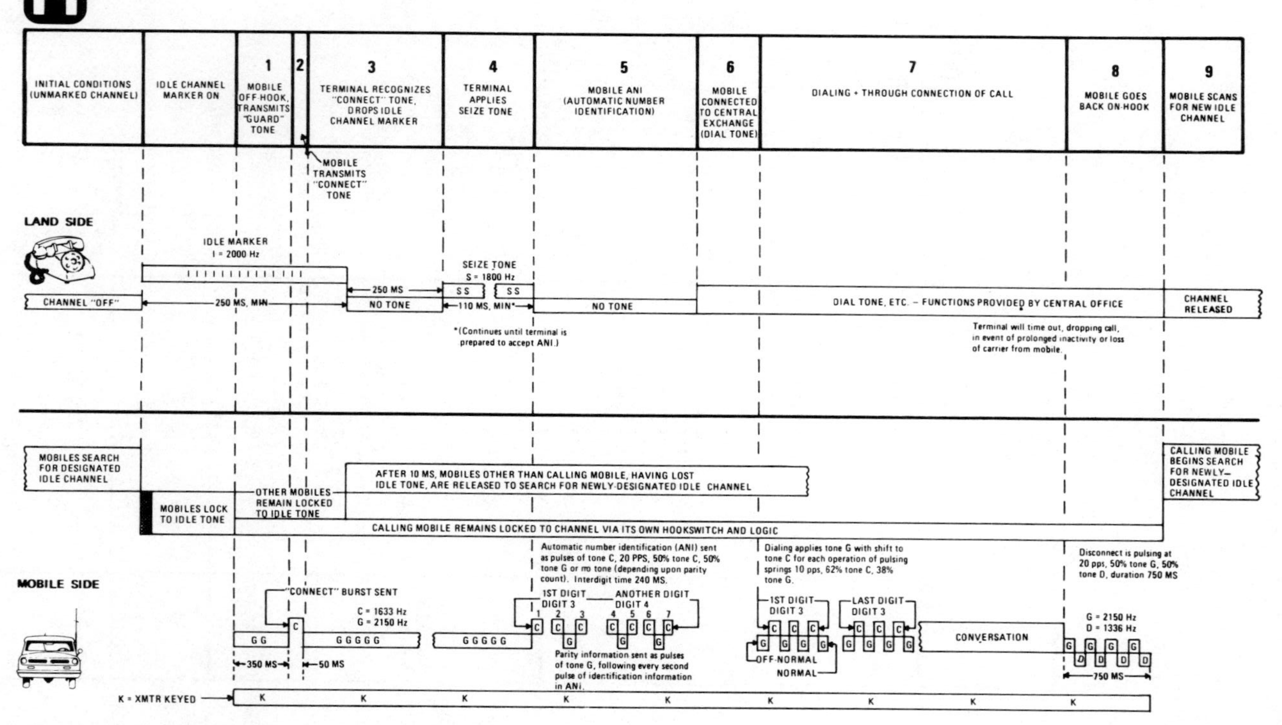

Fig. 11-15. Sequence of events for an IMTS mobile-to-land communciation (courtesy of Harris Corp. RF Communciations Division).

□ If the IMTS terminal does not have a provision for routing roaming operators to a line automatically, the call is routed to the operator for supervision. In this case, the operator will dial the requested number to complete the call.

□ After the mobile unit completes the call and transmits the disconnect signal, the operator is alerted and will disconnect the call and make any necessary records of call length if this is not automatically done in the terminal.

Calls between mobile units in the IMTS system can be handled much the same as a normal call from a mobile subscriber to a wired subscriber. The call is routed from the calling mobile through the IMTS terminal to the central exchange and then back through the IMTS terminal by way of a separate radio channel to the called mobile.

A NEW MOBILE RADIO SERVICE

Recently the FCC has approved a new type of mobile radio service. This new service involves the use of mobile radio "cells." These cells each have their own radio transmitter and a telephone call can be switched from cell to cell and frequency to frequency automatically. The cells would be located adjacent to each other in such a way as to make communication via mobile telephone much easier and available to many more people than it is at present.

SUMMARY

It can be seen that the radiotelephone system of communication is highly complex in design and function. This however, leads to simplistic ease of operation for the user, as all of these complex functions are handled by the internal microprocessor circuits.

It must be remembered that the radiotelephone service is a separate entity and the units installed in the automobiles can be thought of as discrete telephone instruments.

Not all areas offer radiotelephone service to local subscribers. It may be necessary to rent this service from an adjoining city if even this is availalbe. Radiotelephone units are more commonly found in the medium to large cities and in adjoining towns lying within a 20 to 30 mile radius of the central system. Check with your local phone company if you are interested in this service. If they do not offer it locally, they can probably tell you where your closest radiotelephone service is and whether or not it is operational in your desired area of service.

Chapter 12
Homebuilt Telephone Projects

A thorough explanation of telephones, systems, devices, and operation has been provided to this point in the text. It is hoped that the information presented will be of assistance in deciding on which types of phones, phone services, and conveniences the reader may wish to take advantage of. While many of the accessories to the standard telephone instrument can be purchased at very reasonable prices from local outlets, the enterprising experimenter will probably opt to build rather than purchase some of this auxiliary equipment. This practice is highly encouraged by the author who does quite a bit of project design and development when time permits. Through the building and experimentation with electronic projects, the reader may become even more familiar with the principles which lie behind many of the commercially available devices now on the market.

This portion of the text is aimed at the home experimenter and builder. It is assumed that a basic knowledge in electronic kit building has already been gained. However, for the beginner, some basic information will be provided on general construction techniques and some of the component's characteristics which must be known to succeed with a finished project. A normal assortment of components are used for each project, with no one part being highly exotic or very hard to locate at your local electronics or hobby store. The builder will be working with transistors, integrated circuits, diodes, resistors, capacitors, and a few other standardized components.

Each project is presented with a brief explanation of what it is, what it does, and how it accomplishes its function. A schematic diagram and parts list will tell the builder just what he or she needs to complete the project. Often, a layout drawing will be provided to help with the location of parts on the printed circuit board material, which is used as a base for all of these projects. None of the designs incorporate high voltages which might be a safety hazard to inexperienced builders. Most projects are installed externally and do not involve the telephone line directly. This means that FCC licensing and telephone company approval will usually not be necessary. If there is any doubt, however, a call to your local company office should clear up any confusion. Whenever possible, it is good to maintain close relations with your phone company, especially if you are an experimenter. A few of the projects may need phone company approval in some areas. These have been duly noted in the text.

The author urges the builders to experiment with the circuits contained in this chapter. Most are not highly critical regarding many of their components. It may even be possible to improve upon these circuit projects through a bit of experimenting. It is suggested, however, that you do not connect these circuits directly to the phone line or in any other manner not outlined in the text. The aim of these projects is to learn about telephone devices by building working models which serve a useful purpose *and* to always stay within Federal law and the rules of the telephone company.

ELECTRONIC WIRING

When putting together any electronic project, certain skills and tools are necessary to complete the job in a manner which will provide correct and lasting operation. Many projects may be designed for installation on perforated circuit board material, which is available at a low cost from almost any hobby store. This type of construction allows for detailed assembly and provides a sturdy and compact base for all circuit components.

Figure 12-1 shows a piece of perforated circuit board. The wire leads from the components are simply pushed through a convenient hole in the circuit board. Wiring connections are made on the bottom of the board by twisting various wire leads together or connecting them by using single pieces of hookup wire and then soldering. This will also provide for a very neat appearance on the top of the board where the components are located and a sturdiness

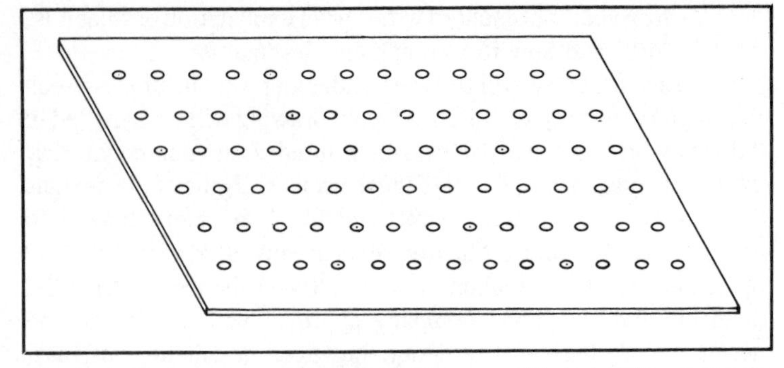

Fig. 12-1. Perforated circuit board provides versatility in building electronic projects.

which is not easily obtained with point-to-point wiring on contact strips.

Several ways of mounting components can be used, and the procedure followed will depend largely upon the amount of space available on the circuit board and also on the type of circuit being built. Figure 12-2 shows examples of vertical and horizontal mounting of various electronic components. Vertical mounting takes up less horizontal space and therefore requires a smaller section of circuit board. Horizontal mounting of the components will require more horizontal space but will not require as much vertical room and provides a flat finished circuit. This latter type of wiring is ideal for circuits which must be housed in a box or container. Vertical mounting will generally take up the least amount of space because the circuit will not be as spread out.

TOOLS

When building the projects outlined in this book, a normal assortment of shop tools will usually be all that is needed. Included in your complement should be needlenose pliers for bending the wire leads of the components and for wrapping them at their contact points. Wire cutters, sometimes known as diags, will also be needed to cut these leads to the correct lengths and for trimming after soldering has been completed. A pocket knife is also desirable and may be used to cut sections of perforated circuit board and for scraping insulation from the surface of painted or enamel-coated wire. A pocket knife will also be useful in clearing away blobs of solder which may accidentally drop in circuit board materials and other components. It is also good to have a wire stripper, electrical tape, and an assortment of Phillips and flathead screwdrivers of the small to miniature variety. Epoxy cement will come in handy for securing the bulkier components to the circuit board, preventing movement from vibration. Other tools which are useful but not absolutely necessary include a set of nutdrivers for installation of the circuit board in a metal case and a de-soldering tool for removal of incorrect solder connections. Alligator clips will be useful for protecting delicate components during the soldering operation. These should be purchased in a general assortment of many different sizes. A special type of insulation known as *heat-shrinkable tubing* can also be used. This material is fitted around the bare leads of components which are in danger of shorting to ground. This loose-fitting insulation is tightened to the conductor by applying heat from a match. Upon heating, the tubing shrinks and molds itself to the conductor.

SOLDERING PROCEDURES

Soldering is the most important part of assembly for any electronic project, regardless of the type of circuit or the components used. One of the largest manufacturers of electronic kits in the world has stated that nearly 90% of the failures involved in putting these kits together has been traced to faulty solder contacts. A hasty job of soldering electronic components will result in a circuit which does not perform dependably or one that may not function at all. Soldering is always a critical part of electronic circuit construction and must be done with care, strictly adhering to proper technique.

The soldering iron most desirable for assembling the telephone projects in this book is the *pencil* type, which has a power

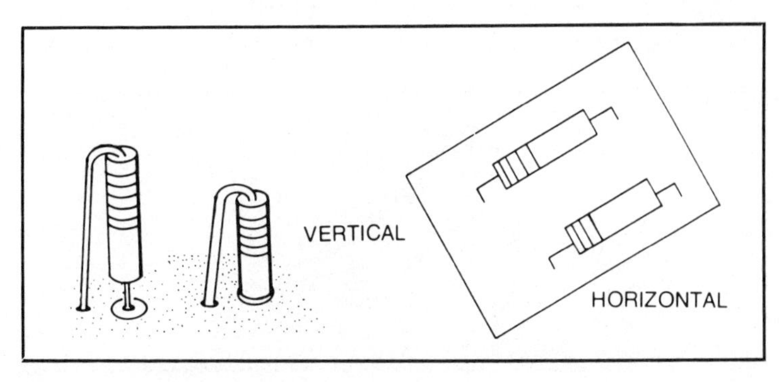

Fig. 12-2. Examples of vertical and horizontal mounting of electronic components.

rating of 25 to 30 watts. This provides adequate heat to get the job done but does not get so hot that the fragile solid-state components are destroyed. Soldering *guns* are very popular for certain types of electronic assembly, but most are rated at more than 75 watts. Soldering guns heat to temperatures which are much higher than required for the assembly of the smaller electronic projects in this chapter. Also, the soldering tips of these guns are overly large for many of the compact applications. Some manufacturers offer expensive soldering stations which include a pencil soldering iron, an insulated holder, and a control box which keeps the temperature of the iron constant at all times. This type of soldering equipment is shown in Fig. 12-3. While these devices are very convenient for electronic applications, they are not necessary, and a simple soldering pencil from a local hobby store may be purchased for less than $10.00 (Fig. 12-4). Make certain you follow the manufacturer's instructions when preparing the tip of a soldering pencil for first use. This is usually when the tinning procedure takes place and involves heating the iron and applying a small amount of solder to the tip. When the tip is covered with a very thin coat of solder, normal soldering functions may be undertaken.

Only one type of solder is suitable for use in the construction of electronic circuits in this book. This is *resin core* solder, which is sold by most electronic and hobby stores and is always identified as such. There is another type of solder which may be sold in hardware stores and plumbing outlets which has an *acid core*. The center of this solder contains an acid which is desirable for plumbing applications but which will ruin electronic circuits and components. The corresive acid core solder usually results in cold solder joints in electronic components leads, and the acid will gradually eat away at the delicate circuit conductors. A *core solder joint* is a connection which has not been made properly and results

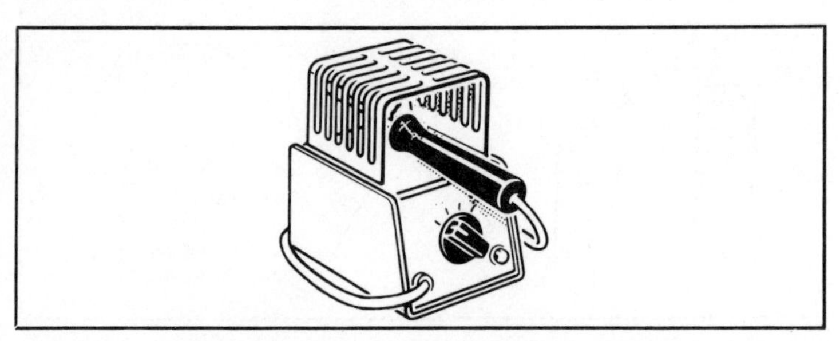

Fig. 12-3. A soldering station consisting of temperature regulated soldering iron, insulated holder, and AC power pack (courtesy of Heathkit).

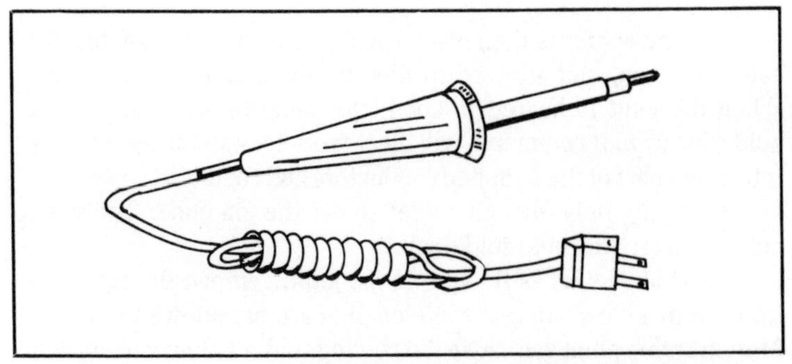

Fig. 12-4. Inexpensive soldering pencil used for low-wattage applications .

in high electrical resistance. High resistance joints present most of the problems in improperly soldered electronic circuits. These bonds do not adequately conduct the flow of electrical current and can sometimes cause rectification in audio circuits. A cold solder joint is a poor or absent electrical connection. It is most often caused by simply dropping the solder onto the joint before the elements have been heated to the proper temperature. This can occur when the tip of the soldering iron is applied to the *solder* rather than to the *joint* to be soldered.

Proper Soldering Techniques

There are several steps involved in forming proper solder connections when building electronic circuits. Each of these must be followed, in order and to the letter, to arrive at a completed product which is electrically stable and dependable. The steps are as follows:

☐ Make certain the elements to be soldered are clear of all foreign matter or debris. Wire conductors, for example, should be scraped clean of all insulation and wiped free of oil, tar or grease.

☐ A firm mechanical joint must be formed from the elements of the joint before soldering is attempted. This is done by tightly wrapping the conductors in such a manner that no physical movement is possible between elements.

☐ The soldering iron should be of an adequate temperature to allow for proper heating. It should be turned on a few minutes before soldering is attempted.

☐ The soldering iron is applied to the joint, *not* to the solder. Once the joint has been properly formed mechanically, the soldering iron tip is placed against it to allow it to heat to the same temperature.

☐ The solder is then placed against the joint, not against the soldering iron, and allowed to flow freely around the elements. When the joint is heated to about the same temperature as the soldering iron, its elements will meet the solder and allow it to flow into every part of the wrapped conductors and contacts.

☐ Apply only enough solder to get the job done. Too much solder can create a cold solder joint.

☐ Once solder is flowing in the joint, remove the tip of the iron and make certain that the elements are not allowed to move. Motion at this point can cause the cooling solder to become cracked or loose in certain areas of the joint.

☐ Allow about 20 seconds for the solder to cool.

☐ Wiggle the protruding elements of the joint to make certain that no physical movement occurs where the solder bond has taken place.

☐ Examine the appearance of the solder joint, looking for any signs of a dull surface or globular solder deposits. A proper solder joint will have a smooth, shiny appearance; while a dull, rough surface indicates a cold solder joint.

While these steps may sound complicated upon first reading them, they will become second nature to you as you complete more and more solder connections. After only a few hours of practice, it will take only seconds to solder each joint in an electronic project. The main trick to soldering is to always apply the tip of the iron to the joint and not to the solder and to allow the solder to *flow* into the crevices of the joint before removing the heat. Remember to use the *least* amount of solder necessary to get the job done. Cold solder joints result when too much solder is used, because the cooling rate is uneven in the different layers of the molten lead which is applied to the joint elements. A soldering iron applied to the outside of a large blob of solder may cause only the outer portion to become molten while the inside remains relatively hard. This latter portion is the part of the solder joint which performs the *electrical* bonding.

A firm *mechanical* joint is stressed because solder is not of adequate mechanical strength to form this physical type of bonding. It serves only as an electrical bond, *not* as a mechanical connection. If solder is used to hold two conductors in place, for example, normal stresses will cause this connection to work loose and the solder contact to crack, since proper mechanical rigidity of the joint was not originally established. Again, solder forms only an *electrical* joint. The elements of the joint itself must form the *mechanical* connection.

Even when using the low-wattage soldering pencils, speed in making the joint is often very important. Some of the solid-state devices used in the telephone projects of this book can be damaged or destroyed when they become heated past their maximum points of endurance. If you are not experienced in proper soldering methods, you would do well to practice upon a more rugged device such as a resistor, capacitor, or even upon two wire conductors wrapped together. Practice proper soldering techniques until they become second nature to you. This will increase the speed with which you're able to make the joints and is important, because the longer the soldering iron is applied ot the leads of a component, the hotter the component gets. A happy medium must be reached wherein adequate time is taken to complete a solder joint without taking so much time that the components become excessively heated.

Heat Sinks

A *heat sink* is often used to aid in the further protection of heat sensitive electronic components when soldering. This is a device which *sinks* or absorbs heat. A pair of needlenose pliers can serve as a very good sink when used to tightly squeeze a lead at a point near the shell of the component, as is shown in Fig. 12-5. Heat will travel up the lead from the point where it is being soldered, but the larger mass of the needlenose pliers will absorb most of it, which prevents a great deal of heat from reaching the case or shell of the component. Alligator clips and special heat sink clips can also be used to form a good source of heat protection. These devices have the advantage of remaining in place after the clip contact has been

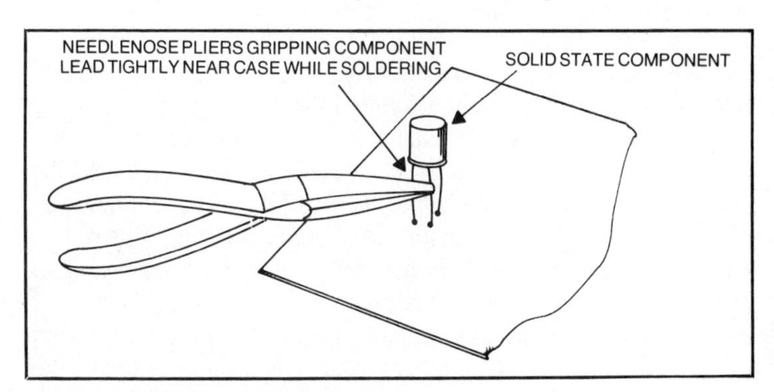

Fig. 12-5. Needlenose pliers can serve as an adequate heat sink.

made and will free the builder's hands for other parts of the soldering procedure. When applying a heat sink, make certain that it is not located too closely to the point on the lead which is being soldered. Placed too closely, the heat sink can pull heat away from the joint and create a cold solder connection. The heat sink is best placed at a point on the lead nearest the component case. Figure 12-6 shows a heat sink attached to a transistor lead.

Again, the proper soldering of electronic circuits is of paramount importance in electronic building. If you take shortcuts when putting together the circuits in this book, you are bound to run into trouble, either when the circuit is first tested, or later, when poor soldering connections break down. A few minutes spent in properly completing a project can save many future hours of troubleshooting, resoldering, and replacing heat-damaged components. Do not attempt to even start on a project until you know the correct methods of soldering.

BUILDING PROCEDURES

In addition to the proper soldering techniques just discussed, there are certain procedures that should be followed when building any type of electronic circuit. These are designed to help the new and inexperienced builder become more proficient at what he is doing, to cause himself the fewest problems, and to successfully complete all of the projects he attempts to build. We have all seen results of half-completed projects. These are the ones which were started long ago and were to be completed as soon as an additional part was obtained, but which just never got finished. Half-completed projects are often subject to breakage and other types of damage because they are usually not installed in a box or covering which provides mechanical protection.

The uncompleted project is not usually the result of lack of interest, lack of ability, or lack of skill. It is often the result of beginning a project before the builder has all of the parts necessary to complete it. This is a cardinal rule of electronic building: never begin a project until all parts, components, connectors, and the housing are on hand to complete the project. When you begin an electronic project with certain components missing, you cannot build the circuit in an orderly manner, as would be the case if all parts were on hand. The builder makes certain mental notes about parts which have been left out and which are to be replaced, and then, at a later time, completely forgets about them. A few of the components which were not on hand originally may be obtained,

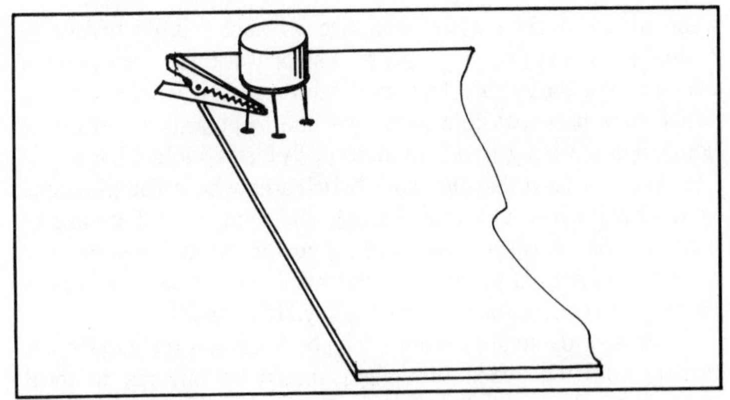

Fig. 12-6. Clip-on heat sink attached to a component lead.

wired into the circuit, and then, assuming that the project is finished, power is applied. Unremembered by the neophyte builder, a component or two was omitted from the circuit, a component which the builder was supposed to have made a mental note of. Since this has been completely forgotten, the builder assumes the circuit is finished and finds that it does not work properly, or at all. He now has to troubleshoot the circuit and will be more apt to look for poor connections or damaged components rather than to seek out *missing* parts. The circuit often winds up a total failure and is tossed into the junk box as a source for spare parts. Here is a good example of a circuit which probably would have worked perfectly if proper techniques had been followed.

While many builders will complete an electronic circuit having all the electrical components on hand, the case or box which is to house the finished product is often saved for last. There is nothing quite so fragile as an electronic circuit on a perforated board which is not mounted in a protective case. These boards have nasty habits of falling from work benches or of breaking when accidentally placed under heavy objects. As soon as your circuit is completed, it should immediately be mounted in a protective case after initial testing.

A major cause of improperly wired circuits is *fatigue*. Experienced builders never work on circuits when they are tired, sleepy, or when their minds are on other things. If you work too long on a small circuit board, the vision will often start to blur and hands may begin to shake from being in one position for too long. When you feel the least bit fatigued, stop what you're doing and take a 10 to 15 minute break until you are refreshed again. Don't set a specific day or time to have your circuit completed. When running behind schedule, you may start to rush or work past your point of adequate concentration. The only result from this will be a circuit which may have possible problems due to polarity reversals of components, wiring errors, or improperly formed solder joints.

Make certain that the work bench area where you assemble your circuits has adequate lighting and ventilation for ease of construction. Arrange your seating so the normal assembly of circuits will not put you in an uncomfortable position, causing you to strain or reach in such a manner that you tire rapidly.

Most of these suggestions for good building techniques are merely common sense ideas and should be obvious to most individuals. It is a good practice to have one specific area where electronic assembly is normally done. This gets the builder accustomed to working under set conditions and makes for a more comfortable and relaxed assembly.

By following these construction suggestions, you should be satisfied each time a project is completed, both with the quality of your electronic circuit and with its operation and dependability. You should also take pride in the fact that a great many electronic projects which are built by other individuals not adhering to these techniques are going unfinished or are causing problems when completed.

Special Integrated Circuit Building Techniques

All of the heating effects which create problems in building with integrated circuits can be overcome by using a socket. The socket is soldered into the circuit before the device is inserted and there is no possibility of any damage occurring due to heat. Adequate time may be taken when soldering these sockets without fear of any type of heat damage. Proper soldering techniques are still dictated, however, as a cold solder joint at a socket will cause just as severe a problem within the circuit and possibly more because of the added resistance created by the friction contact of the device within the socket.

If solid-state devices are to be used with sockets, it is extremely important to make certain the device leads are cleaned of any foreign materials, especially those of an oily nature. A dirty lead can form a high resistance contact within the socket and cause the same types of circuit problems which are most often brought about by cold solder joints. The socket contacts should also be cleaned to make certain that no grit or foreign material has covered

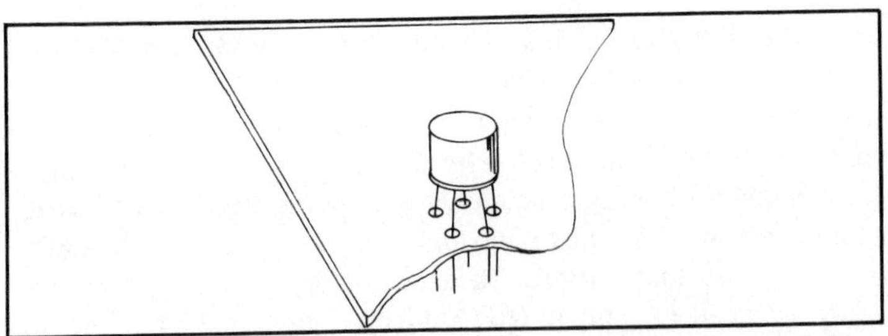

Fig. 12-7. An integrated circuit in a circular can packaged is easily mounted to a section of perforated circuit board by forcing the component leads through the available holes.

the areas which make contact with the leads. Periodic inspection of the socket is necessary, especially if the electronic circuit is used out-of-doors or in an area which is subject to dust and dirt buildup. A circuit which uses sockets is not quite as dependable as one which uses direct solder contact, so if high vibration applications are anticipated, the socket technique may not be practical.

Component Mounting

Integrated circuits used in these projects are normally of two varieties, the *circular can* and the *DIP*, which is an abbreviation for *Dual In-Line Package*. There is a third integrated circuit configuration which is called a flat pack. This last type is most often used for computer applications and is very difficult to work with in a typical home shop due to the extremely close spacing of the circuit leads.

Circular can integrated circuits are the type which look very much like transistors with many leads instead of just three. Often, a small tab will protrude horizontally from the bottom edge of the case to give some means of reference when determining the pin connections of the IC leads. The mounting of this type of integrated circuit to a circuit board is identical to the mounting of transistors, except more device leads must be contended with. This packaging is most conducive to the home builder because it allows for point-to-point wiring and does not necessarily relegate the builder to using printed circuit boards. When using the DIP integrated circuit, circuit boards are always required unless a socket is used which terminates in long wire leads instead of the normal pin connections. Wire lead extensions can be soldered directly to a DIP IC, but the chances are great that this process will damage the component because of heating effects. It is almost impossible to connect any sort of heat sink to the extremely short pins. Also, this packaging is usually accomplished with a plastic case and will melt and become disfigured under conditions of extreme heat. The use of a DIP socket will alleviate all of these problems.

The mounting of integrated circuits and other solid-state components when building a telephone project in this book is best accomplished by using a small piece of perforated circuit board, which is available at most radio and hobby supply stores. It is also recommended that sockets be used for any integrated circuits which are available only in DIP configurations. Figure 12-7 shows how an integrated circuit of the circular can variety can be easily mounted on this type of circuit board by inserting each of the leads through a separate hole and then soldering from beneath. This also creates an attractive finished circuit while being viewed from the top side of the board. Many of the circular can packages for ICs contain a small plastic tube at their centers which acts as a divider and keeps the package slightly above the circuit board. This allows for adequate ventilation on all sides of the device housing. The perforated circuit board method of construction is technically point-to-point wiring, as opposed to circuit board construction, but the perforated board acts as an excellent base or mounting platform for all components.

Figure 12-8 shows how a completed circuit might look. Notice that a vertical mounting technique has been used to conserve space for the resistors involved in the structure. This is accomplished by bending the top lead of the resistor down along the side of the carbon body and clipping both leads so that the ends are even. The same is true of the mounting of small electrolytic capacitors which contain axial leads. These same components could just as easily have been mounted in a horizontal position (flush with the circuit board) if so desired. The vertical construction is intended to conserve horizontal or circuit board space. All connections are made from beneath the board by twisting various leads together and soldering them in the correct fashion.

Integrated circuits of the DIP variety will often fit in a perforated circuit board with closely spaced holes. Point-to-point wiring may be used with this type of IC if it will fit the circuit board properly, but an IC socket would be preferred. It is important to use care when installing a DIP IC in a socket. The pins of the integrated circuit are very delicate and are easily bent or even broken when forced improperly. Correct insertion procedures call for aligning all the pins on one side of the IC with the holes along one side of the socket. Notice that each pin is tapered in a manner which suddenly becomes square at the midway point. Now, start each pin into its own socket hole, but do not seat them all the way. In other words, only the tip of each pin is started into its respective hole in the socket. Next, line up the pins on the other side of the IC into their respective socket holes. Make certain that the pins on the first side have not slipped from their holes while this is being done. It may be necessary to slightly bend some of the pins in order to get them to align properly. This can be accomplished with a toothpick or other small pointed device to gently force the tip of the pin into the correct slot. At this point check all of the IC pins to make certain they are correctly inserted into each of the socket slots. Now, press firmly at the center of the IC in order to cause the remaining portion of each of the pins to snap firmly into place. A slight rocking motion when pressing the IC may cause easier entry. Removal of the device from the socket is much less complicated and is done by simply inserting a small screwdriver under one end and gently pulling upward until the IC snaps out. Practice this procedure with a defective integrated circuit if possible, because if the pins are badly disfigured, the component may be ruined.

BUILDING SUMMARY

While this brief discussion on building technique applied to circuits which use ICs and other solid-state devices, it should be pointed out that heat damage can occur to any electronic component. Small circuit boards generally use tiny, low-wattage components. Resistors, capacitors, and coils can be easily damaged, so solder these as rapidly as possible, while making sure that you don't rush to the point where cold solder joints are formed.

Solid-state components are extremely simple devices which may be used to construct rugged, reliable remote-control systems. If strict attention is given to their selection, ratings, and mounting, the circuits that are built from them should last a lifetime. Take the

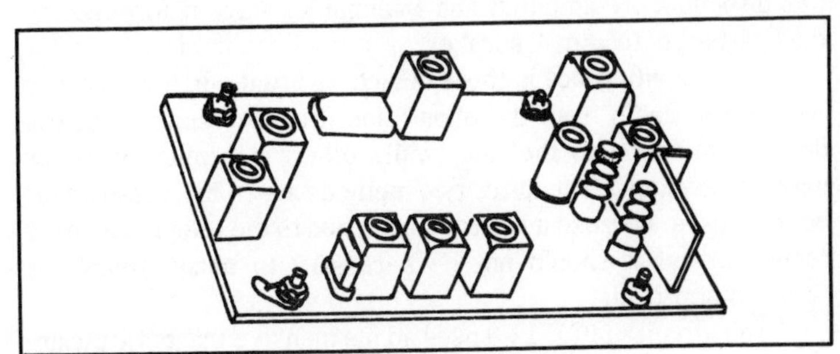

Fig. 12-8. A completed circuit using vertical mounting technique.

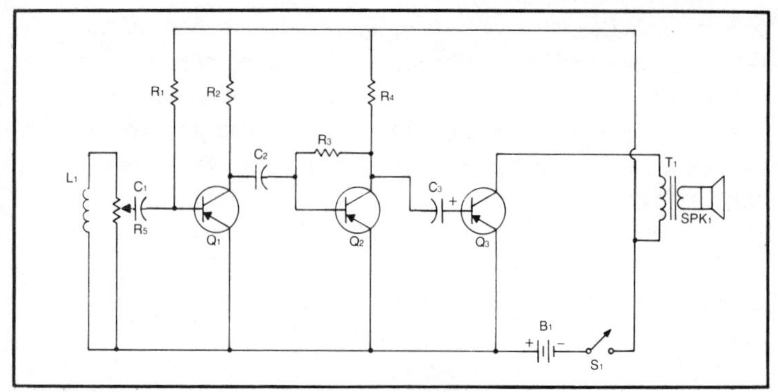

Fig. 12-9. Schematic diagram of magnetic pickup telephone amplifier.

time needed to properly design your circuit and practice proper soldering techniques. Generally, you should make your circuits as simple and uncluttered as possible. This will add to the reliability and subtract from the repair time should a malfunction in the circuitry occur. Take the time to make good solder connections but not so long that the delicate device is damaged from the heat. After soldering is completed, allow adequate time for the devices to cool to room temperature before applying power. Many good solid-state devices have been unnecessarily destroyed by applying operating voltage and current a few seconds before they had cooled from recent soldering.

MAGNETIC PICKUP TELEPHONE AMPLIFIER

Many of the telephone accessory devices available from hobby stores and electronic outlets are designed to amplify a telephone conversation so that it may be heard by other persons in the same room without having to crowd around the receiver portion of the handset. All of these amplifiers work in the same basic manner by sampling the audio information on the phone line and then passing this through a pre-amplifier and an amplifier stage before feeding the final output to a small speaker.

The big difference in these devices is usually in the way they sample the audio from the phone line. Some circuits require a direct connection to the line, while others use microphone or magnetic pickups. The latter two methods are to be preferred for our purposes, because no direct connection to the telephone line is required, and it should not be necessary to obtain telephone company approval.

The circuit in Fig. 12-9 uses an inexpensive magnetic pickup coil which should be available through electronic mail order catalogs or from some hobby stores. This coil is inductively coupled to the telephone line by placing it beneath the base of the telephone instrument. Enough coupling is obtained from this mounting position to cause the current fluctuation which occurs when a conversation is being conducted to drive the first amplification stage of our circuit. Also, see Table 12-1.

Transistor Q1 serves as a voltage amplifier and provides enough drive to feed the base of transistor Q2. This latter device further amplifies the diminutive signal which was provided by the induction coil and feeds this output to the base of Q3, which is the output amplification stage. Q3 is directly coupled to the small audio output transformer, which feeds a miniature 4 ohm speaker.

A crude volume control is included at the output of the induction coil and may be used to effectively vary the volume of the speaker output. This is a low-powered circuit, so booming volume will not be the case, but the speaker output should be adequate enough to be heard within a reasonable distance.

All components, such as resistors and capacitors, should be of the miniature variety with ratings of at least 15 volts. Half-watt carbon resistors were used for this project, although quarter-watt components are also acceptable. These ultra-miniature resistors will allow you to make your circuit even smaller, but they are often difficult to find and much more expensive than their half-watt counterparts. The transformer is a miniature transistor output device. Larger units of the same primary and secondary impedance may be used, but this will probably mean that your circuit will fill a much larger space.

Wiring of this project is fairly non-critical. Make certain that all transistors are connected to the various circuit points in the

Table 12-1. Parts List for Fig. 12-9.

Parts
B1 - 9 volt transistor radio battery
C1, C2 - 0.005 uf ceramic, 25V
C3 - 80 uf electrolytic, 50V
L1 - Telephone pick-up coil
Q1, Q2 - ECG 102A (Sylvania)
Q3 - ECG 104 (Sylvania)
R1, R3 - 1 Mohm, ½ watt carbon
R2 - 6.8 Kohm, ½ watt carbon
R4 - 62 Kohm, ½ watt carbon
R5 - 10 Kohm, ½ watt variable control
S1 - SPST miniature switch
SPK1 - 4 ohm miniature speaker
T1 - 500 ohm primary to 4 ohm secondary (Allied No. 705-0527)

proper manner. It is necessary to first identify the three leads of each as to their emitter, collector, and base connections. A reversal will mean that the circuit will not work and that the transistor may be damaged. Only one electrolytic capacitor is used, which must be connected while observing the correct polarity in the circuit. The polarity markings are clearly shown in the schematic drawing. The only other device which must be connected in a properly polarized manner is the 9-volt battery. A reversal here could damage all three transistors and possibly some of the other components.

Figure 12-10 illustrates a suggested parts layout on a piece of perforated circuit board measuring approximately 4″ square. A much smaller board may be used if you are a more experienced circuit builder, but for the neophyte, the larger construction area may be easier.

Begin this project by inserting the three transistors as shown in the last drawing. Their leads are fitted through the holes in the circuit board material. Once inserted, bend the leads outward to temporarily secure the transistors for further connections and soldering. The remaining components may now be inserted in a like manner as shown. Make sure the electrolytic capacitor, C3, is connected with its positive lead attached to the base of Q3 and the negative lead to the junction of the collector of Q2 and the two carbon resistors.

Starting at the Q1 end of the circuit board, connect one lead of C1 to the base of Q1. Now connect one lead from R1 to the same point. Solder this connection which should include a total of three component leads. Proceed in this same manner throughout the remainder of the circuit, moving from left to right. Before soldering any connection points, count the number of leads which are to be bonded. Compare this with the schematic drawing to make certain that you are not omitting the connection of one component, or possibly connecting a component lead to a wrong contact point.

When you finally arrive at the transformer connection, be careful not to reverse the primary and secondary windings. Different manufacturers use different conductor color coding for their transformer leads, but they also supply a data sheet telling the owner which leads connect to which windings. The 500 ohm primary leads from the transformer connect, respectively, to the collector of Q3 and the miniature toggle switch. A transformer is not a polarized device, so either of the primary leads may be

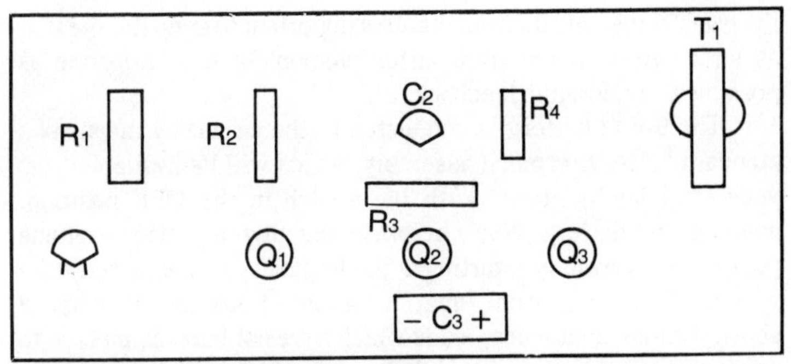

Fig. 12-10. Parts layout for amplifier project.

connected to the collector or the switch. The same applies to the two secondary leads, which are attached to the miniature speaker in any order.

One component has not yet been wired into the circuit if you followed the previous instructions and started by soldering one lead of C1 to the base of Q1. This component is R5, the variable control or potentiometer. This was saved for last because it will require connection to the magnetic pickup coil through a length of cable. If you have not already done so, push the three terminals of the potentiometer through the circuit board and connect the remaining lead of C1 to the center terminal only. Solder this connection. Now, connect one of the remaining potentiometer terminals to the base of Q1 through a short length of hookup wire. Solder the Q1 connection only. The hookup wire connection to the potentiometer terminal will be soldered later.

If the leads coming from the magnetic pickup are long enough, you may now connect one to the bare terminal of the potentiometer and the other to the terminal to which the hookup wire has already been run. Solder both of these points and your circuit is completed. If the pickup leads are not long enough, connect this component to the rest of the circuit by using a short length of shielded audio hookup wire.

In most normal areas, the circuit will work fine just like it is, although it should be installed in some sort of plastic or aluminum box. The type of circuit housing used is left up to the builder, and the description of the building of this project was handled in breadboard fashion. For those of you who wish to mount this circuit in a proper enclosure, it would be best to mount the potentiometer through the front face of this box and make connections to the amplifier circuitry by means of hookup wire. The speaker may be extended a short distance by using a two conductor cable between the transformer and its contacts. It is important to keep the speaker as far away from the transmitter element of the telephone as possible to avoid audio feedback.

The 9-volt battery is connected to the circuit by means of a standard battery terminal assembly, which will be available from your local hobby store. With the switch in the OFF position, connect this battery. Now, before doing anything else, examine your project carefully. Starting at the lefthand side again, trace the wiring connections from input to output. Look for any sign of shorted component leads, leads which have not been connected to any point, and for small globules of solder which may have accidentally dropped across some component leads during the building process. If everything appears to be in proper order, you may begin the operational checkout.

Install the induction coil beneath the phone or in the manner which is recommended by the manufacturer. Locate the speaker as far from the handset as possible. Lift the receiver and throw the toggle switch to the on position. You should hear the dial tone, but if you don't, adjust R5 until the tone is heard clearly. Next, phone a friend and test the circuit while carrying on an actual conversation. If you have proceeded correctly, you should be pleased with the results.

Troubleshooting

If your circuit does not work, immediately throw the switch to the off position. Feel the cases of the transistors with your finger to see if they are warm or hot. A high operating temperature might indicate a circuit malfunction. First of all, check the battery to make certain that it is fresh. Even though it may have been only recently purchased, it could still be bad, especially if it has been stored for a long period of time. It is a good idea to check these batteries where purchased if a battery tester is available. If the battery is good, recheck your wiring. You may have accidentally reversed a polarized component. If the battery leads have been reversed, correct your error and then perform the test procedure again. Circuits which have been connected to reversed polarities often incur damage to their solid-state components. If your circuit doesn't work after correcting the battery reversal, suspect a damaged transistor. The transistors are really the only components which could be damaged in a short period of time. Occasionally, defective transistors may be purchased unknowingly. If you suspect a transistor malfunction, take the device back to its place of purchase and have it checked or replaced.

Project Summary

The magnetic pickup telephone amplifier project is simple and efficient. It can be assembled in a few hours and usually for less than fifteen dollars, including the aluminum case. If the builder shops around, the entire project may be built for as little as five dollars, assuming that a supplier of surplus parts is available. Amateur radio operators usually have a wealth of spare components which may be had for little or nothing. So, if you know a ham operator, this may be an excellent place to look first.

This project is useful as well as interesting. It is convenient for broadcasting phone conversations to persons within the same small room. Since it is battery operated and requires no direct connection to the telephone instrument or line, it may be carried with you and used whenever and wherever you desire.

Some builders have elected to use power supplies which operate from the AC line to avoid the inconvenience of constantly replacing batteries. This would be desirable if the amplifier were to be used on a daily basis and in a permanent location. Later in this chapter, some information will be given on power supplies which may be used to operate many of these projects. An easy way to gain AC power with this circuit is to purchase an AC battery replacement from your local hobby store. One is available from Radio Shack for less than ten dollars which is powered from the AC line and provides from 4.5 to 9 volts output.

ANOTHER TELEPHONE AMPLIFIER

Another type of telephone amplifier differs from the previous project mostly in the method used to sample the audio signals. The electronic circuitry is much different in appearance, in that it is of hybrid design, using both an integrated circuit and a transistor. While the schematic will look different from the previous one, the electronic processes are very similar. They are accomplished in a different manner, however.

Instead of a magnetic induction coil as a phone line sensor, this amplifier circuit uses a contact microphone which is applied to the receiver end of the handset by a rubber suction cup. The component used in the construction of this circuit was purchased from Radio Shack for less than two dollars and is called a telephone pickup. Almost any high impedence contact pickup should work with this circuit, which may also be used with a tape recorder microphone to act as a tiny PA system.

Figure 12-11 shows the schematic diagram which consists of only a few components. The integrated circuit is a common 741 operational amplifier which takes the place of the first two transistorized stages in the previous amplifier project. This is the pre-amplifier section of the project. The pickup is connected to leads 2 and 3 of the IC, which amplifies the microphone output to a level which is usable by the input circuit of the transistor amplifier stage. The single transistor Q1 feeds its output directly to the 8 ohm miniature speaker. A volume control is provided which will allow the circuit to be used with a minimum of audio feedback. Also, see Table 12-2.

The circuit is assembled on a 4″ square section of perforated circuit board, although smaller sections may be used by the more experienced builder. A socket is used to contain the integrated circuit, which is a DIP type having 8 pins. Only five of these 8 pins are used in this circuit. The remaining pins will have no connections. The IC socket is used here because it is very difficult to solder directly to the tiny pins. Integrated circuits are sometimes damaged by the heating effects of a soldering iron, but the use of a socket precludes all possibility of this. All soldering connections are made to the IC socket before the integrated circuit is snapped into place.

Figure 12-12 shows a suggested parts layout for this circuit which may or may not be followed exactly, as component placement is largely non-critical. Since this is an amplifier, all wiring and component leads should be kept as short as possible to prevent possible oscillations from occurring internally.

Begin construction by pushing the leads of the IC socket through the holes in the circuit board. Next, install the transistor in its proper position, using the same technique. The leads of the transistor and the socket should be bent outward to temporarily secure them to the circuit board. Now, install the rest of the components, using Fig. 12-12 as a guide. Starting at the left-hand

Table 12-2. Parts List for Fig. 12-11.

Parts
C1 - 1.5 uf electrolytic, 15V
C2 - 0.12 uf ceramic, 15 V
IC1 - ECG 941M (Sylvania)
Pickup - Radio Shack Telephone Pickup
Q1 - ECG 184 (Sylvania)
R1 - 100 Kohm, ½ watt variable control
R2 - 100 Kohm, ½ watt carbon
R3 - 10 Kohm, ½ watt carbon
SPK1 - 8 ohm miniature speaker

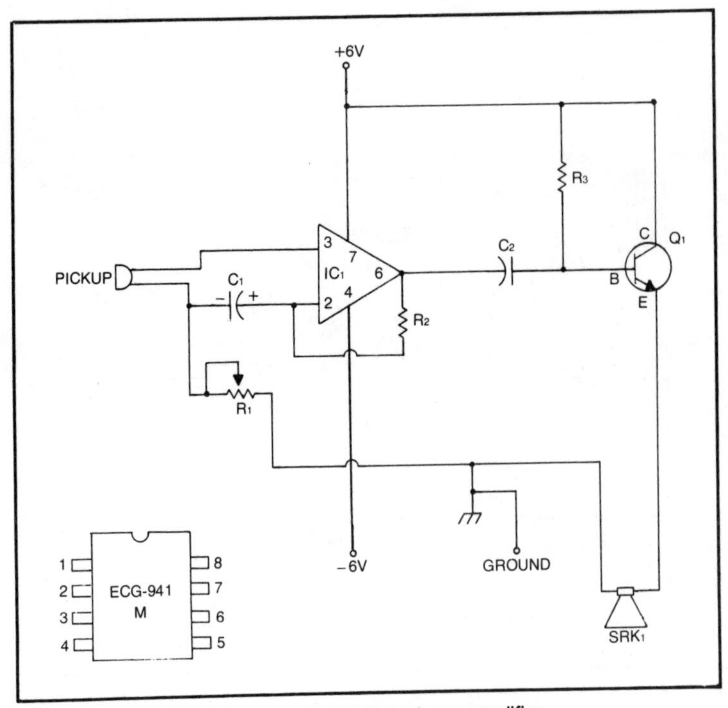

Fig. 12-11. Schematic diagram of hybrid telephone amplifier.

side of the schematic and circuit board, begin connecting the components as indicated. Solder only when you are certain that all leads have been connected. Be especially careful around Q1, as high temperatures for extended periods could cause this component to be damaged. Make certain that you have correctly identified the IC pin numbers and aligned them to the IC socket. Since integrated circuits may often be inserted into their sockets in either direction, you must arrive at the proper direction for placement and then wire the socket pins accordingly. The inset in Fig. 12-11 will show you how to correctly identify the integrated circuit pin numbers.

Other than the power supply, the IC and transistor, the only other polarized device is the electrolytic capacitor labeled C1 in the schematic. Make certain that its positive lead is connected to pin number 2 of the integrated circuit and to R2.

The volume control, R1, may be installed through a plastic or aluminum case if desired, and it is recommended that the telephone pickup element be connected to the circuit by means of a mating plug and receptacle. If the pickup you use is already fitted with a

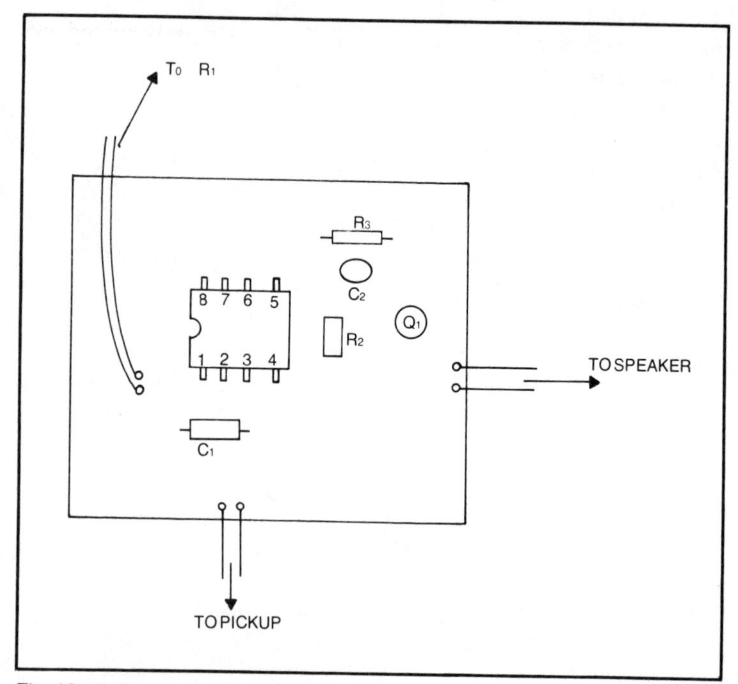

Fig. 12-12. Parts layout for hybrid amplifier.

male plug, purchase a female receptacle which may be wired to pin 3 of the IC and to the negative lead of C1. When this receptacle is mounted in the plastic or aluminum case, the pickup element may be connected in a matter of seconds to the circuit.

Some builders choose to connect the speaker in the same manner as was used for the pickup. This will work well, but should the circuit be activated when the speaker is disconnected, damage could occur to the output transistor. For this reason, it is recommended that the speaker be permanently wired through a length of cable to the remainder of the circuit.

Check all of your wiring connections at this stage of the assembly. Make certain that there are no remaining leads that need to be installed. Bear in mind, however, that only five of the eight leads from the IC socket are used in this circuit. If all wiring seems to be correct, the IC may be inserted into its socket. This is best done by lining up all of the pins on one side into their corresponding socket slots. Do the same for the other side and then double check to make certain that all pins are pushed slightly into their proper position. Now, with a rocking motion press down on the top of the IC case. This component should lock into place easily. Once locked, check to make certain that all pins are properly seated and that none has slipped out of the mating slot.

The circuit may now be tested, but the problem of a power supply must be first taken care of. Two 6-volt batteries may be used when connected as shown in Fig. 12-13. Six-volt batteries are not easy to come by in small sizes and can sometimes be quite expensive. Do not be confused into thinking that these two batteries combine to supply 12 volts to the amplifier circuit. The integrated circuit requires a dual-polarity power supply. This means that one portion of the integrated circuit requires a positive 6-volt supply in relationship to the circuit ground, which is indicated by a chassis connection in Fig. 12-11 and lies between R1 and the speaker. The other portion of the integrated circuit requires a negative 6-volt supply in relationship to the same ground. This is the reason for using two 6-volt batteries for the supply. One is wired to supply a negative potential, the other supplies a positive 6-volts. Both supplies have the same common ground to the amplifier circuit.

The Power Supply

It may be more convenient to build a simple dual-polarity power supply which operates from the AC line. This latter circuit may be built on the same circuit board as the telephone amplifier or may be built separately and connected to the amplifier circuit by a length of cable.

Figure 12-14 shows the schematic drawing of a dual-polarity power supply whose output is 6.3 volts DC, positive and negative,

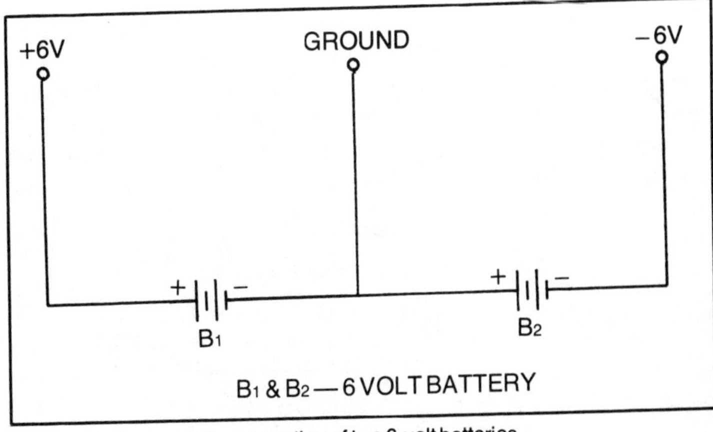

Fig. 12-13. Dual-polarity connection of two 6-volt batteries.

with respect to a common ground. While it is possible to build such a supply using two transformers and two individual circuits, this particular schematic shows that only a single transformer is used. This saves space, time and expense. Two rectifier legs are brought off of the transformer secondary. One provides the negative potential, while the other is positive in relationship to the common ground. Also, see Table 12-3.

Experienced builders will probably think that a standard bridge rectifier circuit is used in this project, but this is not accurate. While the four diodes do resemble a bridge circuit, note that the transformer center tap is grounded. This is really a full-wave center tap DC power supply. The first set of rectifiers on the left-hand side of the schematic form the negative leg, while the pair to the right make up the positive portion of the circuit. The transformer center tap is the common ground for this power supply, so the DC output voltage will be approximately half the value of the total transformer secondary voltage. A 12.6-volt transformer is used with this circuit and provides an output of 6.3 volts (approximately) after rectification and regulation.

A simple electronic regulator circuit is used to stabilize the voltage and consists of a series resistor and 6.3-volt zener diode in each voltage leg. The zener diodes conduct when the output voltage exceeds a value of 6.3 volts. The increased current drain from this conduction causes voltage to be dropped across the series resistor, and the output voltage is dropped back to its desired value. This process takes place constantly and in a matter of a very small fraction of a second, so a voltmeter placed across the output circuit will indicate a constant 6.3 volts.

This circuit is best built by installing the transformer fuse socket and switch in a small aluminum box, as is shown in Fig. 12-15. The remainder of the circuitry from the rectifiers out is placed on a small piece of perforated circuit board, which may be the same piece used for the amplifier circuit. Figure 12-16 shows a suggested component layout for this board or board section.

Table 12-3. Parts List for Fig. 12-14.

C1, C2 - 500 uf electrolytic, 25V
CR1, CR2 - 6.3V, ½ watt zener diode
D1-D4 - 50 PIV, 1 amp diodes
F1 - ½ ampere line fuse
R1, R2 - 100 ohm, ½ watt carbon
S1 - SPST miniature switch
T1 - 12.6V, 1 ampere transformer

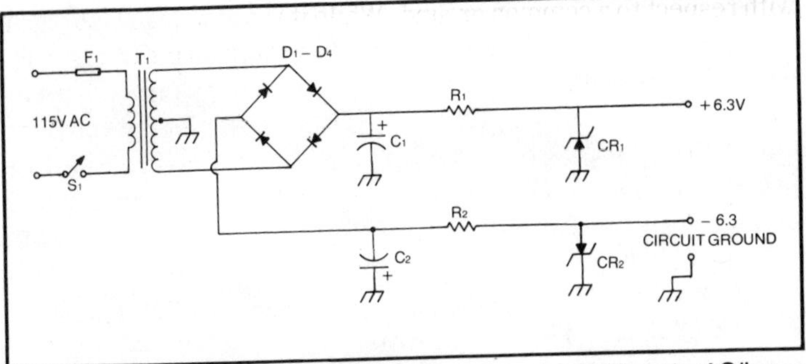

Fig. 12-14. Schematic of dual-polarity power supply operated from the AC line.

Unlike some of the earlier projects, this power supply is composed almost entirely of polarized devices. Only the two series resistors are non-polarized. The rest of the components on the circuit board must be connected in the proper manner. The banded end of a diode is the end in which the schematically-drawn arrow points. This will aid you in properly placing these components on the board for connection. Note that the top leg, or positive portion, of the supply is a mirror image of the bottom. Capacitor C1 is connected with its positive lead to the voltage line, while its negative lead goes to circuit ground. But, C2, which is the equivalent of C1 in the former circuit, has its negative lead attached to the voltage line and its positive terminal grounded. The two zener diodes are also connected in reverse order. Again, the zener diodes will usually have their cathodes indicated by a small band, or the schematic representation may actually be printed on their case sides.

The circuit ground may consist of a length of No. 14 copper wire installed on the bottom of the circuit board from end to end, as shown in Fig. 12-17. The center tap of the transformer and all component grounds may be attached here. Alternately, a single wire may be run from the center tap of the transformer through to the output and all ground connections made here. This is best if the transformer is to be mounted on the circuit board instead of in a metal box.

Do not neglect to include the small fuse which is a one-half ampere in-line type. Should a short-circuit occur within the power supply, this fuse will quickly open up and stop all current flow from the AC line. If the fuse were omitted, a short-circuit could cause high amounts of current to flow into the primary windings of the transformer. This could quickly cause overheating and a possible

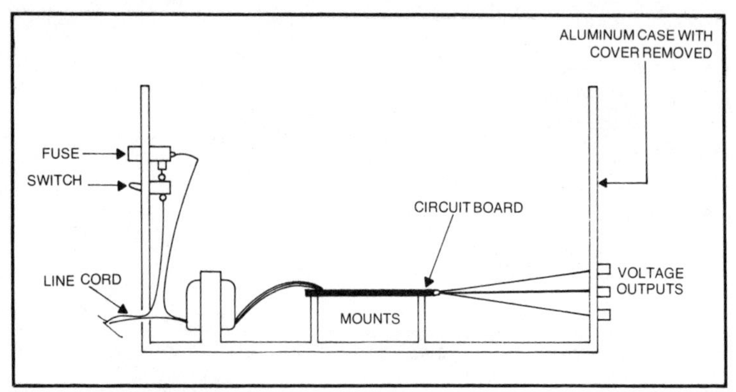

Fig. 12-15. Installation of transformer, fuse socket, switch, and circuit board in an aluminum enclosure.

fire. Through-the-case fuse holders are available which may be quickly installed through an aluminum case and allow fuse replacement from the exterior. Alternately, an in-line fuse holder may be purchased which will be enclosed in the aluminum case. A fuse clip might also be found which will mount directly to the circuit board and provide friction contact to hold a fuse in place.

Once construction is completed, look over your work. A short-circuit here could damage the silicon diodes. Even though the circuit is fused, solid-state devices have a habit of being destroyed due to current surges in a fraction of a second less than the time it takes for the fuse to open up. The fuse can only prevent sustained heavy current drain for fire protection. It cannot protect delicate solid-state components from current surges.

If all seems to be in order, connect the power plug to the wall outlet and place the probes of a DC voltmeter (which will measure a value of at least 7 volts) to the power supply outlet terminals. Start with the positive supply by placing the positive or red voltmeter probe across the +6.3 V terminal and the negative or black probe on the circuit ground. Turn the toggle switch to the on position, and a reading of approximately 6.3 volts should be obtained. If nothing is registered, turn the switch to the off position and remove the line plug from the wall. Check the fuse to see if it is blown. If it is, some serious wiring error is present, and a close inspection should reveal this. Correct this error and replace any components which seem to be burned or damaged.

On the positive side, if 6.3 volts is obtained, remove the probes and reposition them. This time, the black probe should be attached to the −6.3 V terminal and the red probe placed across

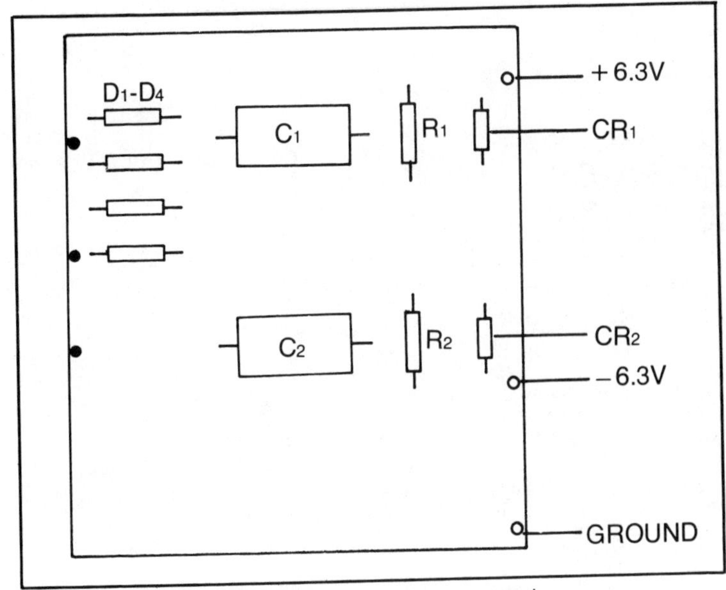

Fig. 12-16. Component layout for dual-polarity power supply.

circuit ground. Another reading of approximately 6.3 volts should be obtained.

If you get a voltage reading at one output and none at the other, this means that there is trouble in the circuit leg which is not delivering output voltage. Turn the supply off, remove the line plugs from the wall, and examine the portion of the circuitry which is not functioning. The cause will most likely be a damaged diode or one which was improperly positioned in the circuit.

Once your power supply is working properly, attach the leads to the power supply points on the amplifier circuit board. The chassis ground from the power supply attaches to the ground connection between R1 and the speaker. The plus and minus voltage outputs connect to the like inputs at the amplifier circuit.

Checkout Procedure

Checkout procedure is very simple. Connect the pickup to the telephone handset. This will probably mean placing the suction cup on top of the receiver end. Turn the power supply switch to the on position and adjust R1 until the dial tone is heard from the speaker output. R1 in this case is the variable control in the amplifier circuit and should not be confused with the series resistor marked in the same manner in Fig. 12-14. The speaker should be placed as far

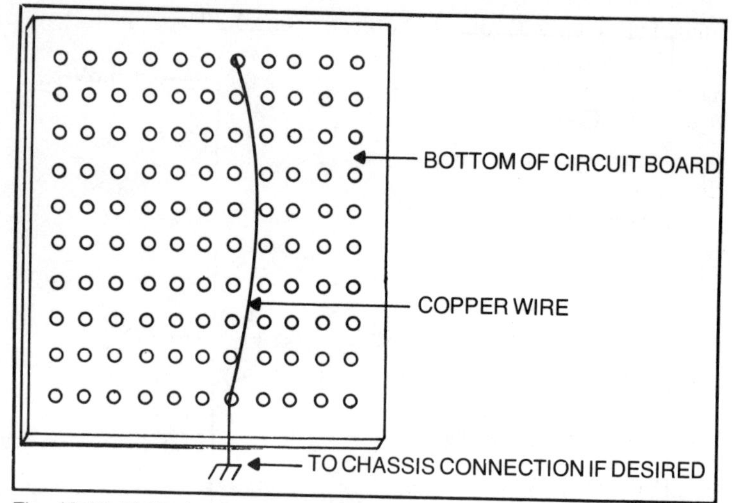

Fig. 12-17. Method of installing a ground buss at the bottom of a piece of perforated circuit board.

from the handset transmitter element as possible to avoid feedback.

If the amplifier won't function, measure the circuit voltage between pin 4 and ground and pin 7 and ground. The positive probe is placed at pin 7 and the negative probe at ground. When measuring the voltage of pin 4, the positive probe now attaches to circuit ground, while the black one is placed across pin 4. If the voltage readings seem to be normal, you probably have a wiring error or faulty component in the amplifier circuit. A thorough check here should provide the solution.

Project Summary

This particular circuit is more complex than the former amplifier, but it is designed for more or less permanent mounting and operates from the AC line. Again, it is not necessary to obtain telephone company approval to use this attachment, as no direct connection has been made to the telephone line. Using this circuit, it is possible to conveniently enlarge the scope of telephone usage for business and conference purposes. In the home, the entire family may be able to listen to the remarks of a distant friend or relative without having to waste time and long distance money by inefficiently passing the handset from person to person. The suction cup pickup attaches to the handset in a matter of a few seconds and the power supply and amplifier circuitry is activated immediately upon throwing the power switch. These is no long period of waiting for the circuit to be connected and operational. This is an important feature of any telephone attachment. The power supply may be left with its power cord attached to the wall outlet at all times, as no power is applied to any of the circuits until the switch is thrown to the on position.

Caution: Unlike other projects, this one operates from the AC line. When performing service, *always* remove the power plug from the wall. Even when the switch is on the off position, the full line voltage is present within the aluminum case or on the power supply circuit board. This is a very dangerous shock hazard and is completely avoided when the supply power cord is disconnected.

ELECTRONIC PHONE PATCH

A phone patch is a circuit or device which enables an audio device to insert its output into the phone line and also enables the signals from the phone line to be fed to the input of an audio device. For example, amateur radio operators have used phone patches for years to allow persons in foreign countries or distant states to talk with residents in the operator's home town. Radio contact was first established. The amateur operator then called the local party on the telephone and the phone patch was used to connect the telephone line to the transmitter and receiver. When the distant party spoke into the transmitter, it was received by the local operator's equipment and automatically sent down the phone line. When the local party spoke into their telephone instrument, the output was fed directly to the microphone input of the transmitter. Figure 12-18 provides a block diagram of the basic functioning of this system.

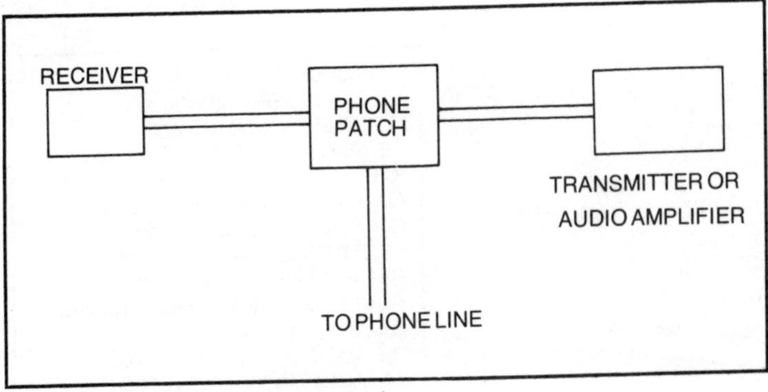

Fig. 12-18. Block diagram of phone patch.

A phone patch circuit requires direct connection to the phone line. While it is possible to simply attach the patch connection to the two wires leading to your telephone instrument connection, most local phone companies will require the installation of a phone coupler which they perform for a service charge. There is also a rental charge on the coupler which is paid on a monthly basis.

Figure 12-19 shows a schematic diagram of the circuit which contains two major parts. To the left, a transistor circuit takes the output from the receiver and feeds it to the matching transformer which attaches to the phone coupler. To the right is an integrated circuit system which receives the output from the phone line, amplifies it and provides a feed to a microphone or other audio input circuit. Also, see Table 12-4.

A gain control is provided in the integrated circuit portion of this system which will control the level of the output to the audio channel. Also, a balance control is present which will provide an adjustment to both the audio input and output sections of the circuit, keeping them at equal levels.

Circuit construction is straightforward. Figure 12-20 shows a suggested parts layout which is designed to keep the two circuits separated. The only common connection points are found at the balance control and its circuit ground.

The integrated circuit is installed in an IC socket which is inserted in the normal manner through the holes in the perforated circuit board. This is an 8-pin DIP package, although only 7 of the pins are connected to the circuit. No connections are made to pin 5, and this portion of the IC socket may be clipped away. There are many polarized devices in this overall circuit. Make certain that all of the capacitors are installed with their positive and negative leads as shown in the schematic diagram. The matching transformer has

Table 12-4. Parts List for Fig. 12-19.

Parts
C1 - 500 uF electrolytic 50V
C2-C6 - 12 uF electrolytic 20V
C7 - 50 pF ceramic, 25 V
C8 - 0.005 ceramic 25V
IC1 - ECG 975 (Sylvania)
Q1 - ECG 123A (Sylvania)
R1 - 22Kohm, ½ watt carbon
R2, R4, R5 - 600 ohm, ½ watt carbon
R3 - 1.2 Kohm, ½ watt carbon
R6 - 10 Kohm, ½ watt variable control
R7 - 1 Kohm, ½ watt carbon
R8 - 100 Kohm, ½ watt variable control
T1 - Allied Electronics No. 705-0545

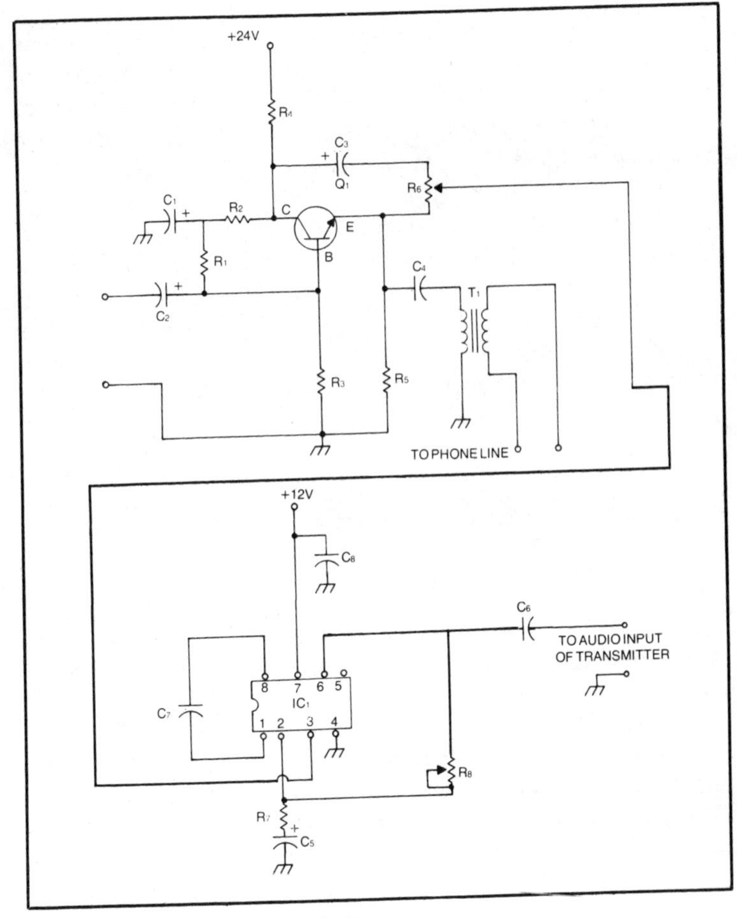

Fig. 12-19. Phone patch schematic diagram.

a primary and secondary impedence of equal value, ideally 600 ohms, although 500-ohm units will also work. Allied Electronics offers such a transformer for less than $8.00, but many attractive buys may be found on the surplus market. Many of the hobby transformers found through local electronics outlets have 500-ohm primary windings and 4 to 8-ohm secondaries. Since these cost only a dollar or so, two of these components may be purchased and wired as shown in Fig. 12-21. From an electronics standpoint, the first transformer exhibits an impedence of about 500 ohms and transforms this to an output impedence of about 8 ohms. The second transformer has been reversed so that its primary is 8 ohms and transforms this amount to 500 ohms. The result of this

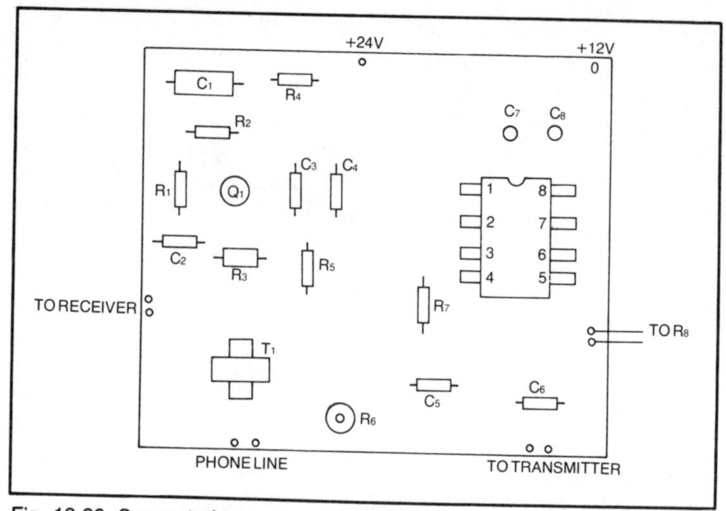

Fig. 12-20. Suggested parts layout designed to keep the two circuit sections separated.

two-transformer circuit is a device which has a 500-ohm input impedence and a 500-ohm output impedence. This is exactly what is required. True, this two-transformer circuit may be a bit larger than a single transformer device, but the former may cost half as much as the latter.

In building this circuit, start with the transistor portion first and complete all of the connections. Next, build the integrated circuit portion, soldering all connections. Do not insert the IC into its socket until later. By connecting a short length of hookup wire between pin 3 of the IC and the center contact arm of the 10 K ohm control, the two circuits are joined into one complex system.

It is best to power this device from an AC-derived source. Note that two voltages are required, 24 and 12 v.d.c. Both of these voltages are positive in relationship to ground. Figure 12-22 provides a schematic of a suggested power supply which is very similar to the one used with the previous amplifier project. This is two supplies derived from one transformer source, but here the similarity ends. This circuit uses a full-wave bridge rectifier for the higher voltage source and a portion of the bridge and the center tap of the transformer for the 12-volt source, which is derived from a full-wave, center tap portion of the circuit. Also, see Table 12-5.

Again, series resistors are used along with zener diodes to provide adequate regulation. Construction of this supply is handled in exactly the same manner as was specified for the earlier supply project. The circuit ground connection of the power supply is attached to the phone patch ground and the two voltage sources connected where indicated.

Test the power supply circuit using the previous procedures, and once this is found to be working properly, attach the remaining circuitry. Connect the output from the radio receiver or other audio device to the capacitor input of transistor Q1 and ground. Connect the output of the transformer to the phone line coupler. The output from the integrated circuit section is attached to the microphone input of a transmitter or audio amplifier by means of a two-conductor shielded cable which terminates in a male jack designed to mate with the input connector. During the testing procedure, the circuit should not be mounted in its permanent enclosure. This can be done after checkout is complete.

Call a friend on the telephone who can help you check the circuit properly. He or she should be someone who is familiar with electronic circuit building and can give you the information you need during this test. After the party is reached, turn on your receiver or other device which is connected to the circuit, and send 30 seconds of audio down the phone line. If the circuit is working properly, the other party should be able to tell you so and to advise as to whether or not the volume is too low, too loud, or just right.

Now, have the party speak for a minute or so while this audio information is fed to the transmitter input or to an audio amplifier. The latter is preferred, since it will be necessary to operate a transmitter into a dummy antenna while monitoring on a separate receiver. This can be rather complicated. The audio amplifier, on the other hand, will allow you to check the level and quality of the telephone line output without going through a lot of inconvenience. The balance control may be adjusted to match the volume levels of the audio fed into the phone line and that which comes from the phone line. Also, a gain control is provided in the IC section with

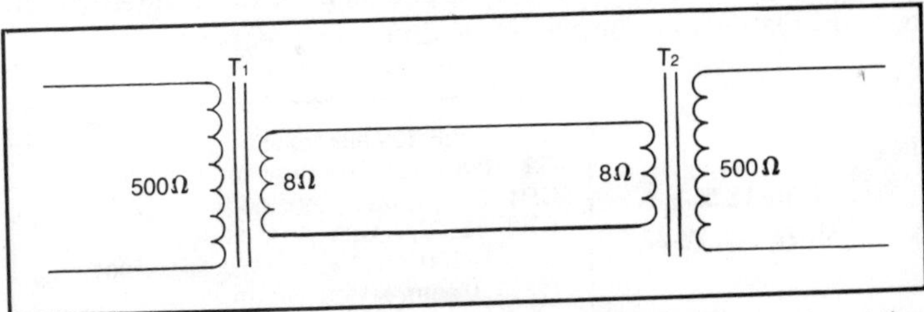

Fig. 12-21. Interconnection of two inexpensive transformers to arrive at the needed input and output impedances.

which you can raise or lower the phone line output to the transmitter input.

This circuit will be very useful for amateur radio operators and for persons who need to receive and process audio information from the phone line while maintaining the capability of feeding audio information into the line. While the assembly and use discussion was tailored mainly for amateur radio applications, this circuit might also be used for high quality tape recording off the phone line or for quality reproduction of the audio information as it is heard by the party at the other end. The carbon elements which serve as microphones in most standard telephone instruments are not high fidelity devices and much quality is lost when accurate reproduction is desirable. These elements perform adequately where voice communications are desired, but they fail miserably when audio frequencies fall below 300 Hz. or rise above 3000 Hz.

Be certain to check with the phone company regarding your intentions to use this device, as they will probably want to install a line coupler to isolate this circuit from their system. In some areas, this may not be necessary, but be safe and legal by contacting your phone company first.

TELEPHONE BROADCASTER

The two amplifier projects were complex because the audio information had to be sampled through pickup coils or elements and then fed to a multi-stage amplification circuit. Every audio device in your home has a similar amplification circuit, and it might be easier to try and tap into one of these circuits than to build one from scratch. Your standard AM transistor or table radio has an amplifier section which samples the detected radio frequency signals and then builds this current up to a level which will drive an audio speaker.

The circuit shown in Fig. 12-23 samples the telephone line audio through a Radio Shack pickup and then feeds this information to an oscillator which broadcasts within the frequency range of your

Table 12-5. Parts List for Fig. 12-22.

C1 - 500 uF 50V electrolytic
C2 - 1000 uF 30V electrolytic
CR1 - 24V, ½ watt zener diode
CR2 - 12.6V, ½ watt zener diode
D1-D4 - 100 PIV, 1 ampere rectifier diodes
R1 - 150 ohm, ½ watt carbon
R2 - 100 ohm, ½ watt carbon
T1 - 25.2V.C.T. transformer, 1 ampere rating

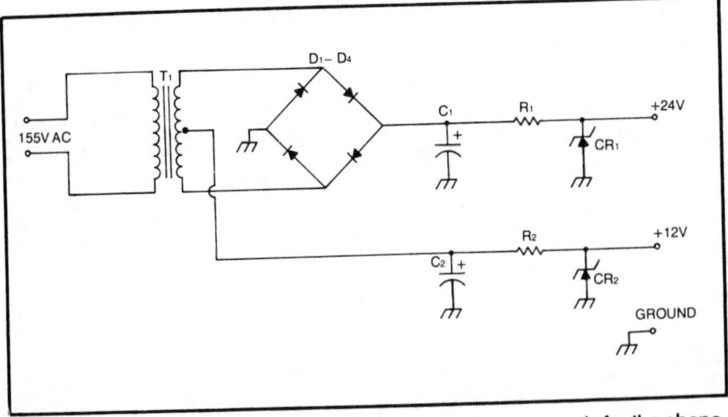

Fig. 12-22. Schematic of a dual voltage, AC-derived power supply for the phone patch circuit.

AM radio dial. All one has to do to recapture the transmitted information is to tune the AM radio to the frequency of the oscillator. The internal detection and amplification circuitry of the radio does the rest. Also, see Table 12-6.

From looking at the schematic, it can be seen that this is a very simple circuit which uses only a single transistor, two resistors, three capacitors, and a loopstick antenna in addition to a 9-volt battery and miniature switch. Loopstick antennas are available from many hobby outlets. They may also be salvaged from old transistor and tube-type AM radio receivers. Some of these inductive devices may have a long winding followed by a short winding. Only the larger winding is used. The shorter winding may be cut away or left in place, as desired. If the larger winding has a center tap lead, this may be cut away. The entire coil is used for this circuit.

Wiring is a little more critical here to prevent frequency drift. As the frequency rate changes, the received signal can simply drift away from the spot where you have your radio tuned. It will be necessary to constantly re-tune if solid construction techniques are not exercised. Keep all component leads and conductor lengths to a minimum.

Figure 12-24 shows a suggested component layout on a perforated circuit board. Do not deviate significantly from this suggested layout, as various problems associated with radio frequency oscillators may develop. Insert the component leads through the circuit board and make all connections. Solder these connections only when you are certain that all leads have been

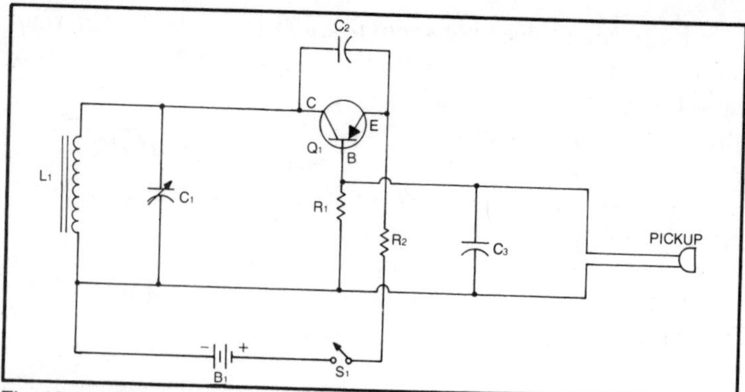

Fig. 12-23. Wireless telephone broadcaster.

properly wrapped and are as short as physically possible. The telephone pickup may be attached to the circuit by means of a plug and receptacle. The cable length between capacitor C3 and the pickup is relatively non-critical and may be as long as is convenient. Examine all circuit wiring before applying power.

The checkout procedure is simple when using a nearby AM radio. Tune the radio to a dead spot near the center of the AM band. Connect the pickup to the telephone handset and turn the switch to the on position. Now, adjust the variable capacitor C1 while speaking into the telephone handset until your voice is heard through the radio speaker. This might occur at different settings of C1, so choose the setting which provides best audio quality. It is best to tune C1 until you are close to the radio receiver frequency and then adjust the receiver's frequency control as a fine tuning method. At this point in construction, your telephone transmitter will be very unstable, and its frequency will change each time your hand touches the capacitor shaft.

Table 12-6. Parts List for Fig. 12-23.

B1 - 9 volt
C1 - 365 pf variable
C2 - 100 pf 15V
C3 - 0.005 uf disk 15V
L1 - Ferrite Loopstick Antenna
Q1 - ECG 126 (Sylvania)
R1 - 1000 ohm ½ watt
R2 - 50 Kohm ½ watt
S1 - SPST miniature toggle
Pickup - Radio Shack Telephone pickup contact microphone

If the circuit does not seem to oscillate, go back over your work, looking for a wiring error. If none is found, you may have a dead battery or transistor Q1 might be defective. Replacement of these components should correct the problem.

Once it has been established that the circuit is working properly, it is time to anchor the components more securely to the circuit board. This is done by using epoxy cement or another type of bonding material. A drop is placed over each of the components and the top of the circuit board. This includes the transistor and L1. The variable capacitor should also be bonded to the circuit board, but do not let any of the bonding compound come in contact with the rotary shaft. It is necessary to maintain the shaft movement for future frequency adjustment. Once the epoxy has set, install the entire circuit in a plastic enclosure through which S1 may be mounted, along with a receptacle for the pickup plug. C1 should ideally possess a screwdriver-adjusted shaft. If this is the case, a small hole may be drilled through the plastic enclosure in line with the shaft. When a frequency adjustment is desired, a non-inductive screwdriver or tuning tool can be inserted for tuning purposes.

It is best to mount this device in some fixed position out of the way of normal human activity. Human body capacitance reacting with the loopstick antenna can cause the frequency to deviate. An ideal mounting position would be on the back of a plastic radio. This is especially convenient since the radio serves as the amplifier.

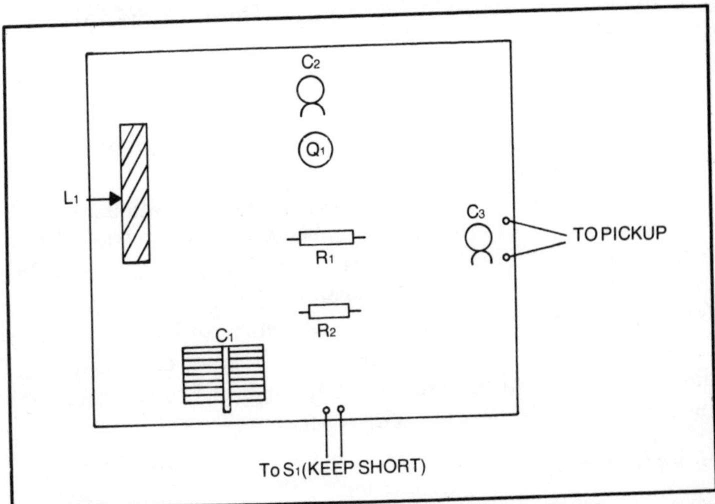

Fig. 12-24. Component layout on perforated circuit board for wireless broadcaster.

The range of this tiny transmitter is diminutive, although distances of 50 feet have been covered by adding an external antenna no longer than 10 feet which consists of a single conductor attached to the top portion of L1. This may be a necessary addition if you cannot conveniently place the AM radio near the telephone transmitter circuit.

Proceeding Further

For the avid experimenter Fig. 12-25 may be of interest. It demonstrates a method whereby a crude wireless telephone system might be designed for limited, short-range communications. Here, the AM radio is placed very near the telephone handset. Another transmitter circuit is built, but instead of feeding it with a contact pickup, a small crystal microphone is substituted. The transmitter at the telephone handset is tuned to the same frequency as the radio receiver at the remote location. The radio receiver at the handset is tuned to another frequency which corresponds with the output frequency of the remote transmitter. The transmitting element in the handset is placed very near the radio speaker.

Now, once the system is set up, a person at the remote location could speak into his tiny transmitter and be heard at the other end of the phone line by transmitting to the radio and then into the telephone through the radio's speaker. Likewise, when the party on the other end spoke, this information would be transmitted through the pickup, into the circuit, and finally to the AM radio at the remote location.

This is not really a practical system, as none of the transmitting circuits are all that stable. Then too, it would be necessary to manually establish the phone connection before the remote operation could begin.

The idea of transmitting telephone audio information can also be brought to fruition by using children's walkie talkies. The reason this hypothetical system has been discussed is to provide the reader with some possible avenues of experimental pursuit. It is important to remember at all times to abide by FCC regulations which govern the use of certain types of transmitting devices. The AM broadcaster does not require a license as long as the antenna length does not exceed 10 feet.

Project Summary

The wireless broadcaster, when properly aligned and mounted in a solid location, can be useful in providing short-range,

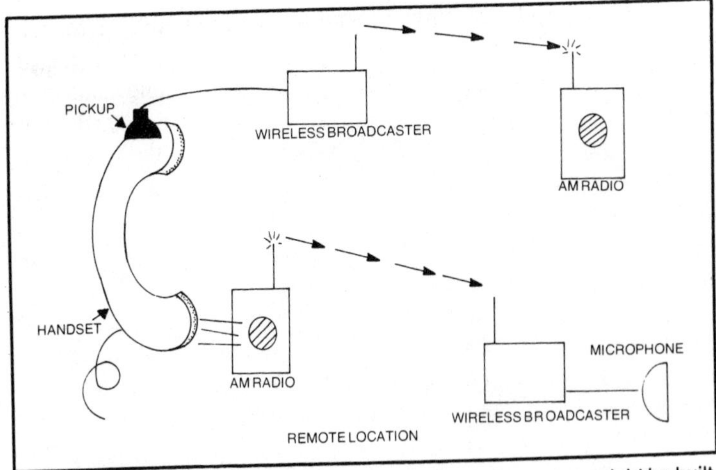

Fig. 12-25. Method by which a crude, wireless telephone system might be built for short range communciations.

wireless telephone amplification for use around conference tables or in the home. The stabilization of all circuit components is a necessity for practical operation. If this step is omitted, the slightest bump or jar to the circuit will result in a frequency shift which will render the circuit totally unreliable. The epoxy cement is available from most hobby stores, although any type of glue will probably work. Some of the "super" adhesives which are advertised on the modern market and sold through discount stores should work very well.

The reader is again cautioned not to exceed the 10-foot maximum allowable antenna length. The antenna should consist of an insulated piece of stranded hookup wire soldered to the top of L1 and extended to the greatest height possible. To exceed the 10-foot length would be a violation of FCC regulations and the broadcast range would probably not be extended any practical distance.

Since the circuit is rather touchy regarding output frequency, remember to mount the finished project at a location which renders it free from all human contact. This location should also be as free from vibration as possible and not subject to great variations in temperature.

CHAPTER SUMMARY

Many persons who have been involved with electronics for the last thirty to forty years have been known to say, not too kindly, that we are living in an appliance age. This refers to what are often known as appliance operators, hobbyists who no longer know how to build their own projects and circuits, but who must purchase everything in a completed, working form.

The term appliance operator does not fit many of the active hobbyists and experimenters here in the United States who would prefer to build their projects rather than buy 100% of the circuits they experiment with. Indeed, the age of the solid-state device has made it much more practical and efficient to be a builder than it ever was three or four decades ago. While it is true that individual circuit components have often been replaced by entire circuits found on a tiny silicon chip and housed in a single component package, it should be noted that the circuits on a chip have been combined in a million different ways by experimenters to arrive at complex circuits which were not even dreamed of ten years ago. This type of building may be more antiseptic than in the days when large vacuum tubes and transformers were involved. However, today's experimenter has the opportunity to delve further into an electronics technology which has advanced a hundred thousand times past that of the era known as "the good old days" of experimenting and building.

More than ever before, the electronics experimenter must not only know that a circuit works properly, but *why* it functions. This was not always true in earlier times; but today, the builder must know why before he can seek to improve upon or come up with an entirely new circuit idea.

All too often, the author has noted that those individuals who cry the loudest about the ineptitude of today's up-and-coming experimenters are the ones who had achieved technical proficiency at one time and then refused to continue to learn and advance with the continuing state of the art. These persons are merely mad at themselves and often seek to build themselves up by attempting to lower others rather than raising themselves.

This last chapter has involved some simple electronic circuits which will serve as a challenge to the newcomer to electronics building and will still be interesting to those with more experience. It is hoped that these circuits will be further experimented with, modified, and improved upon. Telephones and the telephone industry offer many avenues for the electronics experimenter. One might begin by attempting to copy some of the conveniences presently offered, but by using relatively inexpensive, home-built circuits. Commercially manufactured devices and equipment may even be modified and improved upon by making minor circuit changes or by adding new circuit elements to those already in existence.

Let your imagination be a guide. Many of the telephone products available to all of us today are the results of services and conveniences which were dreamed of a few years ago. When experimenters pursue their dreams, most of them become reality.

Index

Index

A comprehensive state-of-the-art sourcebook: terminology, equipment, accessories, installat ion, repair . . . with projects!

The Master Handbook of Telephones

by Robert J. Traister

At last . . . an up-to-the-minute bible of modern telephone equipment and systems covering literally every facet of telephonic communications from standard phones and decorator models to mobile phones, answering services, electronic telephones, *all* kinds of accessories, scramblers, security devices, even wireless systems and homebuilt phones! It's a complete sourcebook on the state-of-the-art from early hand-crank units to today's ultra-sophisticated electronic and laser communications techniques!

You'll find every fact, every technique you need to work with all types of telephone equipment: the range of available units and accessories, how they operate, what you can legally install and repair yourself, how to build a variety of useful gadgets and projects from scratch, *and much, much more*! Clear, concise descriptions and directions, plus loads of detailed illustrations, diagrams, and schematics make it easy for you to grasp every detail . . . there's even a 8-page full-color section!

Beginning with a look at basic telephone operation and the use of dial pulsers and Touch-Tone™ pads, you'll learn about jacks, connectors, outlets, telephone ownership requirements . . . plus ITT rotary and pushbutton phones, dial-in-handset, classic, custom, and wall phones. There's plenty of info on basic answering devices with details on specific models. Memory dialers, call diverters, cordless handsets, conference units, automatic dialers, amplifiers, recorders, scramblers are all included . . . and there's a full explanation (given in non-technical language) of lightware, laser, video facsimile, and satellite communications systems, even a section on interfacing computers with a phone line. The big project section is loaded with step-by-step directions on wiring and building procedures and special integrated circuit techniques. You'll be able to build an amplifier, an electronic phone patch, component mountings . . . even an actual phone booth!

Here's a handbook that's so complete and so detailed that *anyone*, regardless of experience, can easily install, use, *and* repair almost *any* kind of telephone equipment, and do it like a professional!

Robert J. Traister is the author of many books on topics ranging from electronics to sports cars. An avid amateur radio enthusiast, he has an Advanced Class Amateur License and has a special interest in radio teletype and slow-scan TV.